SIGNALS AND COMMUNICATION TECHNOLOGY

For further volumes:
www.springer.com/series/4748

Hu Sheng · YangQuan Chen · TianShuang Qiu

Fractional Processes and Fractional-Order Signal Processing

Techniques and Applications

Springer

Hu Sheng
School of Electronic and Information Engineering
Dalian Jiaotong University
Dalian
People's Republic of China
hu.sheng@djtu.edu.cn

TianShuang Qiu
School of Electronic and Information Engineering
Dalian University of Technology
Dalian
People's Republic of China
qiutsh@dlut.edu.cn

YangQuan Chen
Department of Electrical and Computer Engineering
CSOIS
Utah State University
Old Main Hill 4160
Logan, UT 84322-4120
USA
yangQuan.chen@usu.edu

ISSN 1860-4862 Signals and Communication Technology
ISBN 978-1-4471-2232-6 e-ISBN 978-1-4471-2233-3
DOI 10.1007/978-1-4471-2233-3
Springer London Dordrecht Heidelberg New York

British Library Cataloguing in Publication Data
A catalogue record for this book is available from the British Library

Library of Congress Control Number: 2011940854

Cover design: VTeX UAB, Lithuania

Printed on acid-free paper

Springer is part of Springer Science+Business Media (www.springer.com)

My wife, Shuyuan Zhao, and
My parents, Yuming Sheng and Guiyun Wang

– Hu Sheng

The memory of my father Hanlin Chen, and
My family, Huifang Dou, Duyun, David, and Daniel

– YangQuan Chen

The memory of my father Zhuxian Qiu, and
My family, Lian Wang and Yumeng

– TianShuang Qiu

Foreword

If an engineer wishes to optimize the operation of an electric or hybrid vehicle, the first thing she needs is a mathematical model to control the charging/discharge process. It is an axiom of control theory that the order of the model should match that of the system to be controlled. So, it is not surprising that fractional (non-integer) order control systems are needed to regulate the fractional-order diffusion kinetics inherent in electrochemical cells. What is the origin of such anomalous, and in some cases, non-Gaussian behavior? How do we describe the electrical signals that arise? And, what tools are available today for design engineers to use for fractional-order signal processing? In this new book (Fractional Processes and Fractional-Order Signal Processing) the authors answer these questions in a manner that is easily understood by students, researchers and by practicing engineers.

In order to put this contribution into its proper context, we should first realize that while the tools of fractional calculus are well known—some key elements having been developed by mathematicians in the 18th and 19th centuries—the application of fractional calculus in physics, chemistry, biology and engineering arose only in the latter half of the 20th century. Second, unlike the mature field of "analysis" that forms the foundation for the conventional integral and differential calculus, fractional calculus is still undergoing rapid growth and expansion; driven in part by the same forces that are developing new mathematical theories of chaos, fractals and complex systems. Unfortunately, this dynamic of growth sometimes hinders the transfer and assimilation of new knowledge into the scientific and technical disciplines.

The problem is somewhat akin to the development of new schools of philosophical inquiry in ancient Greece: Stoicism, Epicureanism, and Skepticism. The certainty of the Sophists that human knowledge could be acquired by observation and analysis, was offset by the Epicureans who felt that all such knowledge was useful, while the Skeptics—who saw only the limits of new knowledge—counseled all to suspend judgment. This type of philosophical argument is unfortunately being played out today between the applied mathematics community, the fractional calculus clan, and advocates of nonlinear, chaotic and complex processes (i.e., processes assumed to be beyond our ability to model). The challenge therefore for fractional

Epicureans is to seek aid from mathematical Sophists to place fractional calculus within the canon of college calculus, while helping 'complexity theory' Skeptics to, in the words of Coleridge, "suspend their disbelief".

Given this situation, it is natural that the authors play to all parts of the house in their presentation. This is evident in the four parts of the monograph, which address: basic theory, the theory of fractional-order processes, the analysis of signals generated by fractional-order processes, and the applications of such models to a wide range of physical, mechanical and biological systems. The first part thus provides students with a review (and other readers with a refreshing summary) of random signals and stochastic processes while outlining the basics of fractional-order models, experimentally observed fractional-order behavior and signal processing. The second part focuses on constant-order fractional processes and on multifractional processes. Hence, the purpose of this section is to connect the reader with fractional-order processes on one hand and with the more fundamental concepts of fractional Browning motion and fractal measures of stability (via the Hurst estimators) on the other.

The third section of the book provides the core of the topical analysis with constant-, variable- and distributed-order fractional signal processing described. The presentation spans theory and application (analogue and digital) for the implementation of continuous-time and discrete-time operators. Canonical fractional second order filter models are used to illustrate the expected behavior for the constant-order case. The synthesis of variable-order fractional signal processing is next described for cases of locally stable long memory with an example of the analogue realization in the case of a temperature-dependent variable order fractional integrator and others for variable-order fractional systems. Finally, this section closes with the description of distributed-order integrators and differentiators and a distributed-order low pass filter. Part four of the book presents six separate examples of applications of fractional-order signal processing techniques: two biomedical examples (latency of evoked potentials, multifractional properties of human sleep EEG), one model of molecular motion, an elaboration of optimal fractional-order damping strategies for a mass-spring viscoelastic vibration damper, analysis of electrical noise using fractional signal processing, and a variable-order fractional analysis of changes in water level in the Great Salt Lake.

Given the breadth and depth of the analysis in this monograph the fractional Epicurean and perhaps even the mathematical Stoics should be sated with its range of theory and applications. However, here the Skeptic might recall the philosophical paradox of describing the "state" of a river: the state is never the same as it always varies in time and place, and never repeats. And so it is with fractional calculus, it seems, with its bewildering variety of definitions for fractional-order operators (Riemann-Liouville, Caputo, Weyl, Grunwald-Letnikov, Riesz, etc.). It is notable then that for this text, only the Riemann-Liouville and the Caputo definitions are employed. The emphasis here is not on different observers on the shore of the river, but on different ways for one observer to view the "state", e.g., as a fractional order process with constant-order, variable-order, or distributed-order dynamic measures.

Just as the measurements by Hurst of the flow of the Nile river in Egypt provided the first early evidence for the utility of fractal models of "real-world" phenomena, the fractional-order models of biological, hydrological, electrical and mechanical phenomena presented here give the strongest evidence of the utility of fractional-order analysis. These models are robust, resilient, and adaptive. As physics, engineering, chemistry and biology move toward more complete representations of complex materials and processes—in tissue engineering, for example, the regeneration and regrowth of new organs—scientists and engineers need an expanded set of tools to deal with processes that clearly change, like a river, with time and place. This book provides one clear view of how to develop and apply fractional-order models to contemporary problems in polymer formation, protein folding, and even the optimal charging of lithium-ion batteries for the next generation of electric vehicles.

Distinguished Professor of Bioengineering
Department of Bioengineering
University of Illinois at Chicago, IL

Richard L. Magin

Preface

In ancient times, instead of fractions, the natural numbers were used. However, integers cannot always result when measuring or equally dividing things. As time went by, fractions and then non-integers were gradually understood and applied. With the introduction of fractions and more generally nonintegers, people were able to have a closer look at the beauty of nature around them. For example, people long ago realized that a rectangle of the 'golden ratio' $1.618 : 1$ is most pleasing. The natural exponential $e = 2.71828\ldots$, and the ratio of any circle's circumference to its diameter, $\pi = 3.14159\ldots$ are widely used in mathematics and engineering. The 'beauty' of the fraction was recognized and people came to use the 'fractional view' to observe the world, to use 'fractional thinking' to understand the natural phenomena, and to use 'fractional techniques' to solve the problems at hand.

The term 'fractal' was introduced by Mandelbrot in 1975 [192]. Fractal refers to the self-similar geometric shape, that is, a shape in which is almost identical to the entire shape except in size [91, 102]. Many objects manifest themselves in fractal shape, such as clouds, coastlines and snow flakes. In fractal theory, the fractal dimension was used to characterize the state of nature. Different from the conventional integer dimension, the fractal dimension can be fractional or any non-integer number. Based on the fractal theory, the traditional concept of three-dimensional space can be extended to the fractal (fractional) dimension (FD) which can be applied to characterize complex objects.

Likewise, (integer-order) calculus can be extended to fractional or noninteger order calculus. It should be remarked at this point that due to historical reasons, the term 'fractional' we use here and throughout this monograph should actually be understood as 'non-integer' or 'arbitrary real number' to be precise. Fractional calculus, i.e., fractional-order differentiation and integration, is a part of mathematics dealing with derivatives of arbitrary order [139, 203, 209, 218, 237]. Leibniz raised the question about the possibility of generalizing the operation of differentiation to non-integer-orders in 1695 [237]. Fractional calculus, developed from the field of pure mathematics, has been studied increasingly in various fields [64, 142, 311, 315, 323]. Nowadays, fractional calculus is being applied to many fields of science, engineering, and mathematics [49, 74, 78, 135, 290]. Fractional calculus provides a better description of various natural phenomena, such as viscoelasticity and anomalous diffusion, than integer-order calculus can provide. The most fundamental reason for the superiority of the use is that fractional calculus based models can capture the memory and the heredity of the process. It is safe to say that fractional calculus provides a particularly useful and effective tool for revealing phenomena in nature because nature has memory.

This "fractionalization" idea can go on. Taking the Fourier transform as an example, we can naturally talk about the fractional Fourier transform (FrFT), which is a linear transformation generalizing the Fourier transform, first introduced by Victor Namias in 1980 [213]. FrFT can transform a signal from time domain into a domain between the time and frequency domains. So, the FrFT differs from the conventional Fourier transform by the rotation of fractional times of the $\pi/2$ angle in the time-frequency plane. FrFT is widely used in filter design, signal detection and image recovery [32].

Another important "fractionalization" is fractional low-order moments (FLOM). FLOM is based on the non-Gaussian α-stable distribution, a powerful tool for impulsive random signals. The density functions of α-stable distribution will decay in the tails less rapidly than the Gaussian density function does. So α-stable distribution can be used to characterize signals which exhibit sharp spikes or occasional bursts of outlying observations more frequently than normally distributed signals do. The α-stable distribution based techniques have been applied to describe many natural or man-made phenomena in various fields, such as physics, hydrology, biology, financial and network traffic [43, 70, 253, 270]. The α-stable distribution has a characteristic exponent parameter $\alpha (0 < \alpha \leq 2)$, which controls the heaviness of its tails. For a non-Gaussian stable distribution with characteristic exponent $\alpha < 2$, its second-order moment does not exist. α-stable distribution has only finite moments of order less than α. So the FLOM plays an important role in impulsive processes like the role of mean and variance in the Gaussian processes.

In this monograph, we will introduce some complex random signals which are characterized by the presence of heavy-tailed distribution or non-negligible dependence between distant observations, from the 'fractional' point of view. Furthermore, the analysis techniques for these fractional processes are investigated using 'fractional thinking'. The term 'fractional process' in this monograph refers to some random signals which manifest themselves by heavy-tailed distribution, long range dependence (LRD)/long memory, or local memory. Fractional processes are widely found in science, technology and engineering systems. Typical heavy-tailed distributed signals include underwater acoustic signals, low-frequency atmospheric noises, many types of man-made noises, and so on. Typical LRD/long memory processes and local memory processes can be observed in financial data, communications networks data and biological data. These properties, i.e., heavy-tailed distribution, LRD/long memory, and local memory, always lead to difficulty in correctly obtaining the statistical characteristics and extracting the desired information from these fractional processes. These properties cannot be neglected in time series analysis, because the tail thickness of the distribution, LRD, or local memory properties of the time series are critical in characterizing the essence of the resulting natural or man-made phenomena of the signals. Therefore, some valuable fractional-order signal processing (FOSP) techniques were provided to analyze these fractional processes. FOSP techniques, which are based on the fractional calculus, FLOM and FrFT, include simulation of fractional processes, fractional-order system modeling, fractional-order filtering, realization of fractional systems, etc. So, random signals which exhibit evident 'fractional' properties should be investigated using FOSP techniques to obtain better analysis results.

This monograph includes four parts. The first part is the overview of fractional processes and FOSP techniques. The second part presents fractional processes, which are studied as the output of the fractional order differential systems, including constant-order fractional processes and variable-order fractional processes. The third part introduces the FOSP techniques from the 'fractional signals and fractional systems' point of view. In the last part of the monograph, some application examples of FOSP techniques are presented to help readers understand and appreciate the fractional processes and fractional techniques. We sincerely wish that this monograph will give our readers a novel insight into the complex random signals characterized by 'fractional' properties, and some powerful tools to characterize those signals.

Dalian, Liaoning, China Hu Sheng
Logan, UT, USA YangQuan Chen
Dalian, Liaoning, China TianShuang Qiu

Acknowledgements

The purpose of this monograph is to give a unified presentation of our research on fractional order signal processing (FOSP), based on a series of papers and articles that we have published, and partly, on the Ph.D. dissertation of the first author. It has been necessary to reuse some materials that we previously reported in various papers and publications. Although in most instances such materials have been modified and rewritten for the monograph, copyright permissions from several publishers are acknowledged as follows:

Acknowledgement is given to the Institute of Electrical and Electronic Engineers (IEEE) to reproduce material from the following papers.

©2010 IEEE. Reprinted, with permission, from Yan Li, Hu Sheng, and YangQuan Chen. "On the fractional-order distributed parameter low-pass filter", *Proc. of the 49th IEEE Conference on Decision and Control*, Atlanta, Georgia, USA, Dec. 15–17, 2010, pp. 4984–4989 (material found in Chap. 7). doi:10.1109/CDC.2010.5717957.

©2010 IEEE. Reprinted, with permission, from Yan Li, Hu Sheng, and YangQuan Chen. "On distributed order low pass filter", *Proc. of the 2010 IEEE/ASME International Conference on Mechatronic and Embedded Systems and Applications*", Qingdao, ShanDong, China, 2010, pp. 588–592. doi:10.1109/MESA.2010.5552095 (material found in Chap. 7).

Acknowledgement is given to Elsevier B.V. to reproduce material from the following papers:

©2010 Elsevier B.V. Reprinted, with permission, from Blas M. Vinagre, YangQuan Chen, Ivo Petrás. "Two direct Tustin discretization methods for fractional-order differentiator/integrator", Journal of the Franklin Institute, Volume 340, Issue 5, August 2003, pp. 349–362. doi:10.1016/j.jfranklin.2003.08.001 (material found in Chap. 5).

©2010 Elsevier B.V. Reprinted, with permission, from Yan Li, Hu Sheng, YangQuan Chen. "Analytical impulse response of a fractional second order filter and its impulse response invariant discretization", Signal Processing, Volume 91,

Issue 3, March 2011, pp. 498–507. doi:10.1016/j.sigpro.2010.01.017 (material found in Chap. 5).
©2010 Elsevier B.V. Reprinted, with permission, from Hu Sheng, Hongguang Sun, YangQuan Chen, TianShuang Qiu. "Synthesis of multifractional Gaussian noises based on variable-order fractional operators", Signal Processing, Volume 91, Issue 7, July 2011, pp. 1645–1650. doi:10.1016/j.sigpro.2011.01.010 (material found in Chap. 6).
©2010 Elsevier B.V. Reprinted, with permission, from Yan Li, Hu Sheng, YangQuan Chen. "On distributed order integrator/differentiator", Signal Processing, Volume 91, Issue 5, May 2011, pp. 1079–1084. doi:10.1016/j.sigpro.2010.10.005 (material found in Chap. 7).
©2010 Elsevier B.V. Reprinted, with permission, from Hu Sheng and YangQuan Chen. "FARIMA with stable innovations model of Great Salt Lake elevation time series", Signal Processing, Volume 91, Issue 3, March 2011, pp. 553–561. doi:10.1016/j.sigpro.2010.01.023 (material found in Chap. 8).
©2010 Elsevier B.V. Reprinted, with permission, from Hu Sheng, Yan Li, YangQuan Chen. "Application of numerical inverse Laplace transform algorithms in fractional calculus", Journal of the Franklin Institute, Volume 348, Issue 2, March 2011, pp. 315–330. doi:10.1016/j.jfranklin.2010.11.009 (material found in Appendix B).

Acknowledgement is given to World Scientific Publishing Company to reproduce material from the following paper:

©2010 World Scientific Publishing Company. Reprinted, with permission, from Hu Sheng, YangQuan Chen and Tianshuang Qiu. "Heavy-tailed distribution and local long memory in time series of molecular motion on the cell membrane", Fluctuation and Noise Letters, Volume 10, Issue 1, 2011, pp. 93–119. doi:10.1142/S0219477511000429 (material found in Chap. 11).

Acknowledgement is given to The Institution of Engineering and Technology (IET) to reproduce material from the following paper:

©2011 The Institution of Engineering and Technology. Reprinted, with permission, from H. Sheng, Y.Q. Chen, and T. Qiu. "On the robustness of Hurst estimators", IET Signal Processing, April 2011, Volume 5, Issue 2, pp. 209–225. doi:10.1049/iet-spr.2009.0241 (material found in Chap. 3).

Acknowledgement is given to EDP Sciences, Springer-Verlag to reproduce material from the following paper:

©2011 EDP Sciences, Springer-Verlag. Reprinted, with permission, from H. Sheng, H.G. Sun, C. Coopmans, Y.Q. Chen and G.W. Bohannan. "A physical experimental study of variable-order fractional integrator and differentiator", Published online: 4 April 2011, pp. 93–104. doi:10.1140/epjst/e2011-01384-4 (material found in Chap. 6).

Fractional order signals and systems are emerging as an exciting research area. We have a few sources of motivations and education that we feel we must mention.

We acknowledge Prof. Richard Magin for his wonderful textbook on "Fractional Calculus in Bioengineering" and his series of research papers on using fractional calculus in processing bioengineering signals. Our monograph is partly motivated by his works. Dr. Bruce J. West's visit to CSOIS (Center for Self-Organizing and Intelligent Systems) of Utah State University in April 2009 is another source of motivation. His insightful lecture at there entitled "To be average is exceptional" was particularly motivating. Prof. Manuel Ortigueira has published a series of papers on fundamentals in fractional signals and systems that we feel are educating and clarifying. Prof. Raoul Nigmatullin's cumulative knowledge on new methods of random signal characterization offered insights on fractional processes and FOSP. Professors F. Mainardi and R. Gorenflo helped us understand the link between fractional calculus and stable processes. Prof. Ming Li's overview paper "Fractal time series—a tutorial review", among his many contributions on FOSP, was particularly helpful and motivating to us, too.

It has been an exciting journey writing this book. In the past five years, we have been focusing on FOSP with a better and better understanding of the wonderful world of fractional processes and FOSP. The authors would like to first thank Dr. Yan Li, Shandong University, China for helping to solve many mathematical problems of fractional calculus related to distributed-order filters in Chap. 7. Many MATLAB® codes about the discretization of the distributed-order filters were finished with his help. Secondly, we would like to thank Dr. HongGuang Sun, Hohai University, who provided many valuable numerical algorithms of variable-order fractional calculus. We also wish to thank Professor Anhong Zhou, Utah State University, for providing many valuable bioengineering signals.

We would like to express our sincere thanks to Professor Igor Podlubny, Technical University of Kosice, Slovak Republic, for many helpful comments and suggestions. We are particularly grateful to Professor Dingyü Xue, Northeastern University, China, for the helpful suggestions. We also would like to thank Professor Changpin Li, Shanghai University, China, for his help in the mathematical fundamentals of fractional calculus. We would like to express our gratitude to Dr. Stanly Steinberg and Dr. Martha Zúñiga for sharing the Class I major histocompatibility complex (MHCI) molecular trajectories data as used in [327]. We also would like to express our gratitude to Mr. Rongtao Sun for his initial discussion in data preprocessing of electrochemical noise signals.

Hu Sheng would like to thank the China Scholarship Council (CSC) for the financial support for his two years (2008–2010) exchange Ph.D. studentship at CSOIS of Utah State University. He would like to express his sincere thanks to Dr. Yong Zhu, Department of Electronic Engineering at Dalian University of Technology, for many helpful suggestions. Hu Sheng also would like to thank his wife Shuyuan Zhao and his parents Yuming Sheng and Guiyun Wang for their complete support. He gratefully acknowledges some former and current CSOIS members and thanks for their selfless help during his two years stay at the center: Prof. Chunyang Wang, Dr. Haiyang Chao, Dr. Yongshun Jin, Dr. Wei Sun, Long Di, Prof. Dali Chen, Jinlu Han, Calvin Coopmans, Dr. Christophe Tricaud, Hadi Malek, Prof. Deshun Yin, Jun Pan, Shayok Mukhopadhyay, Tobias Fromm and Dr. Ying Luo.

YangQuan Chen would like to thank his wife Huifang Dou and his sons Duyun, David and Daniel, for their patience, understanding and complete support throughout this work. He also wishes to thank Prof. Dumitru Baleanu for inviting him for a plenary lecture at the IFAC FDA 2008, Ankara, Turkey, entitled "*Fractional Order Signal Processing: Techniques, Applications and Urgency*". This monograph was finalized during his sabbatical semester at University of Pretoria, South Africa in Spring 2011, and he appreciates Professors Xiaohua Xia and Ian Craig for hosting his productive stay. He also wishes to thank Mrs. Sara Dadras for her assistance in carefully proof reading this monograph.

TianShuang Qiu would like to thank his family, Lian Wang and Yumeng.

Finally, thanks go to Oliver Jackson of Springer for his encouragement of this project after his Amazon book search and to Charlotte Cross, Editorial Assistant (Engineering) Springer London for her great assistance. Last but not least, we are thankful to four anonymous reviewers for their detailed comments and suggestions that improved our presentation significantly.

Contents

Acronyms

AC	Alternating current
AR	Autoregressive
ARCH	Autoregressive conditional heteroskedasticity
ARIMA	Autoregressive integrated moving average
ARMA	Autoregression and moving average
CE	Counter electrode
CNS	Central nervous system
DEA	Diffusion entropy algorithm
DFT	Discrete Fourier transform
DLMP	Direct least mean p-norm
DLMS	Direct least mean square
Ece	Potential of counter electrode versus the reference electrode
ECN	Electrochemical noise
EEG	Electroencephalogram
EIS	Electrochemical impedance spectroscopy
EP	Evoked potentials
Ewe	Potential of working electrode versus the reference electrode
FARIMA	Fractional autoregressive integrated moving average
fBm	Fractional Brownian motion
FC	Fractional calculus
FFT	Fast Fourier transform
fGn	Fractional Gaussian noise
FHT	Fractional Hilbert transform
FIGARCH	Fractional integral generalized autoregressive conditional heteroskedasticity
FIR	Finite-duration impulse response
FLDTI	Fractional linear discrete time-invariant
FLOM	Fractional low-order moments
FLTI	Fractional linear continuous time-invariant
FOSP	Fractional-order signal processing
FPSD	Fractional power spectral density

FrFT	Fractional Fourier transform
GARCH	Generalized autoregressive conditional heteroskedasticity
GARMA	Gegenbauer autoregressive moving average
GSL	Great Salt Lake
HFE	Heat-Flow Experiment
Icoupling	Current between working electrode and counter electrode
i.i.d.	Independent and identically distributed
IIR	Infinite-duration impulse response
LASS	Local analysis of self-similarity
LDTI	Linear discrete time-invariant
LFSM	Linear fractional stable motion
LP	Linear polarization
LRD	Long range dependence
LTI	Linear time-invariant
MA	Moving average
mBm	Multifractional Brownian motion
mGn	Multifractional Gaussian noise
MSNR	Mixed Signal to noise ratio
NREM	Non-rapid eye movement
NLST	New adaptive EP latency change detection algorithm
MHCI	Class I major histocompatibility complex
PDF	Probability density function
PSD	Power spectral density
RE	Reference electrode
REM	Rapid eye movement
SαS	Symmetric α-stable
SDA	Signed adaptive algorithm
SNR	Signal to noise ratio
SPT	Single-particle tracking
VMP	Versatile multichannel potentiostat
WE	Working electrode
wGn	White Gaussian noise
ZRA	Zero resistance ammetry

Part I
Overview of Fractional Processes and Fractional-Order Signal Processing Techniques

Chapter 1
Introduction

1.1 An Introduction to Fractional Processes and Analysis Methods

Conventional signal processing techniques simplify the analysis process. Random signals being analyzed are assumed to be independent and identically distributed (i.i.d.), with weak coupling between values at different times, and at least stationary. In most cases, however, the sampled signals exhibit properties of being non-stationary, spiky, or long-range dependent (LRD). The questions, such as "what caused these complex phenomena?", "how do we deal with these signals?", and "how can we extract valuable information from them?", are still puzzling the researchers. In this monograph, we will reveal the secrets of these kinds of complex signals and introduce techniques for analyzing them.

The observed samples of random signals were traditionally assumed to be i.i.d. to simplify the underlying mathematics of many statistical methods. According to the central limit theorem, the probability distribution of the sum (or average) of i.i.d. variables with finite variance approaches a normal distribution. So the Gaussian model has been widely used in signal processing, and the second-order statistics, such as variance and correlation, have been widely employed to characterize the random signals. In practical applications, however, the assumption that the random signals are i.i.d. may not be realistic. Many signals, such as financial data, communications networks data, and many types of man-made noises, belong to non-Gaussian distributed processes. What would the distribution be if variables have infinite variance or variables have non-identical distribution? To solve such a problem, the generalized central limit theorem was provided. It states that if the sum of i.i.d. random variables with or without finite variance converges to a distribution by increasing the number of variables, then the limit distribution must belong to the family of stable laws [36, 94]. For independent, non-identically distributed sequences, similar but weaker results hold. This stable model is a direct generalization of the Gaussian model, so the stable model provides a better description of non-Gaussian and non-stationary random noise.

H. Sheng et al., *Fractional Processes and Fractional-Order Signal Processing*, Signals and Communication Technology,
DOI 10.1007/978-1-4471-2233-3_1,

In time series analysis, there is another traditional assumption—that the coupling between values at different times decreases rapidly as the time difference increases. Based on this assumption, many short-range dependent random process models were built, such as the autoregressive and moving average (ARMA) model, and the autoregressive conditional heteroscedasticity (ARCH) model. Similar to the i.i.d. assumption about the random signals, in practice, the algebraic decay of the autocovariance can be observed in many time series. For example, Hurst spent many years analyzing the records of the rise and fall of the Nile river. He found a strange phenomena: the long-range recording of the elevation of the Nile river has much stronger coupling, and the autocovariance function decays slower than exponentially. He also found that, except for conventional stochastic properties such as mean and variance, there is another very important parameter hiding behind the fluctuation of the time series which can characterize the coupling property. In order to quantify the level of the coupling, the rescaled range (R/S) analysis method was provided to estimate the coupling level, which is now called the Hurst parameter. It was found that the average value of the Hurst parameter for Nile river records was 0.72. Since then the long-range dependence (LRD)/long memory phenomenon has attracted numerous research studies. Many valuable Hurst parameter estimators were provided to more accurately characterize the LRD time series. Based on Hurst's analysis, more suitable models, such as fractional autoregressive integrated moving average (FARIMA) and fractional integral generalized autoregressive conditional heteroscedasticity (FIGARCH) were built to accurately analyze LRD processes. The most distinguishing feature of these generalized models is the long memory character. These generalized models can capture both the short and the long memory nature of the time series. Therefore, using the LRD theory, great achievements have been made in various fields, such as video traffic modeling, econometrics, hydrology and linguistics.

The LRD theory was built on the basis of a stationary process or a process with stationary increments. Therefore, it is unreasonable to characterize the non-stationary signals by a constant long memory parameter H. In 1995, Peltier and Vehel studied the multifractional Brownian motion (mBm) in a research report [232]. In the report, the constant Hurst parameter H was generalized to the local Hölder exponent $H(t)$, where H is a function of the time index of the processes. The local Hölder exponent $H(t)$ can capture the local scaling characteristic of the stochastic processes, and can summarize the time-varying nature of the non-stationary processes. But this extension leads to some difficulties. For example, it is a difficult task to accurately estimate the local Hölder exponent $H(t)$ since the increment process of a multifractional process is no longer a stationary self-similar process. Based on the local memory theory, the generalized random process models were studied by some researchers [21, 243]. The generalized locally stationary long memory process FARIMA model was investigated in [30]. Local memory technique offers a valuable competing framework within which to describe the non-stationary time series.

Power-law is the common characteristic of the stable processes, LRD processes and local memory processes. Stable processes are characterized by the power-law distributions; LRD processes possess power-law decay of correlations; local memory processes have local power-law decay of the correlations. Power-law also has a tight relationship with fractal and self-similarity, since the self-similarity also can be described in a power-law form. So, the power-law is considered as a sign of something interesting and complicated happening. For example, $1/f$ noise is characterized by power-law decay for the correlations, and similarly, in the frequency domain $1/f$ noise manifests itself in a power-law decay of the spectrum. In fact, many naturally occurring phenomena were found to follow power-law form. The power-law relation might be the fundamental relation underlying natural phenomenon. Therefore, power-law related topics, such as its origins and validation, have become an active area of research [206, 277].

Where are the power-law results from? Some researchers found interesting results from fractional-order calculus (FC), which is a part of mathematics dealing with derivatives and integration of arbitrary order. Different from the analytical results of linear integer-order differential equations, which are represented by the combination of exponential functions, the analytical results of the linear fractional-order differential equations are represented by the Mittag-Leffler function, which exhibits a power-law asymptotic behavior [122, 140, 255]. Therefore, most of the results of linear fractional-order differential equations often exhibit a power-law-like property. Researchers are also amazed to find that the stable distribution can be considered as the result of a fractional-order diffusion equation [187, 281], and LRD processes can be generated using fractional-order differential systems. Therefore, FC is being widely used to analyze the random signals with power-law size distributions or power-law decay of correlations [117, 188, 237].

Compared with constant-order fractional calculus, the fractional integration and differentiation of variable-order can provide better characterization of non-stationary locally self-similar signals. Variable-order fractional calculus was first suggested by Samko [251]. Lorenzo and Hartley also suggested the variable-order fractional operators and studied their behavior [180]. The variable-order fractional operator has attracted more and more attention during the past decades [68, 128, 129, 290]. Variable-order fractional calculus is the generalization of FC by replacing the fractional-order α by $\alpha(\tau)$, where τ can be a time variable or other variables. Based on the variable-order fractional calculus, many complex systems can be modeled by variable-order fractional differential equations. Similar to LRD processes, which can be generated using constant-order fractional systems, multifractional processes can be synthesized by variable-order fractional systems [271].

Besides fractional calculus theory, fractional Fourier transform (FrFT) theory also provides some valuable tools to analyze LRD signals. FrFT was first introduced by Victor Namias in 1980 [32]. It is the generalization of the classical Fourier transform. Nowadays, FrFT has become one of the most valuable and frequently used techniques in time-varying signal processing and analysis. FrFT can be defined in several different ways, which leads to various physical interpretations. For signal processing, it can be considered as a rotation by an angle α in the time-frequency plane. FrFT is flexible in application and can be computed in about the same time as the conventional Fourier transform. So, the FrFT has several applications in the areas of optical beam propagation, statistical optics, optical system design and optical signal processing, signal detectors and signal and image recovery [32]. Based on

FrFT, many new operations were defined, such as fractional convolution, fractional correlation and fractional power spectrum. Furthermore, it is found that the FrFT can be used to analyze LRD processes and local memory processes. An improved Hurst parameter estimator which is based on FrFT was provided in [60]. Furthermore, FrFT has valuable application potential in variable-order fractional signal analysis. So, FrFT is studied in this monograph as the basis of the fractional-order signal processing (FOSP) techniques.

Power-law distribution, power-law decay of correlation, and locally power-law decay of correlation, do not always independently present themselves in random processes. Some processes exhibit both the power-law distribution and LRD, or the combination of power-law and local self-similarity. In this monograph, we call these processes, which are discussed above, fractional processes. It is challenging to obtain correctly the statistical characteristics and to extract the desired information from these sampled complex random signals. From the above discussion, FC, stable distribution and FrFT based FOSP techniques provide valuable analysis methods, such as fractional signals generation, fractional filtering, fractional systems modeling and fractional systems realization, to study these kinds of complex signals.

Fractional processes and the FOSP techniques are extended from conventional integer-order processes and integer-order signal processing techniques, respectively. In the following sections, after reviewing conventional stochastic processes and conventional signal processing techniques, the fractional-order processes and the FOSP techniques will be introduced.

1.2 Basis of Stochastic Processes

Definition 1.1 A stochastic (or random) process is a family of random variables $\{X(t)|t \in T\}$ defined on a given probability space, indexed by the time variable t, where t varies over an index set T [301].

Familiar examples of stochastic processes include random movement, medical signals, stock market and exchange rate fluctuations, among others.

1.2.1 Statistics of Stochastic Processes

A stochastic process is a family of time functions, and for a specific t, $X(t)$ is a random variable with distribution

$$F_X(x,t) = P\{X(t) \leq x\}, \tag{1.1}$$

where P is the probability. The function $F_X(x,t)$ is called the first-order probability distribution of the process $X(t)$. If the partial derivative of $F(x,t)$ with respect to x

exists,

$$f_X(x,t) = \frac{\partial F_X(x,t)}{\partial x} \tag{1.2}$$

is the first-order probability density function (PDF) of $X(t)$.

The second-order distribution of the process $X(t)$ is the joint distribution

$$F_X(x_1, x_2; t_1, t_2) = P\{X(t_1) \le x_1, X(t_2) \le x_2\}. \tag{1.3}$$

If the partial derivatives of $F_X(x_1, x_2; t_1, t_2)$ with respect to x_1, x_2 exist, the corresponding joint probability density is defined as

$$f_X(x_1, x_2; t_1, t_2) = \frac{\partial^2 F_X(x_1, x_2; t_1, t_2)}{\partial x_1 \partial x_2}. \tag{1.4}$$

1.2.2 Properties of Stochastic Processes

Mean Function [107, 204, 301]

Definition 1.2 If $X(t)$ is a random process, for every value of t, $X(t)$ is a random variable with mean $E[X(t)]$. We call

$$\mu_X(t) = E[X(t)] = \int_{-\infty}^{\infty} x f_X(x,t) \mathrm{d}x \tag{1.5}$$

the mean function of the random process $X(t)$.

Variance Function [107, 204, 301]

Definition 1.3 If a random variable $X(t)$ has the expected value (mean) $\mu_X(t) = E[X(t)]$, the variance of $X(t)$ is given by

$$\begin{aligned} \sigma_X^2(t) &= \mathrm{var}[X(t)] = E[(X(t) - E[X(t)])^2] \\ &= E[(X^2(t)] - E[X(t)]^2 = \int_{-\infty}^{\infty} [x - \mu_X(t)]^2 f_X(x,t) \mathrm{d}x. \end{aligned} \tag{1.6}$$

Correlation Function [107, 204, 301]

Definition 1.4 If $X(t_1)$ and $X(t_2)$ are two random variables of a random process $X(t)$, their correlation is denoted by

$$\begin{aligned} \rho_X(t_1, t_2) &= \mathrm{corr}[X(t_1), X(t_2)] = E[X(t_1)X(t_2)] \\ &= \int_{-\infty}^{\infty} \int_{-\infty}^{\infty} x_1 x_2 f_X(x_1, x_2; t_1, t_2) \mathrm{d}x_1 \mathrm{d}x_2. \end{aligned} \tag{1.7}$$

Autocovariance Function [107, 204, 301]

Definition 1.5 The autocovariance function is

$$\gamma_X(t_1,t_2)=\mathrm{cov}[X(t_1),X(t_2)]=E[(X(t_1)-\mu_X(t_1))(X(t_2)-\mu_X(t_2))]$$
$$=\int_{-\infty}^{\infty}\int_{-\infty}^{\infty}[x_1-\mu_X(t_1)][x_2-\mu_X(t_2)]\cdot f_X(x_1,x_2;t_1,t_2)\mathrm{d}x_1\mathrm{d}x_2. \tag{1.8}$$

An easy calculation shows that

$$\gamma_X(t_1,t_2)=\mathrm{cov}[X(t_1),X(t_2)]=\rho_X(t_1,t_2)-\mu_X(t_1)\mu_X(t_2). \tag{1.9}$$

Cross-Correlation Function [107, 204, 301]

Definition 1.6 Let $X(t)$ and $Y(t)$ be random processes. Their cross-correlation function is

$$\rho_{XY}(t_1,t_2)=E[X(t_1)Y(t_2)]=\int_{-\infty}^{\infty}\int_{-\infty}^{\infty}xyf_{XY}(x,y;t_1,t_2)\mathrm{d}x\mathrm{d}y. \tag{1.10}$$

Cross-Covariance Function [107, 204, 301]

Definition 1.7 Similarly, the cross-covariance function is

$$\gamma_{XY}(t_1,t_2)=\mathrm{cov}[X(t_1),Y(t_2)]=E[(X(t_1)-\mu_X(t_1))(Y(t_2)-\mu_Y(t_2))]$$
$$=\int_{-\infty}^{\infty}\int_{-\infty}^{\infty}[x-\mu_X(t_1)][y-\mu_Y(t_2)]f_{XY}(x,y;t_1,t_2)\mathrm{d}x\mathrm{d}y. \tag{1.11}$$

It is an easy exercise to show that

$$\gamma_{XY}(t_1,t_2)=\mathrm{cov}[X(t_1),Y(t_2)]=\rho_{XY}(t_1,t_2)-\mu_X(t_1)\mu_Y(t_2). \tag{1.12}$$

Moments [107, 204, 301]

Definition 1.8 The nth moment of a random variable $X(t)$ is given by

$$E[X^n(t)]=\int_{-\infty}^{\infty}x^n f_X(x,t)\mathrm{d}x. \tag{1.13}$$

Definition 1.9 The nth central moment of a random variable $X(t)$ is given by

$$E\left\{(X(t)-E[X(t)])^n\right\}=\int_{-\infty}^{\infty}(x-\mu_X)^n f_X(x,t)\mathrm{d}x. \tag{1.14}$$

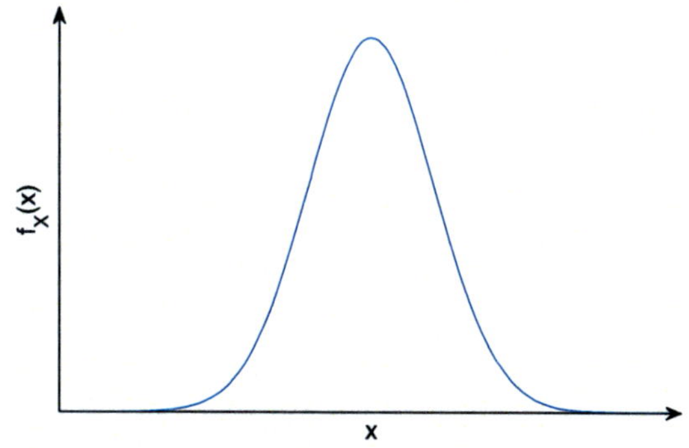

Fig. 1.1 Gaussian distribution

So, the commonly used mean and variance are simply the first-order moment and the second-order central moment, respectively.

1.2.3 Gaussian Distribution and Gaussian Processes

Definition 1.10 Gaussian distribution for a random variable $X(t)$ with mean μ_X and variance σ_X^2 is a statistic distribution with its PDF [107, 114]

$$f_X(x) = \frac{1}{\sigma_X\sqrt{2\pi}} \exp\left[-\frac{(x-\mu_X)^2}{2\sigma_X^2}\right], \quad -\infty < x < \infty. \tag{1.15}$$

Figure 1.1 illustrates the probability density function of Gaussian distribution. The standard normal distribution is given by taking $\mu_X = 0$ and $\sigma_X^2 = 1$ in (1.15). An arbitrary normal distribution can be converted to a standard normal distribution by changing variable $z = (x - \mu_X)/\sigma_X$. Then, new random variable Z has standard normal distribution

$$f_Z(x) = \frac{1}{\sqrt{2\pi}} \exp\left(-\frac{z^2}{2}\right). \tag{1.16}$$

A Gaussian process is a stochastic process for which any finite linear combination of samples will be normally distributed. Assuming that the mean is known, the entire structure of the Gaussian random process is specified once the correlation function or, equivalently, the power spectrum is known. As linear transformations of Gaussian random processes yield another Gaussian process, linear operations such as differentiation, integration, linear filtering, sampling, and summation with other Gaussian processes result in a Gaussian process, too.

1.2.4 Stationary Processes

Definition 1.11 A stochastic process $X(t)$ is called strict-sense stationary if its statistical properties are invariant to a shift of the origin [107, 204, 301].

This means that the processes $X(t)$ and $X(t+\tau)$ have the same statistics for any τ.

Definition 1.12 Two stochastic processes $X(t)$ and $Y(t)$ are called jointly stationary if the joint statistics of $X(t)$ and $Y(t)$ are the same as the joint statistics of $X(t+\tau)$ and $Y(t+\tau)$ for any τ [107, 204, 301].

Definition 1.13 A stochastic process $X(t)$ is called wide-sense stationary if its mean is constant [107, 204, 301]

$$E[X(t)] = \mu_X(t) = \mu_X(t+\tau), \quad \forall\tau \in \mathbb{R} \tag{1.17}$$

and its autocorrelation depends only on $\tau = t_1 - t_2$, that is

$$\rho_X(t, t+\tau) = E[X(t)X(t+\tau)] = \rho_X(\tau, 0). \tag{1.18}$$

1.3 Analysis of Random Signals

A random signal is simply a random function of time. Original random signals have two types: continuous random signals and discrete random signals [120, 204]. If time is continuous, such a random function could be viewed as a continuous random variable. If time is discrete, such a random function could be viewed as a sequence of random variables. Even when time is continuous, we often choose to sample continuous time waveforms in order to work with discrete time sequences rather than continuous time waveforms. In this section, some conventional analysis techniques of random signals are briefly recalled. The random signals being observed are often assumed to be i.i.d. random variables for the purpose of convenient statistical inference. The assumption (or requirement) that observations be i.i.d. simplifies the underlying mathematics of many statistical methods. The assumption is important in the classical form of the central limit theorem, which states that the probability distribution of the sum (or average) of i.i.d. variables with finite variance approaches a normal distribution [93, 278].

1.3.1 Estimation of Properties for Stochastic Signals

In practice, the statistical characteristics of the random signals are always estimated instead of being directly obtained from using the PDF.

Estimation of the Mean Value [107, 204, 301]

Given an i.i.d. random sequence X_n, the unbiased estimation of the mean value is given by

$$\widehat{\mu}_X = \frac{1}{n}\sum_{i=1}^{n} X_i. \tag{1.19}$$

Estimation of the Variance [107, 204, 301]

Given an i.i.d. random sequence X_n, an estimation of the variance is given by

$$\widehat{\sigma}_X^2 = \frac{1}{n}\sum_{i=1}^{n}(X_i - \widehat{\mu}_X)^2. \tag{1.20}$$

Estimation of the Covariance Function [107, 204, 301]

Given an i.i.d. random sequence X_n, the estimation of the covariance function γ_X is given by

$$\widehat{\gamma}_X(j) = \frac{1}{n}\sum_{i=j+1}^{n}(X_i - \widehat{\mu}_X)(X_{i-j} - \widehat{\mu}_X), \quad j = 0, \ldots, n-1. \tag{1.21}$$

Estimation of the Correlation Function [107, 204, 301]

Given an i.i.d. random sequence X_n, the estimation of the correlation function ρ_X is given by

$$\widehat{\rho}_X(j) = \frac{\widehat{\gamma}_X(j)}{\widehat{\gamma}_X(0)}, \quad j = 0, \ldots, n-1. \tag{1.22}$$

Estimation of the Cross-Covariance Function [107, 204, 301]

Given two i.i.d. random sequences X_n and Y_n, the estimation of the cross-covariance function γ_{XY} is given by

$$\widehat{\gamma}_{XY}(j) = \frac{1}{n}\sum_{i=j+1}^{n}(X_i - \widehat{\mu}_X)(Y_{i-j} - \widehat{\mu}_Y), \quad j = 0, \ldots, n-1. \tag{1.23}$$

Estimation of the Cross-Correlation [107, 204, 301]

Given two i.i.d. random sequences X_n and Y_n, the estimation of the cross-correlation function ρ_{XY} is given by

$$\widehat{\rho}_{XY}(j) = \frac{\widehat{\gamma}_{XY}(j)}{\sqrt{\widehat{\gamma}_X(0)\widehat{\gamma}_Y(0)}}, \quad j = 0, \ldots, n-1. \tag{1.24}$$

Estimation of the Moments [107, 204, 301]

Given an i.i.d. random sequences X_n, the estimation of the nth moment function ξ^n is given by

$$\widehat{\xi^n} = \frac{1}{n}\sum_{i=1}^{n} X_i^n, \tag{1.25}$$

and the nth central moment function m^n is given by

$$\widehat{m^n} = \frac{1}{n}\sum_{i=1}^{n}(X_i - \widehat{\mu}_X)^n. \tag{1.26}$$

1.3.2 Simulation of Random Signals

A simple method to generate a random signal is to perform time domain filtering on a white Gaussian noise (wGn) [114, 204]. A traditional stationary continuous random signal $Y(t)$ with a prescribed power spectral density (PSD) $S_{YY}(f)$ can be generated by filtering a white Gaussian noise using a linear time-invariant (LTI) filter. Note that the definition of PSD of a signal $X(t)$ is Fourier transform of its autocorrelation function

$$S_{XX}(f) = \int_{-\infty}^{\infty} \rho_X(\tau) e^{-j\omega\tau} \mathrm{d}\tau, \tag{1.27}$$

where $\rho_X(\tau)$ is the autocorrelation function of $X(t)$. For example, let a wGn $X(t)$ which has PSD $S_{XX}(f) = 1$, pass through an appropriately designed filter having its transfer function $H(f)$, then, the PSD of the output stochastic process is simply $S_{YY}(f) = |H(f)|^2$. Figure 1.2 illustrates the time domain filtering based simulation method of a colored random process. Other time-domain and frequency-domain random signal simulation methods can be used to simulate the random signals and more details can be found in [204].

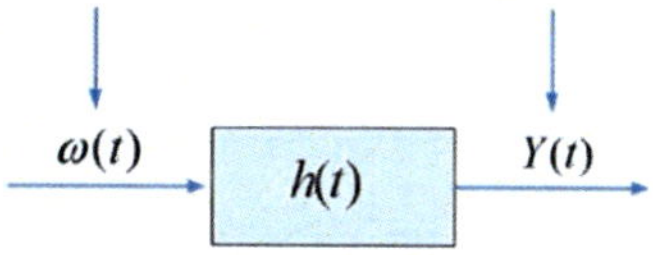

Fig. 1.2 Time domain filtering based simulation method of colored Gaussian random process

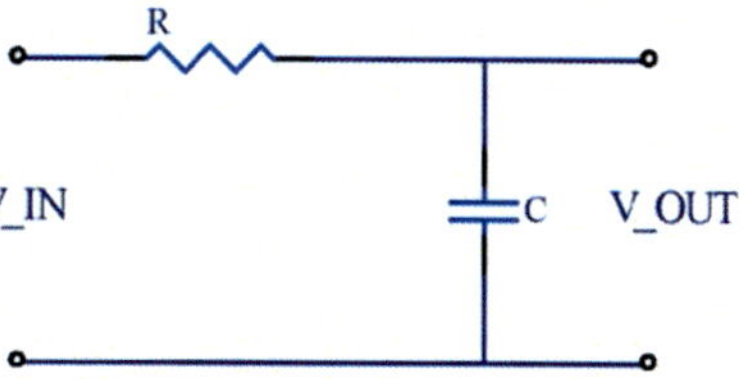

Fig. 1.3 Passive RC first-order low-pass filter

1.3.3 Signal Filtering

In signal processing, filtering technique can be used to remove unwanted contents of the signal, such as band-limited random noise, and to extract useful contents of the signal [5, 259]. Filters may be classified into analogue filters and digital filters [321].

Analogue Filters

An analogue filter is a filter which operates on continuous-time signals. In the real world, analog filters are often electronic circuits, or "analogues", of dynamic systems working in continuous time. Analogue filters have the advantages of high frequency operations without the problem of quantization, and with a smoother transition of low-level analog signals [259, 321]. There are two types of analogue filters: passive and active. Passive implementations of linear filters are based on combinations of resistors (R), inductors (L) and capacitors (C). These types of filters are collectively known as passive filters, because they do not depend upon an external power supply and/or they do not contain any active component. Figure 1.3 illustrates a simple passive RC low pass filter. The cutoff frequency of the RC low-pass filter is

$$f_c = \frac{1}{2\pi RC}. \tag{1.28}$$

The cutoff frequency is a boundary in a system's frequency response at which the energy flowing through the system begins to be reduced below 50%. The impulse response for the capacitor voltage is

$$h_c(t) = \frac{1}{RC} e^{-t/RC} \mu(t), \tag{1.29}$$

where $\mu(t)$ denotes the Heaviside unit step signal

$$\mu(t) = \begin{cases} 1, & t \geq 0 \\ 0, & t < 0. \end{cases} \tag{1.30}$$

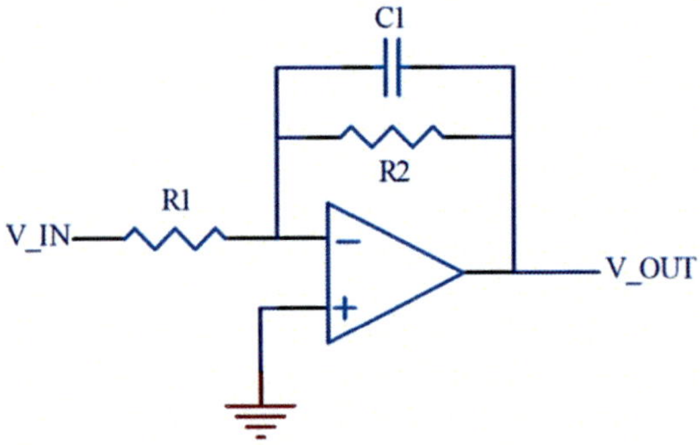

Fig. 1.4 Active first-order low-pass filter

The high pass, band pass filters can also be designed using the combination of capacitors and resistors.

Active filters are distinguished by the use of one or more active components, requiring an external power source. Active filters can be implemented using a combination of passive and active (amplifying) components [259, 321]. Operational amplifiers (OpAmps) are frequently used in active filter designs. These kinds of filters are of high quality, and can achieve resonance without the use of inductors. However, their upper frequency limit is constrained by the bandwidth of the operational amplifiers used. Figure 1.4 illustrates a first-order active low-pass filter. The cutoff frequency of the active low-pass filter is

$$f_c = \frac{1}{2\pi R_2 C_1}. \quad (1.31)$$

Digital Filter

In contrast to the analogue filter's processing of a continuous-time signal, the digital filter performs numerical calculations on a sampled, discrete-time input signal to reduce or enhance certain aspects of the input signal [5]. A digital filter can be characterized by a difference equation or by its Z-transfer function. Digital filters can be classified as either finite-duration impulse response (FIR) or infinite-duration impulse response (IIR) filters. FIR filters, characterized by a finite length impulse response, can be described by a Z-transfer function in a form of a polynomial as follows

$$H(z) = a_0 + a_1 z^{-1} + \cdots + a_m z^{-m} = \sum_{i=0}^{m} a_i z^{-i}. \quad (1.32)$$

IIR filters are characterized by Z-transfer functions in the form of a rational function or ratio of polynomials ($m \leq n$)

$$H(z) = \frac{\sum_{i=0}^{m} a_i z^{-i}}{1 + \sum_{j=1}^{n} b_j z^{-j}}. \quad (1.33)$$

Digital filters have many advantages over analogue filters [5]:

- Digital filters can be easily designed, tested, and implemented in software. To realize analogue filters, however, hardware circuits have to be designed and setup for them.
- Digital filters are programmable, so they can be easily changed without changing the hardware. But the analogue filters can only be changed by redesigning the circuit.
- Digital filters are much more versatile and flexible in their ability to process signals in a variety of ways.

1.3.4 Modeling Random Processes

In order to accurately analyze and characterize random processes, processes with the same nature may be classified together by using a generating model. Many stationary random signals can be generated by LTI systems with white noise as the input driving signal. LTI systems can be modeled by linear differential or difference equations which are described as ARMA models in the discrete case. Therefore, ARMA models are frequently used in discrete time series analysis to characterize stationary time series. An ARMA model is a combination of an autoregressive (AR) model and a moving average (MA) model. The ARMA model is given by a difference equation

$$x(n) + \sum_{i=1}^{p} a_i x(n-i) = e(n) + \sum_{j=1}^{q} b_j e(n-j), \tag{1.34}$$

where $e(n)$ is a discrete wGn. p and q are the orders of the AR-part and the MA-part of the ARMA model, respectively. $a_1, \ldots, a_p$ are the coefficients of the AR-part of the model; $b_1, \ldots, b_q$ are the coefficients of the MA-part of the model.

The generalized autoregressive conditional heteroscedasticity (GARCH) model may be used to characterize and model some observed time series. GARCH model, an extension of the ARCH model commonly used in modeling financial time series, is given by

$$x_t = \mu_t + \epsilon_t, \tag{1.35}$$

$$\epsilon_t = \sigma_t e_t, \tag{1.36}$$

$$\begin{aligned} \sigma_t^2 &= a_0 + a_1 \epsilon_{t-1}^2 + \cdots + a_q \epsilon_{t-q}^2 + b_1 \sigma_{t-1}^2 + \cdots + a_p \sigma_{t-p}^2 \\ &= a_0 + \sum_{i=1}^{q} a_i \epsilon_{t-i}^2 + \sum_{j=1}^{p} a_j \sigma_{t-j}^2, \end{aligned} \tag{1.37}$$

where μ_t represents the regression function for the conditional mean, ϵ_t denotes the error terms, e_t is a white noise with zero mean and unit variance.

1.3.5 Transform Domain Analysis

Transform-domain analysis techniques are often used to map a problem from one domain into another, so the problem may be solved more easily in the transformed domain. For example, the Fourier transform can transform the random signals from time domain to frequency domain. So, the spectrum of the random signal can be easily analyzed in the new domain.

Fourier Transform [32]

The Fourier transform can transform the random signals from time domain to frequency domain.

Definition 1.14 The definition of Fourier transform is

$$\mathcal{F}[f(t)] = \frac{1}{\sqrt{2\pi}} \int_{-\infty}^{\infty} f(t) e^{-j\omega t} \mathrm{d}t = F(\omega). \tag{1.38}$$

Definition 1.15 The inverse Fourier transform is defined as

$$f(t) = \mathcal{F}^{-1}[F(\omega)] = \frac{1}{\sqrt{2\pi}} \int_{-\infty}^{\infty} F(\omega) e^{j\omega t} \mathrm{d}\omega. \tag{1.39}$$

Definition 1.16 For discrete random signal $X(n)$, the discrete Fourier transform is

$$F_X(k) = \sum_{n=0}^{N-1} X(n) e^{-i2\pi k \frac{n}{N}}, \quad k = 0, \ldots, N-1. \tag{1.40}$$

The Fourier transform can be efficiently computed through a fast Fourier transform (FFT) which is essential for high-speed computing.

Laplace Transform [260]

The Laplace transform is very important for simplifying an LTI system model from its linear differential equation with constant coefficients to an algebraic equation which can be solved more easily.

Definition 1.17 The one-sided Laplace transform of an original function $f(t)$ of a real variable t, for $t \geq 0$, is defined by the integral (if it exists)

$$F(s) = \mathcal{L}\{f(t)\} = \int_{0}^{\infty} f(t) e^{-st} \mathrm{d}t, \tag{1.41}$$

where parameter s is a complex number $s = \sigma + i\omega$.

Definition 1.18 The inverse Laplace transform is given by the following complex integral

$$f(t) = \mathcal{L}^{-1}\{F(s)\} = \frac{1}{2\pi i} \lim_{T\to\infty} \int_{\gamma-iT}^{\gamma+iT} e^{st} F(s)\mathrm{d}s, \tag{1.42}$$

where the integration is done along the vertical line $\mathrm{Re}(s) = \gamma$ in the complex plane such that γ is greater than the real part of all singularities of the complex function $F(s)$.

Z-Transform [136]

In signal processing, the Z-transform converts a discrete time-domain signal, which is a sequence of real or complex numbers, into a complex frequency-domain representation. It can be considered as a discrete equivalent of the Laplace transform.

Definition 1.19 The Z-transform of a discrete sequence $f(k)$, $k = 1, 2, \ldots$ is defined as

$$\mathcal{Z}[f(k)] = F(z) = \sum_{k=0}^{\infty} f(k)z^{-k}. \tag{1.43}$$

Definition 1.20 For a Z-transform function $F(z)$, its inverse Z-transform is defined as

$$f(k) = \mathcal{Z}^{-1}[F(z)] = \frac{1}{j2\pi} \oint_C F(z)z^{k-1}\mathrm{d}z, \tag{1.44}$$

where C is a counterclockwise closed path encircling the origin and entirely in the region of convergence. The contour or path C must encircle all of the poles of $F(z)$.

Wavelet Transform [106]

The wavelet transform is a powerful mathematical tool for the analysis of transient, non-stationary, or time-varying phenomena.

Definition 1.21 The Morlet-Grossmann definition of the continuous wavelet transform for a signal $f(x) \in L^2(R)$ is

$$W(a, b) = \frac{1}{\sqrt{a}} \int_{-\infty}^{+\infty} f(x)\psi^*\left(\frac{x-b}{a}\right)\mathrm{d}x, \tag{1.45}$$

where z^* denotes the complex conjugate of z, $\psi^*(x)$ is the analyzing wavelet, a $(a > 0)$ is the scale parameter and b is the position parameter.

From $W(a,b)$ which is following the wavelet transform of a function $f(x)$, $f(x)$ can be recovered using the following formula

$$f(x)=\frac{1}{C_\chi}\int_0^{+\infty}\int_{-\infty}^{+\infty}\frac{1}{\sqrt{a}}W(a,b)\chi\left(\frac{x-b}{a}\right)\frac{\mathrm{d}a\,\mathrm{d}b}{a^2}, \tag{1.46}$$

where

$$C_\chi=\int_0^{+\infty}\frac{\hat{\psi}^*(\nu)\hat{\chi}(\nu)}{\nu}\mathrm{d}\nu=\int_{-\infty}^{0}\frac{\hat{\psi}^*(\nu)\hat{\chi}(\nu)}{\nu}\mathrm{d}\nu. \tag{1.47}$$

Generally speaking, $\chi(x)=\psi(x)$, but other choices can enhance certain features for some applications.

Hilbert Transform [109]

The Hilbert transform is a linear operator which takes a function $u(t)$, and produces a function $H(u)(t)$, in the same domain.

Definition 1.22 Hilbert transform of a function (or signal) $f(t)$ is given by

$$\widehat{g}(t)=\mathcal{H}[f(t)]=\frac{1}{\pi}\int_{-\infty}^{\infty}\frac{f(\tau)}{t-\tau}\mathrm{d}\tau. \tag{1.48}$$

Definition 1.23 For a Hilbert transform function $\mathcal{H}[f(t)]$, the inverse Hilbert transform is defined as

$$f(t)=\mathcal{H}^{-1}[\widehat{g}(t)]=-\frac{1}{\pi}\int_{-\infty}^{\infty}\frac{\widehat{g}(t)}{t-\tau}\mathrm{d}\tau. \tag{1.49}$$

Mellin Transform [230]

The Mellin transform, an integral transform, is closely related to the Laplace transform, the Fourier transform, and allied special functions.

Definition 1.24 The Mellin transform of a function $f(x)$ is defined as

$$\{\mathcal{M}f\}(z)=\varphi(z)=\int_0^{\infty}x^{z-1}f(x)\mathrm{d}x. \tag{1.50}$$

Definition 1.25 The inverse Mellin transform is defined as

$$\{\mathcal{M}^{-1}\varphi\}(x)=f(x)=\frac{1}{2\pi i}\int_{c-i\infty}^{c+i\infty}x^{-z}\varphi(z)\mathrm{d}z. \tag{1.51}$$

1.3.6 Other Analysis Methods

Besides the above random signal analysis techniques, there are also some other random signal analysis methods, such as continuous B-spline based interpolation, entropy analysis techniques, weak signal detection, blind source separation, signal coding and digital data compression, and so on [201, 309, 317].

1.4 Research Motivation

The observed samples of random signals were traditionally assumed to be stationary i.i.d., where the coupling between values at different time instants decreases rapidly as the distance between different time instants increases. In practical applications, however, these assumptions may not be realistic. Many signals, such as financial data, communications networks data, as well as many types of man-made noises, do not satisfy the assumptions. The non-Gaussian, LRD, and non-stationary properties in real world signals always lead to difficulties in signal analysis tasks.

1.4.1 Heavy Tailed Distributions

In signal processing, the assumption of i.i.d. is often needed to apply the central limit theorem. Meanwhile the Gaussian model is commonly used to describe various complex phenomena. However, in practical applications, many analyzed signals, such as financial data, man-made noises and communications networks data, were found to be impulsive, so they cannot be characterized well using the Gaussian model. The estimated variance of these kinds of signals is not convergent, and the density functions decay more slowly than the Gaussian density function. Compared with Gaussian noise, the α-stable model provides a much better description of the impulsive noise [253]. α-stable distribution based on the generalized central limit theorem has been introduced in Sect. 1.1. The density of the α-stable distribution with $\alpha < 2$ is heavy-tailed, and the α-stable processes exhibit sharp spikes. So the α-stable model is especially suitable for the processes without finite variance [215]. The analysis techniques for α-stable signals are based on the fractional low-order moments (FLOM). For an α-stable random variable X with $0 < \alpha < 2$,

$$E|X|^p = \infty, \quad \text{if } p \geq \alpha, \tag{1.52}$$

and

$$E|X|^p < \infty, \quad \text{if } p < \alpha. \tag{1.53}$$

When $0 < \alpha \leq 1$, α-stable processes have no finite first or higher-order moments; when $1 < \alpha < 2$, α-stable processes have the first-order moment and all the FLOMs. The infinite second-order moment of the stable processes with $0 < \alpha < 2$ makes the techniques based on second-order and higher-order moments questionable if not meaningless.

For a non-Gaussian α-stable random variable X with zero location parameter and dispersion γ, the probability P has the following property

$$\lim_{t\to\infty} t^{\alpha} P(|X| > t) = \gamma C(\alpha), \tag{1.54}$$

where $C(\alpha)$ is a positive constant dependent on α. Clearly, stable distributions have inverse power-law tails.

Because the analytical results of linear fractional-order differential equations are represented by Mittag-Leffler functions, which exhibit a power-law asymptotic behavior [122, 140, 255], many efforts have been devoted to investigating the relationship between fractional calculus and α-stable distributions. A relationship between stable distributions in probability theory and the fractional integral was studied in [281]. The relationship between time-fractional and space-fractional diffusion equations and α-stable distributions were studied in [187]. For the time-fractional diffusion equation below

$$\frac{\partial^{\alpha}\mu}{\partial t^{\alpha}} = \gamma \frac{\partial^{2}\mu}{\partial x^{2}}, \quad \mu = \mu(x,t;\alpha), 0 < \alpha < 2, \tag{1.55}$$

where γ is a positive constant, its Green function is a one-sided stable probability density function. For the symmetric space-fractional diffusion equation

$$\frac{\partial\mu}{\partial t} = \gamma \frac{\partial^{\alpha}\mu}{\partial x^{\alpha}}, \quad \mu = \mu(x,t;\alpha), 0 < \alpha < 2, \tag{1.56}$$

where γ is a positive constant, the Fourier transform of its Green function corresponds to the canonic form of a SαS distribution [215].

The relationship between fractional calculus and FLOM was investigated in [71, 72]. For Fourier pair $p(x)$ and $\varphi(\mu)$, complex FLOM can have complex fractional lower orders [71, 72], and

$$\left(D^{\gamma}\varphi\right)(0) = -E\left[|X|^{\gamma}\right], \tag{1.57}$$

$$\left(D^{-\gamma}\varphi\right)(0) = E\left[|X|^{-\gamma}\right], \tag{1.58}$$

where $\gamma \in \mathbb{C}$, $\mathrm{Re}\,\gamma > 0$.

1.4.2 Long Range Dependence

In time series analysis, another traditional assumption is that the coupling between values at different time instants decreases rapidly as the time difference or distance increases. Based on this assumption, many random process models were built, such as ARMA model and ARCH model. Similar to the i.i.d. assumption of the random signals, in practice, algebraic decay of the autocovariance are often observed in many time series.

The above mentioned strong coupling leads to some different features compared to those due to weak dependence, since the dependence between distant observations in a time series can no longer be neglected. The statistical properties of a long memory process can be quite different from those of a set of i.i.d. observations [22]. The correlations function $\rho(n)$ of the stationary short range dependent stochastic models, such as the ARMA processes and Markov processes, is absolutely summable, that is

$$\sum_{n=0}^{\infty} |\rho(n)| = Constant < \infty. \tag{1.59}$$

However, for the processes with long-range dependence, the correlations function $\rho(n)$ is not absolutely summable, i.e.,

$$\sum_{n=0}^{\infty} |\rho(n)| = \infty. \tag{1.60}$$

Besides, the familiar variability properties of sample averages of i.i.d. observations are far from valid for a long memory process [22]. The variance of the sample mean of sample averages for n i.i.d. observations is

$$\mathrm{var}(\overline{X}) = \frac{\sigma^2}{n}. \tag{1.61}$$

However, when the observations are correlated, $\sum_{i\neq j} \rho(i,j) \neq 0$, and the variance of the sample mean of sample averages is

$$\mathrm{var}(\overline{X}) = \frac{\sigma^2}{n}\left[1 + \frac{\sum_{i\neq j} \rho(i,j)}{n}\right]. \tag{1.62}$$

Therefore, the short-range dependent model, such as ARMA and ARCH models cannot be used to characterize these LRD processes with power-law decaying correlations.

Signals with long-range correlations, which are characterized by power-law decaying autocorrelation function, occur ubiquitously in nature and many man-made systems. Because of the strong coupling and the slow decaying autocorrelation, these processes are also said to be long memory processes. Some self-similar processes may exhibit long-range dependence. Typical examples of LRD signals include financial time series, electronic device noises, electroencephalography (EEG) signal, etc. The level of the dependence or coupling of LRD processes can be indicated or measured by the estimated Hurst parameter $H \in (0, 1)$ [22]. If $0 < H < 0.5$, the time series is a negatively correlated process, or an anti-persistent process. If $0.5 < H < 1$, the time series is a positively correlated process. If $H = 0.5$, the time series has no statistical dependence.

Similar to the α-stable distribution, the LRD processes are also closely related to fractional calculus. An LRD process having its autocorrelation function with power-law decay can actually be realized by passing short-range dependent signals through a constant-order fractional-order differential system [221, 271]. In order to capture the property of coupling or power-law decaying autocorrelation, fractional calculus based LRD models have been suggested, such as FARIMA and FIGARCH models [110, 166, 266].

1.4.3 Local Memory

Time-varying long memory characteristics have been observed in telecommunications networks, physiological signals, seismic measurements, etc. [320]. Neither the short-range dependent models, nor the long memory models can capture the local scaling characteristic and the time-varying nature of the non-stationary local memory processes. As a suitable analysis model for non-stationary local memory processes, a multifractional processes model was proposed in [232]. In this generalized local memory model, the constant Hurst parameter H was extended to the case where H is indexed by a time-dependent local Hölder exponent $H(t)$. The extension of the long memory process leads to non-stationary and non-self-similar new stochastic processes, which can capture the complex behavior of non-stationary, nonlinear dynamic systems. Two typical examples of local memory time processes are multifractional Gaussian noise (mGn) and multifractional Brownian motion (mBm) [67].

Different from the LRD processes, which can be generated by constant-order fractional systems, local memory processes cannot be synthesized by constant-order fractional systems. Instead, variable-order fractional calculus should be adopted as a suitable analysis technique. It has been studied in [271] that multifractional processes can be synthesized with the help of the variable-order fractional operator by studying the relationships of wGn, mGn and mBm. Therefore, variable-order fractional calculus is the basic tool for understanding and analyzing variable-order fractional processes characterized by local power-law decay of autocorrelations.

The properties of heavy-tailed distribution, LRD, or local memory are not mutually exclusive. Sometimes, the joint presence of heavy-tailed distribution and long memory, or the joint presence of heavy-tailed distribution and local memory can be found in financial data, communications networks data, and some man-made signals. It is difficult to study these fractional processes with those joint power-law properties. Effective advanced techniques should be developed to analyze these fractional processes. In the next section, some basic FOSP techniques will be introduced.

1.5 Basics of Fractional-Order Signal Processing

The monograph focuses on fractional processes and their analysis techniques. Fractional processes are characterized by heavy-tailed distribution, power-law decay of autocorrelation, or local memory. These properties i.e., heavy-tailed distribution, long-range dependence, and local memory always lead to certain trouble in correctly obtaining the statistical characteristics and extracting desired information from these fractional processes. Therefore, these properties cannot be neglected in time series analysis, because the tail thickness of the distribution and the strong coupling of the time series are critical in characterizing the essence of the resulting natural or man-made phenomenon of the time series. Suitable analysis techniques are needed to study these fractional processes. It has been introduced in the previous sections that the basic theory of the heavy-tailed distribution is α-stable distribution. Fractional calculus, on the other hand, including the constant-order and variable-order fractional calculus, offers a unified basic tool for both long memory and local memory process analysis. Moreover, FrFT based techniques also provide some valuable analysis methods for long memory processes [60]. All FOSP techniques in this monograph are based on these basic theories, i.e., α-stable distribution, fractional calculus and FrFT, which offer many powerful tools, such as fractional filter, simulation of fractional processes, and fractional modeling, to analyze fractional processes in many fields of science and engineering. In this section, some basics of fractional-order signal processing, mainly, fractional calculus, α-stable distribution and FrFT, are introduced.

1.5.1 Fractional Calculus

Fractional calculus is a mathematical discipline which deals with derivatives and integrals of arbitrary real or complex orders [139, 203, 218, 237, 252]. It was proposed more than 300 years ago, and the theory was developed mainly in the 19th century. Several books [139, 203, 218, 237, 252] provide a good source of references on fractional calculus. It has been shown that there are a growing number of physical systems whose behavior can be compactly described using fractional-order systems (or systems containing fractional derivatives and integrals) theory [111]. Moreover, fractional calculus is being applied in almost every current line of inquiry into control theory and its applications [139, 207, 226, 237, 238, 250].

Constant-Order Fractional Calculus

The fractional-order integral of the integrable function $f(t)$ with $\alpha \in \mathbb{R}_+$ is defined as [237]

$$ {}_aD_t^{-\alpha} f(t) = \frac{1}{\Gamma(\alpha)} \int_a^t \frac{f(\tau)}{(t-\tau)^{1-\alpha}} \mathrm{d}\tau, \tag{1.63} $$

where Γ is the Gamma function, ${}_aD_t^{-\alpha}$ is the fractional integral of order α in $[a,t]$. The αth Riemann-Liouville fractional-order derivative of function $f(t)$ is defined by

$$ {}_aD_t^{\alpha} f(t) = \frac{1}{\Gamma(n-\alpha)} \left(\frac{\mathrm{d}}{\mathrm{d}t}\right)^n \int_a^t (t-\tau)^{n-\alpha-1} f(\tau)\mathrm{d}\tau, \tag{1.64} $$

where $n=[\alpha]+1$ and $[\alpha]$ denotes the integer part of α. The Caputo fractional-order derivative of order α of $f(t)$ is defined by [237]

$$ {}_a^C D_t^{\alpha} f(t) = \frac{1}{\Gamma(n-\alpha)} \int_a^t (t-\tau)^{n-\alpha-1} f^{(n)}(\tau)\mathrm{d}\tau, \tag{1.65} $$

where $n=[\alpha]+1$.

Distributed-Order Fractional Calculus

Caputo proposed the idea of the distributed-order differential equation [45], and discussed the solution of differential equations of fractional-order when the fractional-order derivatives are integrated with respect to the order of differentiation [46]. Lorenzo and Hartley explored the two approaches of distributed-order operators: direct approach and independent variable approach [180]. In brief, the distributed-order fractional calculus deals with the following integral form

$$ \int_a^b A(\alpha)_0 D_t^{\alpha} f(t)\mathrm{d}\alpha, \tag{1.66} $$

where $A(\alpha)$ is the order-weighting distribution and a and b are constants.

Variable-Order Fractional Calculus

Fractional integration and differentiation of variable-order were studied by Samko [251]. Variable-order fractional operators were suggested and the behaviors of the operators were also studied by Lorenzo and Hartley [180]. The subject of variable-order fractional operator has attracted more and more attention during the past decades [68, 128, 129, 290]. There are different ways of defining a variable-order differential operator. The definition of variable-order integration based on the Riemann-Liouville fractional integral is defined as

$$ \begin{aligned} {}_cD_t^{-q(t)} f(t) = {} & \frac{1}{\Gamma(q(t))} \int_c^t (t-\tau)^{q(t)-1} f(\tau)\mathrm{d}\tau \\ & + \psi(f, -q(t), a, c, t), \quad t > c, \end{aligned} \tag{1.67} $$

where $\psi(f,-q(t),a,c,t)$ is the so-called "initialization function" and $0 \le q(t) < 1$. Another important definition of the variable-order derivative operators based on

Caputo's differential operator defined by Coimbra is [68]

$$D_t^{\alpha(t)} f(t) = \frac{1}{\Gamma(1-\alpha(t))} \int_{0+}^{t} \frac{f'(\tau)\mathrm{d}\tau}{(t-\tau)^{\alpha(t)}} + \frac{(f(0^+) - f(0^-))t^{-\alpha(t)}}{\Gamma(1-\alpha(t))}, \quad 0 < \alpha(t) < 1. \tag{1.68}$$

Based on the definitions of variable-order differential operators, many complex dynamic systems with complex memory behaviors can be more properly described by variable-order differential equations.

1.5.2 α-Stable Distribution

Non-Gaussian signals and noises tend to produce large-amplitude fluctuations from the average value more frequently than Gaussian ones do. Non-Gaussian signals and noises are more likely to exhibit sharp spikes or occasional bursts of outlying observations than one would expect from normally distributed signals. Underwater acoustic signals, low-frequency atmospheric noise, and many types of man-made noises have all been found to belong to this class. Stable distributions provide a useful theoretical tool for this type of signals and noises [215].

A univariate distribution function $F(x)$ is stable if and only if its characteristic function has the form

$$\varphi(t) = \exp\{jat - \gamma |t|^{\alpha} [1 + j\beta \operatorname{sign}(t)\omega(t,\alpha)]\}, \tag{1.69}$$

where

$$\omega(t,\alpha) = \begin{cases} \tan\frac{\alpha\pi}{2}, & \text{if } \alpha \neq 1 \\ \frac{2}{\pi}\log|t|, & \text{if } \alpha = 1, \end{cases} \tag{1.70}$$

$$\operatorname{sign}(t) = \begin{cases} 1, & \text{if } t > 0 \\ 0, & \text{if } t = 0 \\ -1, & \text{if } t < 0, \end{cases} \tag{1.71}$$

and

$$-\infty < a < \infty, \quad \gamma > 0, \quad 0 < \alpha \leq 2, \quad -1 \leq \beta \leq 1. \tag{1.72}$$

An α stable characteristic function (or distribution) is determined by four parameters: α, a, β *and* γ. α is called the characteristic exponent. A small value of α will imply considerable probability mass in the tails of the distribution. $\alpha = 2$ corresponds to the Gaussian distribution (for any β). γ is a scaling parameter called the dispersion. It is similar to the variance of the Gaussian distribution. β is a symmetry parameter. $\beta = 0$ indicates a distribution symmetric about a. In this case, the distribution is called symmetric α-stable ($S\alpha S$). a is a location parameter [215]. For $S\alpha S$ distribution, a is the mean when $1 < \alpha \leq 2$ and the median when $0 < \alpha < 1$.

Fig. 1.5 α-stable probability density functions

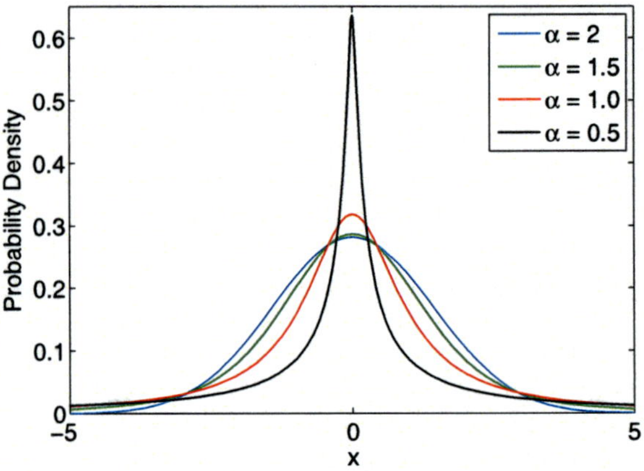

(a) Symmetric α-stable densities, $\beta = 0$, $\gamma = 1$, $a = 0$

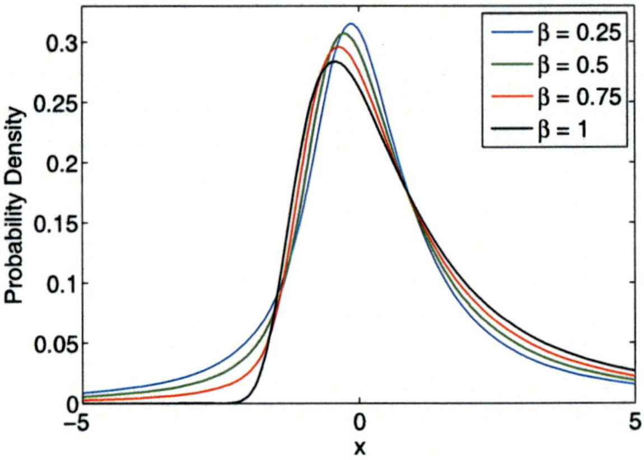

(b) Skewed α-stable densities, $\alpha = 1$, $\gamma = 1$, $a = 0$

Figure 1.5 shows the influence of the parameters on α-stable probability density functions (PDFs). Figure 1.5(a) displays a set of symmetric α-stable densities with different characteristic exponent parameters α, and Fig. 1.5(b) shows a set of skewed α-stable densities with different symmetry parameters β.

1.5.3 Fractional Fourier Transform

FrFT was first introduced by Namias in 1980 [32]. It is the generalization of the classical Fourier transform. Nowadays, FrFT has become one of the most valuable and frequently used techniques in time-varying signal processing and analysis [262]. FrFT is flexible in application and can be computed in about the same time as the ordinary Fourier transform. FrFT has wide applications in the areas of optical beam

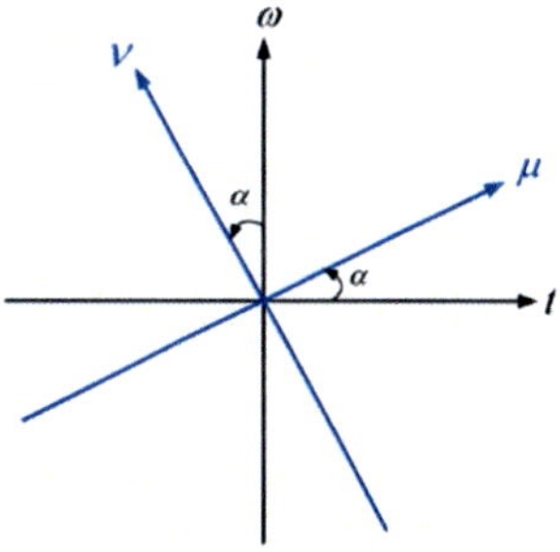

Fig. 1.6 Rotation in the time-frequency plane

propagation, statistical optics, optical system design and optical signal processing, signal detectors and signal and image recovery [32].

FrFT, the generalization of the conventional Fourier transform, has significant practical value due to the wide application of the classical Fourier transform and frequency domain concepts. The continuous FrFT of a function $x(t)$ is given as

$$X_a(\mu) = F^a[x(t)](\mu) = \int_{-\infty}^{\infty} K_p(\mu, t)x(t)\mathrm{d}t, \quad \alpha = a\pi/2, \tag{1.73}$$

where the transform kernel is given by

$$K_p(\mu, t) = \begin{cases} A_\alpha e^{j((\mu^2+t^2)/2)\cot\alpha - j\mu t\csc\alpha}, & \alpha \neq k\pi, \\ \delta(\mu - t), & \alpha = 2k\pi, \\ \delta(\mu + t), & \alpha = (2k+1)\pi, \end{cases}$$

where $\delta(\cdot)$ is the Dirac δ function, and

$$A_\alpha = \sqrt{(1 - j\cot\alpha)/2\pi}. \tag{1.74}$$

Obviously, when $a = 1$ ($\alpha = \pi/2$), X_a is the conventional Fourier transform.

An important property of FrFT is that it can be interpreted as a rotation in the time-frequency plane with an angle α (Fig. 1.6), which offers unique advantages in filtering; signal recovery, reconstruction, and synthesis; image recovery, restoration and enhancement; pattern recognition and so on [32, 262].

The discrete FrFT (DFrFT) has been studied by many researchers [41, 87, 257]. A definition of DFrFT is [87]

$$\begin{aligned} X_\alpha(\mu) &= \tilde{F}^\alpha\left[x(nT)\right](\mu) \\ &= A_\alpha e^{(j\mu^2/2)}\cot\alpha \sum_{n=-\infty}^{\infty} x(nT)e^{j(1/2)\cot\alpha n^2T^2 - jn\mu T\csc\alpha}, \end{aligned} \tag{1.75}$$

where T is the sampling period and other parameters are similar to those in (1.73).

1.6 Brief Summary of Contributions of the Monograph

This monograph introduces some typical fractional processes and fractional signal processing techniques in detail. Some real world applications of FOSP techniques are also presented. The main contributions of the monograph are briefly summarized as follows:

- A summary of the fractional processes and multifractional processes was presented in a systematic way.
- The fractional-order signal processing techniques were summarized in the monograph.
- The relationship between wGn, fGn and fBm, and the relationship between wGn, mGn and mBm were both investigated.
- Numerical synthesis methods of mGn and multifractional stable noises were presented.
- The distributed-order integrator and differentiator, distributed-order low-pass filter, and distributed parameter low-pass filter were studied.
- Fractor based analogue realization of fractional-order integrator and differentiator, and the realization of variable-order fractional integrator and differentiator were presented.
- Robustness analysis of twelve Hurst estimators for constant-order fractional processes, and the tracking performance and the robustness of local Hölder estimators for multifractional processes were studied in detail.
- Several real world application examples of fractional-order signal processing techniques were provided.
- Impulse response invariant discretization of the constant-order fractional filters and the distributed-order fractional filters were presented.
- The validity of applying numerical inverse Laplace transform algorithms in fractional calculus was investigated.

The MATLAB®/Simulink® codes resulted from the above original research efforts are available from MatlabCentral of the MathWorks Inc. website. With these codes, all results in this monograph are reproducible with minimum efforts.

1.7 Structure of the Monograph

Chapter 2 is a collection of several basic concepts and results which will be employed throughout the monograph. It establishes notations and consolidates several concepts of fractional processes and FOSP techniques. All the concepts in this chapter are based on the theory of fractional calculus, α-stable distribution and FrFT briefly introduced in Chap. 1. This chapter aims to provide a relatively broad conceptual framework of the monograph.

Chapter 3 is an introduction to the basic concepts of constant-order fractional processes, including fBm, fGn, fractional stable motion, and fractional stable noise. Another critical part of this chapter is the accuracy and robustness analysis of some existing Hurst parameter estimators. The analysis results provide guidance for properly choosing Hurst estimators for constant-order fractional processes. FOSP applications in Chaps. 8, 9, 11, 12 are all based on the results in this chapter. Chapter 4 may be considered to be an extension of Chap. 3. It deals with the so-called multifractional processes or variable-order fractional processes, which are based on generalizing the constant Hurst parameter H to the case where H is indexed by a time-dependent local Hölder exponent $H(t)$. The tracking performance and robustness of the local Hölder exponent for multifractional processes are studied in this chapter. FOSP applications in Chaps. 9, 11, 12 are all based on the results in Chap. 4.

Chapters 5, 6 and 7 deal with the details of FOSP techniques. Chapter 5 introduces the constant-order fractional signal processing methods to analyze random signals discussed in Chap. 3. Chapter 6 deals with multifractional signal processing methods to analyze random signals studied in Chap. 4. Chapter 7 studies some types of distributed-order fractional filters. Chapters 5, 6 and 7 are necessary for understanding the later chapters.

Chapters 8, 9, 10, 11, 12 and 13 provide several application examples of FOSP techniques in geography, control and biomedical signals. Each of these chapters can be read before or after others. These chapters provide several examples on how to generalize the conventional signal processing methods to FOSP techniques, and how to obtain more valuable information by using these techniques. We hope that these interesting application examples can stimulate the development of new applications by the readers.

Chapter 2
An Overview of Fractional Processes and Fractional-Order Signal Processing Techniques

2.1 Fractional Processes

In this monograph, the term fractional processes refers to the following random processes:

- Random processes with long range dependence (LRD);
- Multifractional processes which exhibit local memory/locally self-similar property;
- Random processes with heavy-tailed distributions;
- Random processes which exhibit both LRD and heavy-tailed distribution properties;
- Random processes which exhibit both local memory and heavy-tailed distribution properties.

It is known that a conventional (integer-order) random signal can be considered as the solution of an integer-order differential equation with the white noise as the input excitation. From the perspective of "signals and systems", a conventional (integer-order) random signal can be regarded as the output of an integer-order differential system or integer-order filter with the white noise as the input signal [114, 204]. Similarly, other studies show in [164, 221, 271] that the fractional signals can be taken as the solutions of constant-order fractional or variable-order fractional differential equations. Therefore, fractional signals can be synthesized by constant-order fractional systems, or variable-order fractional systems with a wGn or a white stable noise as the input signal, where the white stable noise is a cylindrical Wiener processes on Hilbert spaces subordinated by a stable process [38, 121]. In this chapter, fractional processes and FOSP techniques are introduced from the perspective of fractional signals and fractional-order systems.

H. Sheng et al., *Fractional Processes and Fractional-Order Signal Processing*, Signals and Communication Technology,
DOI 10.1007/978-1-4471-2233-3_2, © Springer-Verlag London Limited 2012

2.1.1 Fractional Processes and Fractional-Order Systems

Review of Conventional Random Processes and Integer-Order Systems

A continuous-time LTI (linear time invariant) system can be represented by an integer-order ordinary differential equation in the general form [114, 204]

$$\sum_{j=0}^{N} a_j y^{(j)}(t) = \sum_{i=0}^{M} b_i f^{(i)}(t), \tag{2.1}$$

where $f(t)$ is the input signal, and $y(t)$ is the output signal of the LTI system with proper initial conditions and $N \geq M$. The transfer function of the continuous LTI system under zero initial conditions is

$$H(s) = \frac{\sum_{i=0}^{M} b_i s^i}{\sum_{j=0}^{N} a_j s^j}. \tag{2.2}$$

The output signal $y(t)$ of the LTI system (2.1) can be written as

$$y(t) = \int_0^t h(t-\tau) f(\tau) \mathrm{d}\tau, \tag{2.3}$$

under a zero state condition, where $h(t)$ is the impulse response of the LTI system. (2.3) is also called "zero-state response" of (2.1) under input or driving signal $f(t)$. In this monograph, all responses are in the sense of "zero-state response" unless otherwise indicated. A traditional stationary continuous random signal can be expressed as the output of an LTI system with wGn (white Gaussian noise) as the driving input signal,

$$y(t) = \int_0^t h(t-\tau) \omega(\tau) \mathrm{d}\tau, \tag{2.4}$$

where $\omega(t)$ is wGn, $h(t)$ is the inverse Laplace transform of transfer function $H(s)$, that is $h(t) = \mathcal{L}^{-1}[H(s)]$. In the same way, a stationary stable continuous random signal with heavy-tailed distribution can be considered as the output of an LTI system with white stable noise as the input

$$y(t) = \int_0^t h(t-\tau) \omega_\alpha(\tau) \mathrm{d}\tau, \tag{2.5}$$

where $\omega_\alpha(t)$ is a white stable noise, which will be introduced in Chap. 3.

A linear discrete time-invariant (LDTI) system can be represented by a difference equation of the following general form [114, 204]

$$\sum_{j=0}^{N} a_j y(n-j) = \sum_{i=0}^{M} b_i f(n-i), \tag{2.6}$$

where $f(n)$ is the input sequence, and the $y(n)$ is the output sequence of the LDTI system with $m \le n$. The Z-transfer function of the LDTI system is

$$H(z) = \frac{\sum_{i=0}^{M} b_i z^{-i}}{\sum_{j=0}^{N} a_j z^{-j}}. \tag{2.7}$$

A traditional stationary discrete random signal can be expressed as the output of an LDTI system with the discrete wGn as the input,

$$y(n) = \omega(n) * h(n), \tag{2.8}$$

where $\omega(n)$ is a discrete wGn, '$*$' is the convolution, and $h(n)$ is the inverse Z-transform of $H(z)$, that is $h(n) = \mathcal{Z}^{-1}[H(z)]$.

Similarly, a stationary stable discrete random signal with heavy-tailed distribution can be considered as the output of a discrete LTI system with discrete white stable noise as the input,

$$y(n) = \omega_\alpha(n) * h(n), \tag{2.9}$$

where $\omega_\alpha(n)$ is the discrete white stable noise [215, 253].

Constant-Order Fractional Processes and Constant-Order Fractional Systems

Similar to the integer-order continuous-time LTI system, a constant-order fractional linear continuous time-invariant (FLTI) system can be described by a fractional-order differential equation of the general form [164, 221]

$$\sum_{j=0}^{N} a_j D^{\nu_j} y(t) = \sum_{i=0}^{M} b_i D^{\mu_i} f(t), \tag{2.10}$$

where $f(t)$ is the input, $y(t)$ is the output of the FLTI system, and D^α denotes the fractional derivative of order α. The transfer function of the continuous FLTI system under zero initial conditions is [164, 221]

$$H(s) = \frac{\sum_{i=0}^{M} b_i s^{\nu_i}}{\sum_{j=0}^{N} a_j s^{\mu_j}}, \quad \mathrm{Re}(s) > 0. \tag{2.11}$$

The output $y(t)$ of an FLTI system can also be described as

$$y(t) = \int_0^t h(t-\tau) f(\tau) \mathrm{d}\tau, \tag{2.12}$$

where $h(t)$ is the impulse response of the FLTI system (2.11), and $f(t)$ is the input. A constant-order fractional stationary continuous random signal can be regarded as the output of an FLTI system with wGn as the input,

$$y(t) = \int_0^t h(t-\tau)\omega(\tau)\mathrm{d}\tau, \tag{2.13}$$

where $\omega(t)$ is the wGn, $h(t)$ is the inverse Laplace transform of $H(s)$ in (2.11). In the same way, a constant-order fractional stable continuous random signal can be considered as the output of an FLTI system with the white stable noise as the input,

$$y(t) = \int_0^t h(t-\tau)\omega_\alpha(\tau)\mathrm{d}\tau, \tag{2.14}$$

where $\omega_\alpha(t)$ is the white stable noise.

Similar to the LDTI system, a constant-order fractional linear discrete time-invariant (FLDTI) system can be represented by a constant-order fractional difference equation with the general form [164, 220]

$$\sum_{j=0}^{N} a_j D^{\nu_j} y(n) = \sum_{i=0}^{M} b_i D^{\mu_i} f(n), \tag{2.15}$$

where $f(n)$ is the input, $y(n)$ is the output of the FLDTI system, and D^α denotes the fractional difference operator (delay) of order α, that is $D^\alpha y(n) = y(n-\alpha)$. The transfer function of the FLDTI system is [164, 220]

$$H(z) = \frac{\sum_{i=0}^{M} b_i z^{-\nu_i}}{\sum_{j=0}^{N} a_j z^{-\mu_j}}, \quad |z| = 1. \tag{2.16}$$

A constant-order fractional discrete random signal can be considered as the output of an FLDTI system with discrete wGn as the input,

$$y(n) = \omega(n) * h(n), \tag{2.17}$$

where $\omega(n)$ is the discrete wGn, '$*$' is the convolution, and $h(n)$ is the inverse Z-transform of $H(z)$. A constant-order fractional stable discrete random signal can be considered as the output of a discrete FLDTI system with discrete white stable noise as the input,

$$y(n) = \omega_\alpha(n) * h(n), \tag{2.18}$$

where $\omega_\alpha(n)$ is the discrete white stable noise.

Compared with the constant-order fractional processes, the distributed-order fractional processes and multifractional processes are more complex. Distributed-order fractional processes can be considered as the output of the combination of the constant-order fractional-order systems [180]. Multifractional processes can be considered as the output of a variable-order fractional system which can be represented by a variable-order fractional differential equation. Different from the constant-order fractional systems which can be simply described by transfer functions, the variable-order fractional systems cannot be simply expressed using the Laplace transform, because it is difficult to calculate the Laplace transformation of variable-order fractional differential equations. Variable-order fractional processes will be discussed in Chap. 4.

2.1.2 Stable Processes

Definition 2.1 A random variable X is stable or stable in the broad sense, if for X_1 and X_2 independent copies of X and any positive constants a and b,

$$aX_1 + bX_2 \stackrel{d}{=} cX + d, \tag{2.19}$$

for some positive c and $d \in \mathbb{R}$. The random variable is strictly stable or stable in the narrow sense if (2.19) holds with $d = 0$, for all choices of a and b.

A random variable is symmetric stable if it is stable and symmetrically distributed around 0, e.g. $X \stackrel{d}{=} -X$. Here $\stackrel{d}{=}$ means the equivalence in distribution.

Definition 2.2 A real random variable X is $S\alpha S$, if its characteristic function is of the form

$$\varphi(t) = \exp\{jat - \gamma|t|^\alpha\}, \tag{2.20}$$

where $0 < \alpha \le 2$ is the characteristic exponent, $\gamma > 0$ the dispersion, and $-\infty < a < \infty$ the location parameter.

When $\alpha = 2$, X is Gaussian.

The problem of estimating the parameters of an α-stable distribution is difficult, because majority of the stable family lacks any known closed-form density functions. Since most of the conventional methods in mathematical statistics depend on an explicit form for the density function, these methods cannot be used in estimating the parameters of the α-stable distributions. Fortunately, some numerical methods can be used in the literature for the parameter estimation of symmetric α-stable distributions [215]. The most frequently used method for estimating the parameters of the $S\alpha S$ law with $1 \le \alpha \le 2$ is suggested in [92]. Let $F(\cdot)$ be a distribution function. Then, its f fractile x_f is defined by

$$F(x_f) = f, \tag{2.21}$$

where f is restricted to be $0 < f < 1$. The order statistics $X_{(1)}, \dots, X_{(N)}$ of a random sequence $X_1, \dots, X_N$ satisfy $X_{(1)} \le \cdots \le X_{(N)}$.

Let $X_1, \dots, X_N$ be a random sample sequence from an unknown distribution $F(x)$, whose order statistics are $X_{(1)}, \dots, X_{(N)}$. Specifically, assuming that $0 \le i \le$

N and $\frac{2i-1}{2N} \le f < \frac{2i+1}{2N}$, then

$$\widehat{x}_f = X_{(i)} + [X_{(i+1)} - X_{(i)}]\frac{f - q(i)}{q(i+1) - q(i)}, \tag{2.22}$$

where

$$q(i) = \frac{2i-1}{2N}. \tag{2.23}$$

If $i = 0$ or $i = N$, then $\widehat{x}_f = X(1)$ and $\widehat{x}_f = X(N)$, respectively.

McCulloch generalized the above method to provide consistent estimates for α and c [199]. He also eliminated the asymptotic bias in the Fama-Roll estimators of α and c. Specifically, for the symmetric stable law, the fractile estimate $\widehat{\upsilon}_\alpha$ is that

$$\widehat{\upsilon}_\alpha = \frac{\widehat{x}_{0.95} - \widehat{x}_{0.05}}{\widehat{x}_{0.75} - \widehat{x}_{0.25}}. \tag{2.24}$$

Thus, a consistent estimate $\widehat{\alpha}$ can be found by searching tables in [199], with a matched value of $\widehat{\upsilon}_\alpha$. For fixed α, the following quantity

$$\upsilon_c = \frac{\widehat{x}_{0.75} - \widehat{x}_{0.25}}{c}, \tag{2.25}$$

is independent of α. $\widehat{x}_{0.75}$ and $\widehat{x}_{0.25}$ are all consistent estimators, with the following parameter a consistent estimator of c

$$\widehat{c} = \frac{\widehat{x}_{0.75} - \widehat{x}_{0.25}}{\upsilon_c(\widehat{\alpha})}. \tag{2.26}$$

McCulloch's method provides consistent estimators for all four parameters, with $-1 \le \beta \le 1$ and $\alpha \ge 0.6$ [199].

2.1.3 Fractional Brownian Motion

The definition of 'one-sided' fBm based on the Riemann-Liouville fractional integral, was introduced in [20].

Definition 2.3 The 'one-sided' fBm is defined as

$$B_H(t) = \frac{1}{\Gamma(H + 1/2)} \int_0^t (t - \tau)^{H-1/2} \omega(\tau) \mathrm{d}\tau, \quad 1/2 < H < 1, \tag{2.27}$$

where $\omega(t)$ is wGn.

According to the definition of Riemann-Liouville fractional integral, the fBm can be considered as the $(\alpha + 1)$th integration of wGn.

$$B_H(t) = {}_0D_t^{-1-\alpha} \omega(t). \tag{2.28}$$

So, from the perspective of fractional signals and fractional-order systems, fBm can be generated by $(\alpha + 1)$th integrator with wGn as the input. Besides the above 'one-sided' fBm definition, another frequently used stochastic integral form definition of fBm with index H $(0 < H < 1)$ [144, 193] will be introduced in Chap. 3.

The index H is the Hurst parameter which determines the type of fBm. When $H = 0.5$, fBm is the conventional Brownian motion; when $H > 0.5$ the increments of the fBm process are positively correlated [22].

2.1.4 Fractional Gaussian Noise

fGn is the derivative of fBm [193]. So, the fGn can be expressed as the αth order integration of wGn

$$Y_H(t) = {}_0D_t^{-\alpha}\omega(t), \tag{2.29}$$

where $\omega(t)$ is the wGn. The Hurst parameter of fGn is related to α by $H = 1/2 + \alpha$. Therefore, from the perspective of fractional signals and fractional-order systems the fGn can be simulated by the αth integrator with wGn as the input.

fGn has some distinctive properties. The power spectrum of fGn has an inverse power-law form, and the autocorrelation function of fGn has the power-law decay. Different from the i.i.d. random signals characterized by mean, variance or other high-order statistic properties, fGn is mainly characterized by the Hurst parameter (Hurst exponent) $H \in (0, 1)$, which was named after the hydrologist Hurst who pioneered the field of research in the fifties [123]. There are a number of practical methods which can be used to estimate the Hurst parameter. The best known Hurst exponent estimator is the Rescaled Range method (R/S), which was first proposed by Hurst in the hydrological context. A variety of other estimation techniques exist, such as the Aggregated Variance method [22], the Absolute Value method [297], the Periodogram method [97], the fractional Fourier transform (FrFT) based method [60], Koutsoyiannis' method [153], and so on. A comprehensive evaluation of these Hurst estimators is provided in Chap. 3.

2.1.5 Fractional Stable Motion

The fractional stable motion, which exhibits both the LRD and heavy-tailed distribution properties, is a generalization of fBm. The linear fractional stable motion (LFSM) was studied in [253]. From the perspective of fractional signals and fractional-order systems, the fractional stable motion can be expressed as the output of an $(\alpha + 1)$th fractional integrator with white stable noise as the input,

$$Y_{\alpha,H}(t) = {}_0D_t^{-1-\lambda}\omega_\alpha(t), \quad 0 < \lambda < 1/2, \tag{2.30}$$

where $H = 1/\alpha + \lambda$, and $\omega_\alpha(t)$ is the α-stable noise [253].

2.1.6 Fractional Stable Noise

The fractional stable noise provides the increments of fractional stable motion. So, the fractional stable noise can be constructed as the output of an αth fractional integrator with white stable noise as the input

$$Y_{\alpha,H}(t) = {}_0D_t^{-\lambda}\omega_\alpha(t), \quad 0 < \lambda < 1/2, \tag{2.31}$$

where $H = 1/\alpha + \lambda$, and $\omega_\alpha(t)$ is the α-stable noise [253]. The fractional stable noise will be introduced in detail in Chap. 3.

2.1.7 Multifractional Brownian Motion

Based on the definition of 'one side' fBm, Lim provided the definition of the Riemann-Liouville fractional integral based mBm in [172].

Definition 2.4 The Riemann-Liouville fractional integral based mBm can be described as

$$B_{H(t)}(t) = \frac{1}{\Gamma(H(t)+1/2)}\int_0^t (t-\tau)^{H(t)-1/2}\omega(\tau)\mathrm{d}\tau, \quad 1/2 < H(t) < 1, \tag{2.32}$$

where $\omega(t)$ is wGn.

Therefore, we can consider mGn as the output of $[\alpha(t)+1]$th variable-order fractional integrator with wGn as the input.

$$B_{H(t)}(t) = {}_0D_t^{-1-\alpha(t)}\omega(t). \tag{2.33}$$

Time-dependent local Hölder exponent $H(t)$ is the generalization of the constant Hurst parameter H [232]. Obviously, fBm is a special case of the mBm with a constant Hölder exponent $H(t) = H$. The properties of the mBm will be introduced in Chap. 4.

2.1.8 Multifractional Gaussian Noise

mGn is defined as the derivative of mBm. Therefore, we can consider mGn as the output of $\alpha(t)$th variable-order fractional integrator with wGn as the input. The mGn $Y_{H(t)}(t)$ can be described as

$$Y_{H(t)}(t) = {}_0D_t^{-\alpha(t)}\omega(t), \tag{2.34}$$

where $\omega(t)$ is wGn. The local Hölder exponent $H(t)$ of mBm is related to $\alpha(t)$ by $H(t) = 1/2 + \alpha(t)$. Similar to the mBm which is the generalization of fBm, mGn is the generalization of fGn, and fGn is a special case of the mGn with a constant local Hölder exponent $H(t) = H$.

2.1.9 *Multifractional Stable Motion*

The multifractional stable motion, which exhibits both the local self-similarity and heavy-tailed distribution properties, is a generalization of mBm. The multifractional stable motion $Y_{\alpha,H(t)}(t)$ is presented as

$$Y_{\alpha,H(t)}(t) = {}_0D_t^{-1-\lambda(t)}\omega_\alpha(t), \quad 0 < \lambda(t) < 1/2, \tag{2.35}$$

where $\omega_\alpha(t)$ is α-stable noise [253]. The local Hölder exponent $H(t)$ of multifractional stable motion is related to α and $\lambda(t)$ by $H(t) = 1/\alpha + \lambda(t)$. mBm is the special case of the multifractional stable motion with stable distribution parameter $\alpha = 2$.

2.1.10 *Multifractional Stable Noise*

In the same way, a multifractional stable noise can be considered as the $\lambda(t)$th integration of an α-stable process. The multifractional stable noise is presented as

$$Y_{\alpha,H(t)}(t) = {}_0D_t^{-\lambda(t)}\omega_\alpha(t), \quad 0 < \lambda(t) < 1/2, \tag{2.36}$$

where $\omega_\alpha(t)$ is α-stable noise [253]. Multifractional stable noise exhibits local self-similarity and heavy-tailed distribution. mGn is the special case of the multifractional stable noise with stable distribution parameter $\alpha = 2$.

2.2 Fractional-Order Signal Processing Techniques

In this monograph, like the conventional signal processing methods, FOSP techniques include fractional random signals simulation, fractional filter, fractional systems modeling, and so on. The FOSP techniques are briefly summarized in this section.

2.2.1 *Simulation of Fractional Random Processes*

As stated above, random processes can be generated by performing time domain integer-order filtering on a white Gaussian process [107, 204]. Similarly, the frac-

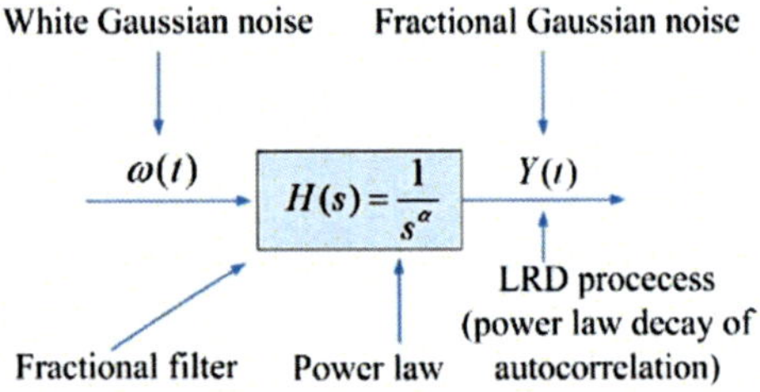

Fig. 2.1 Fractional Gaussian noise simulation

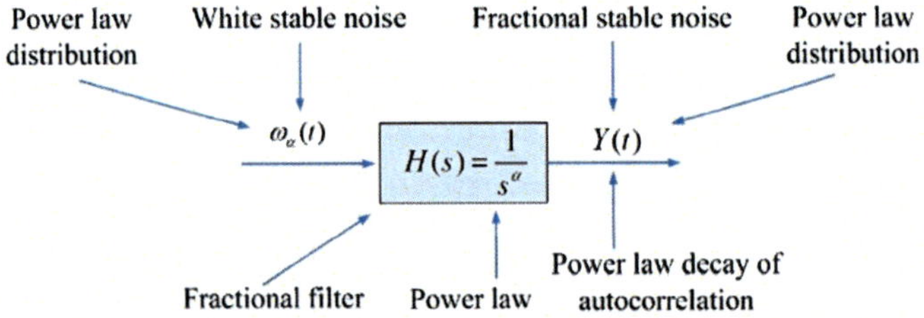

Fig. 2.2 Fractional stable noise simulation

tional random processes can be simulated by performing the time domain fractional-order filtering on a white Gaussian process or a white α-stable process. Different types of fractional filters generate different fractional random signals. For example, fractional Gaussian noise and fractional stable noise can be simulated by a constant-order fractional filter. Figures 2.1 and 2.2 illustrate the simulations of fractional Gaussian noise and fractional stable noise, respectively. The constant-order fractional integrated or filtered signals exhibit the LRD property, that is, the power-law decay of the autocorrelation. Similarly, multifractional Gaussian signals and multifractional stable signals can be simulated by variable-order fractional filters. The output signals of the variable-order fractional filters exhibit the local memory property.

2.2.2 Fractional Filter

It has been introduced in the above subsection that the fractional filters can be used to generate the fractional random signals. Similar to the classification of the fractional signals in this monograph, the fractional filters can also be classified into three types: constant-order fractional filters, distributed-order fractional filters, and variable-order fractional filters. Fractional-order filters are different from the integer-order filters. Integer-order filters generate the short-range dependence on the input signal; constant-order fractional filters generate the LRD property; variable-order fractional filters generate the local memory property. The distributed-order filters can be considered as the summation of the constant-order fractional filters. In this monograph, the constant-order and distributed-order fractional filters are studied. The constant-order fractional filters will be introduced in Chap. 5, and the distributed-order fractional filters will be studied in Chap. 7.

2.2.3 Fractional-Order Systems Modeling

It has been introduced in Sect. 2.1 that a traditional stationary integer-order random signal can be considered as the output of an LTI system with wGn as the input. The continuous-time LTI system can be characterized by a linear difference equation known as an ARMA model in the discrete case. An ARMA(p, q) process X_t is defined as

$$\Phi(B)X_t = \Theta(B)\epsilon_t, \tag{2.37}$$

where ϵ_t is a wGn, and B is the backshift operator. However, the ARMA model can only capture the short-range dependence property of the system. In order to capture the LRD property of the fractional system, the FARIMA(p, d, q) model was proposed [37]. An FARIMA(p, d, q) process X_t is defined as [37]

$$\Phi(B)(1 - B)^d X_t = \Theta(B)\epsilon_t, \tag{2.38}$$

where $d \in (-0.5, 0.5)$, and $(1 - B)^d$ is the fractional differencing operator.

Furthermore, the locally stationary long memory FARIMA(p, d_t, q) model

$$\Phi(B)(1 - B)^{d_t} X_t = \Theta(B)\epsilon_t, \tag{2.39}$$

was suggested in [30], where $\{\epsilon_t\}$ is a wGn and d_t is a time-varying parameter. The locally stationary long memory FARIMA(p, d_t, q) model can capture the local self-similarity of the systems. Besides the above mentioned fractional system models, other fractional models will be introduced in Chaps. 5 and 6.

2.2.4 Realization of Fractional Systems

Realization of fractional systems includes the realization of analogue fractional systems and the realization of digital fractional systems.

Analogue Realization of Fractional Systems

Analogue fractional systems, such as the fractional controllers and fractional filters, can be used widely in engineering. All fractional systems rely on the fractional-order integrator and the fractional-order differentiator as basic elements. Many efforts have been made to design analogue fractional-order integrators and differentiators. Most of these analogue realization methods are based on networks of resistors,

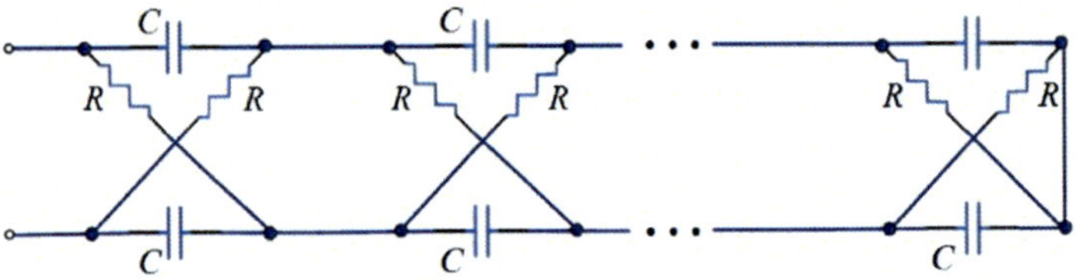

Fig. 2.3 RC transmission line circuit

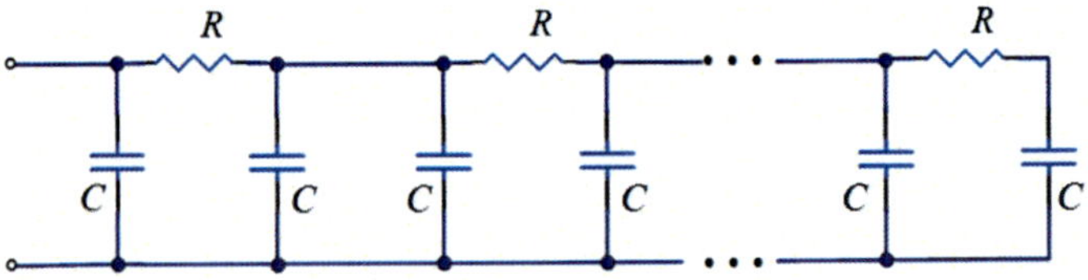

Fig. 2.4 RC domino ladder circuit

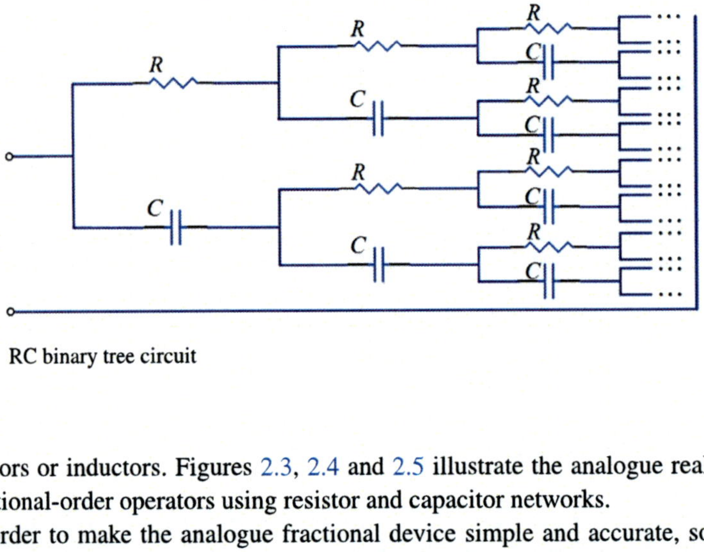

Fig. 2.5 RC binary tree circuit

capacitors or inductors. Figures 2.3, 2.4 and 2.5 illustrate the analogue realization of fractional-order operators using resistor and capacitor networks.

In order to make the analogue fractional device simple and accurate, some researchers have concentrated on smart materials which exhibit realistic fractional behavior. In this monograph, the analogue realization of constant-order fractional-order differentiator/integrator and variable-order fractional differentiator/integrator was based on an electrical element named 'Fractor' (Fig. 2.6), manufactured by Bohannan [27, 28]. The Fractor was originally made from Lithium Hydrazinium Sulfate ($LiN_2H_5SO_4$) which exhibits realistic fractional behavior $1/(j\omega C)^{\lambda}$ over a large range of frequency, where $\alpha \approx 1/2$ [261]. Now, the Fractor is being made from Lithium salts. The analogue realization of fractional systems will be introduced in Chaps. 5 and 6.

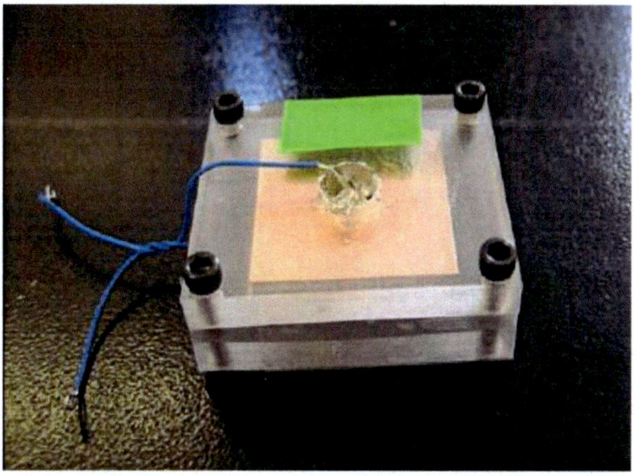

Fig. 2.6 Fractor

Digital Realization of Fractional Systems

Based on the definition of fractional calculus, the calculation of the output of a fractional system depends on the long-range history of the input. Because of the limitation of calculation speed and storage space, the digital realization of fractional systems is difficult. The commonly used methods of approximate digital realization of fractional systems are frequency domain methods and time domain methods. Currently both methods offer limited success in fitting the fractional system.

Frequency domain methods include Oustaloup method [227], Carlson method [237], Matsuda method [237], and so on. Frequency-domain fitting techniques can fit the magnitude of the frequency response very well, but cannot guarantee the stable minimum-phase fitting. Time domain methods are mainly based on fitting the impulse response or the step response of the system. An effective time domain impulse response invariant discretization method was discussed in [59, 62, 63, 182]. There, a technique for designing discrete-time infinite impulse response (IIR) filters to approximate the continuous-time fractional-order filters is proposed, keeping the impulse response of the continuous-time fractional-order filter and the impulse response of the approximate discrete-time filter almost the same.

2.2.5 *Other Fractional Tools*

Besides the above FOSP techniques, there are other FOSP techniques too, such as fractional Hilbert transform, fractional spectrum analysis, fractional B-spline, and so on. These FOSP techniques provide new options for analyzing complex signals.

Fractional Hilbert Transform

The fractional Hilbert transform (FHT), the generalization of the conventional Hilbert transform, was proposed in [176]. FHT has been successfully used in digital

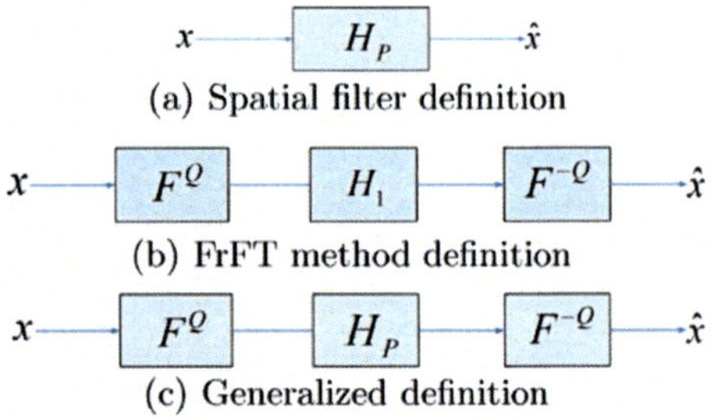

Fig. 2.7 Block diagrams for the different implementations of the fractional Hilbert transform. **(a)** Spatial filter. **(b)** FrFT method. **(c)** Generalized definition

image processing. There are three commonly used definitions of FHT. The first definition is based on modifying the spatial filter with a fractional parameter, and the second one is based upon the fractional Fourier plane for filtering. The third definition is the combination of these two definitions. The transfer function of the first definition is [176]

$$\tilde{H}_P(\nu) = \exp(+i\phi)u(\nu) + \exp(-i\phi)u(-\nu), \tag{2.40}$$

where P is the fractional order, $u(\nu)$ is a step function, and $\phi = P\pi/2$. The second type FHT is defined as [176]

$$V_Q = F^{-Q} H_1 F^Q, \tag{2.41}$$

where F^α is the FrFT operation of order α, Q is a fractional parameter, and

$$\tilde{H}_1(\nu) = \exp\left(+i\frac{\pi}{2}\right)u(\nu) + \exp\left(-i\frac{\pi}{2}\right)u(-\nu). \tag{2.42}$$

The third definition of FHT is [176]

$$V_Q = F^{-Q} H_P F^Q. \tag{2.43}$$

Figure 2.7 illustrates the three definitions of FHT.

Fractional Power Spectrum Density

Definitions of fractional spectrum density (FPSD) fall into two types: FrFT based and FLOM based. FrFT based FPSD was developed from combining the conventional PSD and the FrFT method. FPSD exhibits distinctive superiority to non-stationary signals. FrFT based fractional power spectrum is defined as

$$P^\alpha_{\varepsilon\varepsilon}(\mu) = \lim_{T\to\infty} \frac{E|\xi_{\alpha,T}(\mu)|^2}{2T}, \tag{2.44}$$

where $\xi_{\alpha,T}(\mu)$ is the αth FrFT of $\varepsilon_T(t)$, and $\varepsilon_T(t)$ is the truncation function in $[-T,T]$ of the sample function of the random process $\varepsilon(t)$.

FLOM based fractional power spectra include the covariation spectrum and the fractional low-order covariance spectrum [184].

Definition 2.5 The covariations spectrum is defined as [215]

$$\begin{aligned}\tilde{\phi}^c_{xx}(\omega) = \mathcal{F}[R^c_{xx}(\tau)] &= \int_{-\infty}^{\infty} R^c_{xx}(\tau)e^{-j\omega\tau}\mathrm{d}\tau \\ &= \int_{-\infty}^{\infty} [x(t), x(t-\tau)]_\alpha e^{-j\omega\tau}\mathrm{d}\tau,\end{aligned} \tag{2.45}$$

where $[x(t), x(t-\tau)]_\alpha$ is the covariation defined as

$$[x(t), x(t-\tau)]_\alpha = \frac{E[x(t)(x(t-\tau))^{\langle p-1\rangle}]}{E(|x(t-\tau)^p|)}\gamma_{x(t-\tau)}, \quad 1 \le p < \alpha, \tag{2.46}$$

where γ_y is the scale parameter of Y, and $z^{\langle\alpha\rangle} = |z|^\alpha \operatorname{sgn}(z)$.

Definition 2.6 Fractional low-order covariance spectrum is defined as

$$\tilde{\phi}^d_{xx}(\omega) = \mathcal{F}(R^d_{xx}(\tau)) = \int_{-\infty}^{\infty} R^d_{xx}(\tau)e^{-j\omega\tau}\mathrm{d}\tau, \tag{2.47}$$

where

$$R^d_{xx}(\tau) = E[x(t)^{\langle A\rangle}x(t-\tau)^{\langle B\rangle}], \quad 0 \le A < \frac{\alpha}{2}, \; 0 \le B < \frac{\alpha}{2}. \tag{2.48}$$

FLOM based fractional power spectrum techniques have been successfully used in time delay estimation [184].

Fractional Splines

Fractional B-splines can be considered as the generalization of the usual integer-order B-splines. There are three commonly used definitions of fractional B-splines, they are causal fractional B-splines, anti-causal fractional B-splines, and non-causal symmetric fractional B-splines [25, 222, 308]. Fractional causal B-splines are defined by taking the $(a+1)$th fractional difference of the one-sided power function.

Definition 2.7 The fractional causal B-splines are specified in the Fourier domain

$$\hat{\beta}^\alpha_+(\omega) = \left(\frac{1-e^{j\omega}}{j\omega}\right)^{\alpha+1}. \tag{2.49}$$

Definition 2.8 The anti-causal B-splines of degree α is defined in Fourier domain as

$$\hat{\beta}^\alpha_-(\omega) = \left(\frac{e^{j\omega}-1}{j\omega}\right)^{\alpha+1}. \tag{2.50}$$

Definition 2.9 The non-causal symmetric fractional B-splines of degree α is defined in Fourier domain as

$$\hat{\beta}_*^\alpha(\omega) = \left| \frac{\sin \omega/2}{\omega/2} \right|^{\alpha+1}. \tag{2.51}$$

2.3 Chapter Summary

This chapter provides an overview of basic concepts of fractional processes and FOSP techniques from the perspective of fractional signals and fractional-order systems. Section 2.1 deals with the constant-order fractional-order processes and variable-order fractional processes. All these fractional processes can be generated by fractional-order systems driven by white Gaussian noise. Section 2.2 briefly introduced some FOSP techniques including fractional processes simulation, fractional filter, fractional systems modeling, analogue/digital realization of fractional systems, and other fractional tools. All discussions on FOSP techniques are centered around fractional calculus, FrFT and α-stable distribution. A detailed introduction of constant-order fractional processes and multifractional-processes will be provided in the following two chapters, respectively. The constant-order fractional signal processing techniques, variable-order fractional signal processing techniques and distributed-order filters will be introduced in Chaps. 5, 6 and 7, respectively.

Part II
Fractional Processes

Chapter 3
Constant-Order Fractional Processes

Fractional processes with constant long memory parameter are increasingly involved in areas ranging from financial science to computer networking. In Chap. 2, we discussed a new perspective on fractional signals and fractional-order systems. Specifically, a fractional process with a constant long memory parameter can be regarded as the output signal of a fractional-order system driven by wGn. Since the main property of constant-order fractional processes is LRD, we often call this kind of processes LRD processes. Due to the requirements of accurate modeling and forecasting of LRD time series, the subject of their self-similarity and the estimation of their statistical parameters are becoming more and more important. An LRD process can be characterized by its long memory parameter H, the Hurst parameter or Hurst exponent [123]. The Hurst exponent has close relationship with power-law, long memory, fractals, fractional calculus and even chaos theory. Therefore, Hurst exponent estimation is crucial to fractional system modeling and forecasting. In this chapter, LRD processes and Hurst parameter estimators are introduced. Furthermore, the robustness and the accuracy of twelve Hurst parameter estimators are extensively studied.

3.1 Introduction of Constant-Order Fractional Processes

3.1.1 Long-Range Dependent Processes

LRD phenomenon occurs in many fields of study, such as economics, agronomy, chemistry, astronomy, engineering, geosciences, hydrology, mathematics, physics and statistics.

Definition 3.1 A stationary process is said to have long-range correlations if its covariance function $C(n)$ (assume that the process has finite second-order statistics) decays slowly as $n \to \infty$, i.e. for $0 < \alpha < 1$,

H. Sheng et al., *Fractional Processes and Fractional-Order Signal Processing*,
Signals and Communication Technology,
DOI 10.1007/978-1-4471-2233-3_3,

$$\lim_{n\to\infty} \frac{C(n)}{n^{-\alpha}} = c, \tag{3.1}$$

where c is a finite, positive constant.

That is to say, for large n, $C(n)$ looks like c/n^{α} [107]. The parameter α is related to the Hurst parameter via the equation $\alpha = 2 - 2H$. We can also define the LRD using the spectral density.

Definition 3.2 The weakly-stationary time-series $X(t)$ is said to be long-range dependent if its spectral density obeys

$$f(\lambda) \sim C_f |\lambda|^{-\beta}, \tag{3.2}$$

as $\lambda \to 0$, for some $C_f > 0$ and some real parameter $\beta \in (0, 1)$.

The parameter β is related to the Hurst parameter by $H = (1 + \beta)/2$ [65]. Hurst parameter H is a simple parameter which can characterize the level or degree of LRD. The Hurst exponent, which was proposed for the analysis of long-term storage capacity of reservoirs more than a half century ago, is used today to measure the intensity of LRD. Indeed a lot of time series are often described in the model with LRD characters. The case when $0 < H < 0.5$ indicates the time serial is a negatively correlated process, or anti-persistent process; the case when $0.5 < H < 1$ indicates that it is a positively correlated process; the case when $H = 0.5$ indicates that the process is almost not a correlated process.

From the above definition we can see that LRD manifests as a high degree of correlation between data points distantly separated in time. In the frequency domain, LRD manifests a significant level of power at frequencies near zero. However, the restriction we discussed above is that the process has finite second-order statistics. If the process has infinite second-order statistics, we cannot describe the LRD with the covariance function. There are well-defined self-similar processes with stationary increments, infinite second-order moments [22]. Typical examples of LRD processes with infinite second-order moments are stable self-similar processes and FARIMA with infinite variance [37, 40, 150], which is introduced in Chap. 5.

The LRD processes have some important properties. The variances of the sample mean are decaying more slowly than the reciprocal of the sample size. As a consequence, classical statistical tests and confidence intervals lead to the wrong results. The rescaled adjusted range statistic is characterized by a power-law: $E[R(m)/S(m)] \sim a_1 m^H$ as $m \to \infty$ with $0.5 < H < 1$. The fact that the autocorrelations decay hyperbolically rather than exponentially implies non-summable autocorrelation function, that is, $\sum_k r(k) = \infty$. In this case, even though the $r(k)$s are individually small for large lags, their cumulative effect is important. The spectral density $f(\cdot)$ obeys a power-law near the origin [247]. Two commonly seen LRD processes are fBm and fGn, respectively.

3.1.2 Fractional Brownian Motion and Fractional Gaussian Noise

Fractional Brownian Motion

Definition 3.3 The fBm with Hurst index H $(0 < H < 1)$ is defined as the stochastic integral, for $t \geq 0$

$$B_H(t) = \frac{1}{\Gamma(H+1/2)} \left\{ \int_{-\infty}^{0} [(t-s)^{H-1/2} - (-s)^{H-1/2}] \mathrm{d}W(s) + \int_{0}^{t} (t-s)^{H-1/2} \mathrm{d}W(s) \right\}, \tag{3.3}$$

where W denotes a Wiener process defined on $(-\infty, \infty)$.

Figure 3.1 illustrates 1000 points of fBm with different Hurst parameters. The index H is the Hurst parameter indicating the degree of self-similarity $(0.5 \leq H < 1)$. When $H = 0.5$, fBm is the usual Brownian motion. The fBm process $B_H(t)$ has the covariance function [296]

$$\mathrm{cov}(B_H(s), B_H(t)) = \frac{\sigma^2}{2} \left(|t|^{2H} + |s|^{2H} - |t-s|^{2H} \right). \tag{3.4}$$

Mean value of fBm is [134]

$$E(B_H(t)) = 0. \tag{3.5}$$

Variance function of fBm is [134]

$$\mathrm{var}[B_H(t)] = \frac{\sigma^2}{2} |t|^{2H}. \tag{3.6}$$

For $\sigma > 0$ and $0 < H < 1$:

$$E(B_H^2(t)) = \sigma^2 |t|^{2H}. \tag{3.7}$$

The fBm has the following properties,

- Stationary increments

$$B_H(t) - B_H(s) \sim B_H(t-s), \tag{3.8}$$

where "$\sim$" means equal in distribution.
- Long-range dependence: For $H > 1/2$, the fBm is an LRD process

$$\sum_{n=1}^{\infty} E[B_H(1)(B_H(n+1) - B_H(n))] = \infty. \tag{3.9}$$

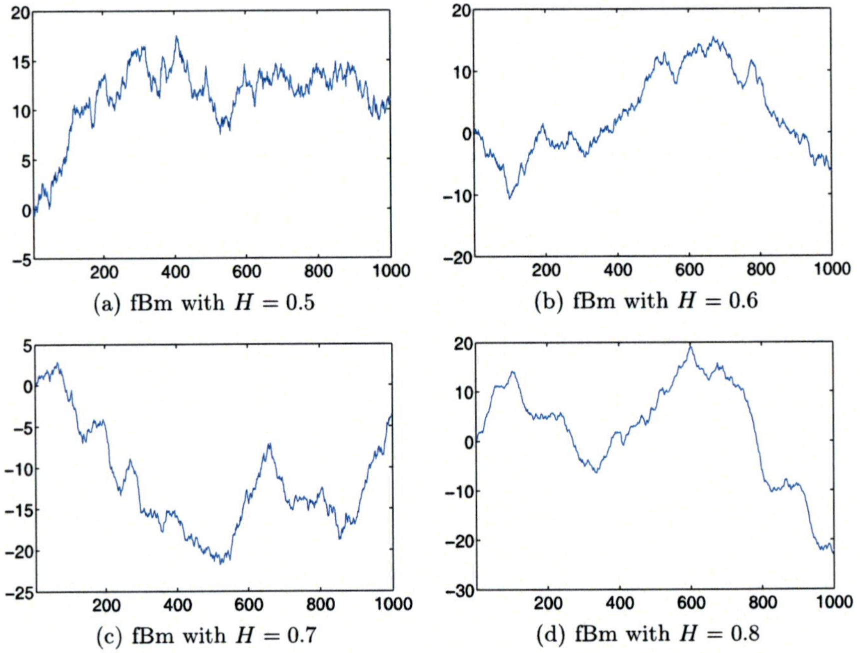

Fig. 3.1 Fractional Brownian motion: **(a)** $H = 0.5$; **(b)** $H = 0.6$; **(c)** $H = 0.7$; **(d)** $H = 0.8$

- Self-similarity

$$B_H(at) \sim |a|^H B_H(t). \tag{3.10}$$

A Riemann-Liouville fractional integral based definition of fBm is

$$B_H(t) = \frac{1}{\Gamma(H+1/2)} \int_0^t (t-\tau)^{H-1/2} \omega(\tau) \mathrm{d}\tau, \tag{3.11}$$

where $H > 0$ and $\omega(t)$, $t > 0$ is the one-sided white Gaussian noise.

Fractional Gaussian Noise

fGN is the increment sequence of fBm

$$X_k = Y(K+1) - Y(k), \quad k \in \mathbb{N}, \tag{3.12}$$

where $Y(k)$ is an fBm [223]. The fGn X_k is a stationary process. The mean value of an fGn is

$$E(X_k) = 0. \tag{3.13}$$

The second-order moment of fGn is

$$E(X_k^2) = \sigma^2 = E(Y(1)^2). \tag{3.14}$$

Its autocovariance function is given by [134, 163]

$$\gamma(k,H)=\frac{\sigma^2}{2}\left[|k+1|^{2H}-2|k|^{2H}+|k-1|^{2H}\right], \tag{3.15}$$

where $k\in\mathbb{N}$, and

$$\sigma^2=\frac{\Gamma(2-H)\cos(\pi H)}{\pi H(2H-1)}, \tag{3.16}$$

where Γ is the Gamma function. Its correlation function is given by [134, 162]

$$\rho(k,H)=\frac{\sigma^2}{2}\left[|k+1|^{2H}-2|k|^{2H}+|k-1|^{2H}\right]. \tag{3.17}$$

Hurst parameter H also plays an important role in fGn process. Fractional Gaussian noise with $H<0.5$ demonstrates negatively autocorrelated behavior; fGn with $H>0.5$ demonstrates positively correlated or long memory or persistent behavior; and the special case of fGn with $H=0.5$ corresponds to classical Gaussian white noise. The spectral density of an fGn with $H>0.5$ has the following inverse power-law form as $\xi\to 0$

$$S(\xi)\sim C_s|\xi|^{-\beta},\quad 0<\beta<1, \tag{3.18}$$

where C_s is a constant and $C_s>0$. This type of signal is also called $1/f$ noise. The $1/f$ fluctuations are widely found in practice. Figure 3.2 illustrates 1000 points fGn with different Hurst parameters.

3.1.3 Linear Fractional Stable Motion and Fractional Stable Noise

Linear Fractional Stable Motion (LFSM)

The fractional stable processes, which display both LRD and heavy-tailed distribution, were studied in detail in [253]. The most commonly used fractional stable processes are LFSM and fractional stable noise. An LFSM process $L_{\alpha,H}$ is defined as [253]

$$\begin{aligned}L_{\alpha,H}=\int_{-\infty}^{\infty}&\left\{a\left[(t-s)_+{}^{H-1/\alpha}-(-s)_+{}^{H-1/\alpha}\right]\right.\\&\left.+b\left[(t-s)_-{}^{H-1/\alpha}-(-s)_-{}^{H-1/\alpha}\right]\right\}M_{\alpha,\beta}(\mathrm{d}x),\end{aligned} \tag{3.19}$$

where a, b are real constants, $M_{\alpha,\beta}$ is an α-stable random measure, $|a|+|b|>0$, $0<\alpha<2$, $0<H<1$ and for $x\in\mathbb{R}$

$$(x)_+^a=\begin{cases}x_+^a, & \text{when } x>0\\ 0, & \text{when } x\le 0,\end{cases} \tag{3.20}$$

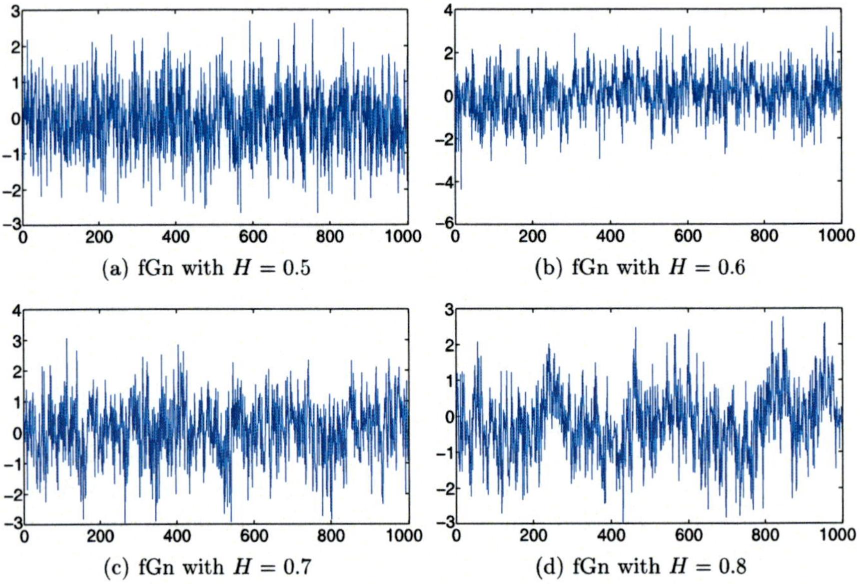

Fig. 3.2 Fractional Gaussian noises: **(a)** $H = 0.5$; **(b)** $H = 0.6$; **(c)** $H = 0.7$; **(d)** $H = 0.8$

and $x_-^a = (-x)_+^a$.

The representation of LFSM is based on the representation of fBm, where the exponent $H - 1/2$ is substituted by $H - 1/\alpha$. When $\alpha = 2$, the LFSM reduces to fBm. Figure 3.3 illustrates the 1000 points LFSMs with different Hurst parameters H and stable parameters α.

Fractional Stable Noise

Fractional stable noise is the stationary increments of LFSM.

$$\begin{aligned} X_j &= L_{\alpha,H}(a,b;\, j+1) - L_{\alpha,H}(a,b;\, j) \\ &= \int_{-\infty}^{\infty} \Big\{ a\Big[(j+1-x)_+{}^{H-1/\alpha} - (j-x)_+{}^{H-1/\alpha} \Big] \\ &\quad + b\Big[(j+1-x)_-{}^{H-1/\alpha} - (j-x)_-{}^{H-1/\alpha} \Big] \Big\} M_{\alpha,\beta}(\mathrm{d}x), \qquad (3.21) \end{aligned}$$

where a, b are real constants, $M_{\alpha,\beta}(\mathrm{d}x)$ is an α-stable random measure, $|a| + |b| > 0$, $0 < \alpha < 2$, $0 < H < 1$ and for $x \in \mathbb{R}$. Fractional stable noise can also be constructed as functionals of white α-stable noises by using a transformation induced from fractional integral operators [121]. When $H > 1/\alpha$, the fractional stable noise has long-range dependence; when $H < 1/\alpha$, the fractional stable noise has negative dependence. There is no long-range dependence when $0 < \alpha < 1$ because

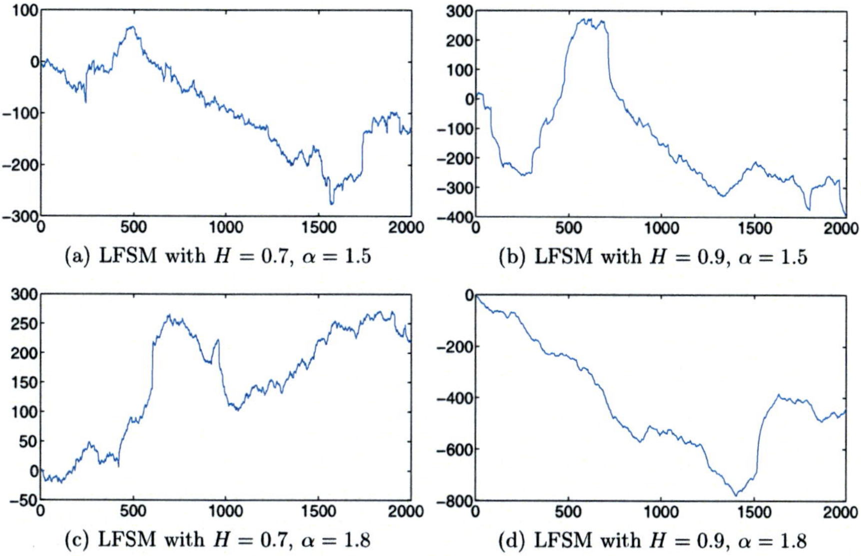

(a) LFSM with $H = 0.7$, $\alpha = 1.5$ (b) LFSM with $H = 0.9$, $\alpha = 1.5$

(c) LFSM with $H = 0.7$, $\alpha = 1.8$ (d) LFSM with $H = 0.9$, $\alpha = 1.8$

Fig. 3.3 Linear fractional stable motion: **(a)** $H = 0.7, \alpha = 1.5$; **(b)** $H = 0.9, \alpha = 1.5$; **(c)** $H = 0.7$, $\alpha = 1.8$; **(d)** $H = 0.9$, $\alpha = 1.8$

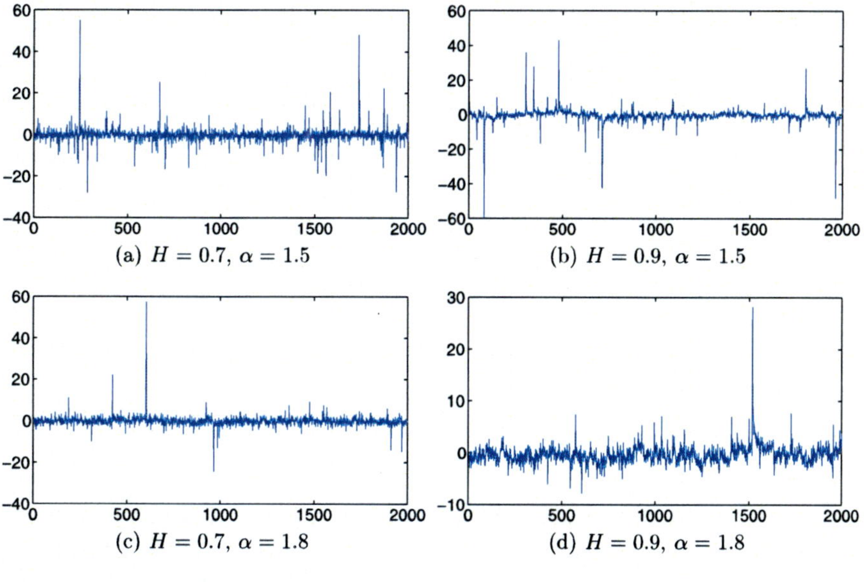

(a) $H = 0.7$, $\alpha = 1.5$ (b) $H = 0.9$, $\alpha = 1.5$

(c) $H = 0.7$, $\alpha = 1.8$ (d) $H = 0.9$, $\alpha = 1.8$

Fig. 3.4 Fractional stable noises

$H \in (0, 1)$. The value $H = 1/\alpha$ lies on the boundary between long-range and negative dependence. Figure 3.4 illustrates the 1000 points fractional stable noises with different Hurst parameters H and stable parameters α.

3.2 Hurst Estimators: A Brief Summary

Many Hurst estimators have been proposed to analyze the LRD time series [293]. These methods are mainly useful as simple diagnostic tools for LRD time series.

3.2.1 *R/S Method [194]*

R/S method is one of the most well-known estimators. It is described briefly as follows. For a given time series of length N, $R(n)$ is the range of the data aggregated over blocks of length n, and $S^2(n)$ is the sample variance. For fGN or FARIMA processes,

$$E[R(n)/S(n)] \sim C_H n^H, \tag{3.22}$$

as $n \to \infty$, where C_H is a positive, finite constant independent of n. The Hurst parameter can be estimated by fitting a line to a log-log plot of $R(n)/S(n)$ versus n. Both the number of blocks and the number of points of the time series should not be chosen too small.

3.2.2 *Aggregated Variance Method [22]*

For a given time series X_i of length N, the corresponding aggregated series is defined by

$$X^{(m)}(k) = \frac{1}{m} \sum_{i=(k-1)m+1}^{km} X(i), \quad k = 1, 2, \ldots, \tag{3.23}$$

for successive values of m. Its sample variance is

$$\widehat{\mathrm{var}} X^{(m)} = \frac{1}{N/m} \sum_{k=1}^{N/m} (X^{(m)}(k) - EX^{(m)})^2. \tag{3.24}$$

For fGN and FARIMA processes, $\mathrm{var}^{(m)} \sim \sigma^2 m^\beta$ as $m \to \infty$ where σ is the scale parameter and $\beta = 2H - 2 < 0$. The sample variance $\widehat{\mathrm{var}}(X^{(m)})$ should be asymptotically proportional to m^{2H-2} for large N/m and m, and the resulting points should form a straight line with slope $\beta = 2H - 2$ and $-1 \le \beta < 0$.

3.2.3 Absolute Value Method [297]

The Absolute Value method is very similar to the Aggregated Variance method. The data is divided in the same way as (3.23) to form aggregated series. The Absolute Value method calculates the sum of the Absolute Values of the aggregated series, $\frac{1}{N/m}\sum_{k=1}^{N/m}|X^{(m)}(k)|$. For fGN and FARIMA processes with parameter H, the result should be a straight line with slope $H-1$.

3.2.4 Variance of Residuals Method [298]

Variance of Residuals method was proposed in [233]. The time series is divided into blocks of size m. Within each of the blocks, the partial sums of the series are calculated and then fit a least-squares line to the partial sums within each block and compute the sample variance of the residuals,

$$\frac{1}{m}\sum_{t=1}^{m}(Y(t)-a-bt)^2. \tag{3.25}$$

The procedure is repeated for each of the blocks, and the resulted sample variances are averaged. By doing so, we can get a straight line with slope of $2H$ if the result is plotted on a log-log scale versus m.

3.2.5 Periodogram Method and the Modified Periodogram Method [97, 113]

The Periodogram method is defined by

$$I(\xi)=\frac{1}{2\pi N}\left|\sum_{j=1}^{N}X_je^{ij\xi}\right|^2, \tag{3.26}$$

where ξ is the frequency, and X_j is the data. For a series with finite variance, $I(\xi)$ is an estimate of the spectral density of the series. A process with LRD should have a periodogram which is proportional to $|\lambda|^{1-2H}$ close to the origin, so the log-log plot of $I(\xi)$ should have a slope of $1-2H$ close to the origin. The periodogram method can be modified using Modified Periodogram method to obtain better estimation results. For Modified Periodogram method, the frequency axis is divided into logarithmically equally spaced boxes, and the periodogram values inside each box are estimated and averaged. Compared with the Periodogram method, the Modified Periodogram method can reduce the variance of the estimates and the bias in the estimation of the Hurst parameter H.

3.2.6 Whittle Estimator [298]

The Whittle estimator is also based on the periodogram. It involves the function

$$Q(\eta) := \int_{-\pi}^{\pi} \frac{I(\lambda)}{f(\lambda;\eta)} \mathrm{d}\lambda, \tag{3.27}$$

where $I(\lambda)$ is the periodogram, $f(\lambda;\eta)$ is the spectral density at the frequency λ, and η denotes the vector of unknown parameters. The Whittle estimator is the value of η which minimizes the function Q. When dealing with fGN or FARIMA processes, η is simply the parameter H or d. If the series is assumed to be FARIMA(p, d, q), then η includes also the unknown coefficients in the autoregressive and moving average parts.

3.2.7 Diffusion Entropy Method [105]

The purpose of the Diffusion Entropy algorithm is to establish the possible existence of scaling without altering the data with any form of detrending. The existence of scaling implies the existence of a PDF $p(x,t)$ that scales according to the equation

$$p(x,t) = \frac{1}{t^{\delta}} \cdot F\left(\frac{x}{t^{\delta}}\right), \tag{3.28}$$

where δ is the PDF scaling exponent. From the Shannon entropy

$$S(t) = -\int_{-\infty}^{\infty} p(x,t) \cdot \ln p(x,t) \mathrm{d}x, \tag{3.29}$$

after simple algebra, we get the following fundamental relation:

$$S(\tau) = A + \delta\tau, \tag{3.30}$$

where

$$A \equiv -\int_{-\infty}^{\infty} F(y) \cdot \ln[F(y)] \mathrm{d}y, \tag{3.31}$$

and

$$\tau = \ln(t/t_0). \tag{3.32}$$

Equation (3.30) is the key relation for understanding how the DEA is used for detecting the PDF scaling exponent δ.

3.2.8 Kettani and Gubner's Method [138]

An approach to estimating the LRD parameter was proposed by Kettani and Gubner [138]. Let $X_1, X_2, \dots, X_n$ be a realization of a Gaussian second-order self-similar process. The estimated Hurst parameter can be calculated by

$$\hat{H}_n = \frac{1}{2}[1 + \log_2(1 + \hat{\rho}_n(1))], \tag{3.33}$$

where $\hat{\rho}_n(k)$ denotes the sample autocorrelation. The 95% confidence interval of H is centered around the estimate $\hat{H}_n$. For an FARIMA$(0, d, 0)$ process,

$$\hat{d}_n = \frac{\hat{\rho}_n(1)}{1 + \hat{\rho}_n(1)}. \tag{3.34}$$

The 95% confidence interval of d is centered around the estimate $\hat{d}_n$.

3.2.9 Abry and Veitch's Method [1]

A wavelet-based tool for the analysis of LRD and a related semi-parametric estimator of the Hurst parameter was introduced by Abry and Veitch [1]. The scale behavior in data can be estimated from the plot of the

$$\log_2\left(\frac{1}{n_j}\sum_k |d_x(j,k)|^2\right) \tag{3.35}$$

versus j, where $d_x(j,k) = \langle x, \psi_{j,k}\rangle$, and $\psi_{j,k}$ is the dual mother wavelet. This wavelet-based method can be implemented very efficiently allowing the direct analysis of very large data sets, and is robust against the presence of deterministic trends.

3.2.10 Koutsoyiannis' Method [153]

Koutsoyiannis proposed an iterative method to determine the standard deviation σ and the Hurst exponent H that minimize the error $e^2(\sigma, H)$ [153]:

$$\begin{aligned} e^2(\sigma, H) :=& \sum_{k=1}^{k'} \frac{[\ln \sigma^{(k)} - \ln s^{(k)}]^2}{k^p} \\ =& \sum_{k=1}^{k'} \frac{[\ln \sigma + H \ln k + \ln c_k(H) - \ln s^{(k)}]^2}{k^p}, \end{aligned} \tag{3.36}$$

where $s^{(k)} \approx c_k(H)k^H\sigma$ and $c_k(H) := \sqrt{\frac{n/k-(n/k)^{2H-1}}{n/k-1/2}}$. The minimization of $e^2(\sigma, H)$ can be done numerically. The algorithm for estimation of both H and σ has been successfully applied in the analysis of climate change [153].

3.2.11 Higuchi's Method [116]

Higuchi provided an approach to estimating the Hurst parameter of the LRD time series on the basis of the fractal theory [116]. For a given time series $X(1), X(2), \ldots, X(N)$, a newly constructed time series X_k^m is defined as

$$X(m), X(m+k), \ldots, X(m+\lfloor (N-m)/k \rfloor \cdot k) \quad (m=1,2,\ldots,k), \tag{3.37}$$

where $\lfloor\ \rfloor$ denotes the greatest integer function. The normalized length of the series, X_k^m is

$$L_m(k) = \frac{N-1}{[(N-m)/k]\cdot k^2} \sum_{i=1}^{[(N-m)/k]} |X(m+ik) - X(m+(i-1)\cdot k)|. \tag{3.38}$$

Then $\langle L(k) \rangle \propto k^{-D}$, where $D = 2 - H$.

3.3 Robustness of Hurst Estimators

Different estimation methods have different scopes of application, with different accuracy and robustness. Some efforts have been made to compare these estimators both in theory and practice [130, 244, 296, 297]. However, up to now, the quantitative evaluation of the robustness of these estimators is still short of reasonable for the noisy LRD time series and the LRD time series with infinite variances. Therefore, it is important to thoroughly evaluate these Hurst estimators. The intention of this section is to provide some basic information for these robustness questions.

The most well-known models of LRD processes are fractional Gaussian noise (fGN) and fractional ARIMA (FARIMA) processes. In this section, the fGN and the FARIMA with stable innovations time series, introduced in detail in Chap. 4, are used to evaluate the robustness of several different Hurst parameter estimators. Two types of noises are added when we analyze the Hurst parameter estimators for the LRD time series. The first type of added noise is Gaussian, because generally, the Gaussian model effectively characterizes the added noise. The second type of noise added to the LRD time series is non-Gaussian. Many noises, in practice, are non-Gaussian, such as low-frequency atmospheric noise and many types of man-made noises [202]. Most non-Gaussian noises are impulsive in nature, so they provide another basis for evaluating the robustness of Hurst parameter estimators. Furthermore, the studied LRD time series itself has infinite variance, such as that found in computer networks signals and stock returns time series. So the robustness of these estimators to the LRD time series with infinite variance is also evaluated. All twelve Hurst estimators introduced in the previous section are analyzed.

3.3.1 Test Signal Generation and Estimation Procedures

Most of the Hurst parameter estimators function under the assumptions that the observed process is stationary and Gaussian or at least a linear process [22]. But, in reality, these assumptions are usually not satisfied. Most estimators are vulnerable to trends, periodicity and other sources of measurement noises. So, the robustness properties of estimators in real world applications are crucial.

The first step in evaluating the robustness of these estimators is to generate the processes which exhibit the self-similar properties. Taqqu et al. [18] described some methods for generating discretized sample paths of LRD processes such as fBm, fGN and FARIMA processes. Among them, the fGN and the class of FARIMA processes are the most popular models to exactly simulate self-similar processes. fGN, itself a self-similar process, is used to model phenomena in many disciplines, e.g. in computer networks signal processing, economics and queueing systems. FARIMA processes can be described using an autocorrelation function, which decays hyperbolically. It differs significantly from the related traditional, short-range dependent stochastic processes such as Markov, Poisson or ARMA processes.

In order to accurately evaluate the Hurst parameters we generate the exact fGN processes using the embedding method (for $1/2 \le H < 1$) and Lowen's method (for $0 < H < 1/2$) [284]. We generate 100 replications of simulated fGNs at every different Hurst value between 0.01 and 0.99 in steps of 0.01. The data length for each Hurst value is 10000. For each of the estimation methods, the Hurst parameters are estimated for 100 times at each Hurst value. Furthermore, for every estimator, the mean values at each Hurst value are calculated. Then we can analyze the difference between the estimations and the true Hurst value for $0 < H < 1$.

In order to analyze the robustness of these twelve estimators for noise corrupted LRD time series, two types of noises are added. In most situations, to simplify the implementation and analysis of the estimators, we usually assume that the additive noise is Gaussian. So, firstly, 30 dB signal to noise ratio (SNR) white Gaussian noise is added. All the methods mentioned above are evaluated using 30 dB SNR white Gaussian noise corrupted LRD process. Furthermore, there are a number of important situations where dominant interferences are impulsive and should be characterized more accurately as an $S\alpha S$ noise. Here, $S\alpha S$ ($\alpha = 0.9$) noise is added. The same analysis methods are used to analyze the robustness of the above twelve Hurst estimators.

Some processes have both infinite variances and LRD properties. Because some Hurst estimators are based on the second-order statistics, the accuracy of these estimators is intensely affected. The two best models are linear fractional stable motion (LFSM) process and the FARIMA time series with $S\alpha S$ innovations. Here, the FARIMA time series with $S\alpha S$ innovations model is used to analyze the robustness of Hurst parameter estimators. For FARIMA time series with $S\alpha S$ innovations model, the Hurst parameter H can be described as $H = d + 1/\alpha$ [253], where d is the fractional differencing exponent. For convenience, we concentrate on $\alpha \in (1, 2)$ and $d \in (0, 1 - 1/\alpha)$, so $H \in (1/\alpha, 1)$. We use the Stilian Stoev's simulation methods [285] to generate the FARIMA time series with $S\alpha S$ innovations, where $\alpha = 1.5$ and $H \in (2/3, 1)$. Most Hurst estimators assume that the FARIMA time series to be analyzed has no heavy-tailed distributions, so the Hurst parameter H can be calculated by $H = d + 1/2$. However, for FARIMA with infinite variance time series, the Hurst parameter H should be $H = d + 1/\alpha$. For the estimation of the Hurst value for FARIMA with heavy-tailed distributions time series, the error $1/\alpha - 1/2$ should be considered cautiously. Therefore, it is very important to estimate the parameter α of the time series with heavy-tailed distributions before we estimate the LRD parameter H. So, it is to be noted that the following estimation results of FARIMA with stable innovations time series for all twelve estimators have been corrected considering the error $1/\alpha - 1/2$. In practice, for a heavy-tailed distribution LRD time series with unknown parameter α, the Absolute Value method, the Diffusion Entropy method, and Higuchi's method can provide better estimation results, although they are not perfectly accurate for time series with known parameter α.

3.3.2 Comparative Results and Robustness Assessment

In this subsection we analyze the robustness of the twelve estimators [51]. As already mentioned above, the twelve estimators are applied to four types of different LRD processes, namely,

- (LRD1) LRD processes alone;
- (LRD2) LRD processes corrupted by 30 dB SNR white Gaussian noise;
- (LRD3) LRD processes corrupted by 30 dB SNR α-stable noise ($\alpha = 0.9$);
- (LRD4) FARIMA with α-stable innovations ($\alpha = 1.5$ and $H \in (2/3, 1)$).

Figures 3.5 through 3.16 show the estimation results of all these twelve estimators. In all the figures, the blue lines are the true Hurst values H and the red dots are the estimation values $\hat{H}$.

Results of R/S method

The robustness analysis results of R/S method for four LRD processes are presented in Fig. 3.5. From Fig. 3.5(a) we can see that the R/S method is biased for almost all Hurst values ($0 < H < 1$) of LRD1. The estimated Hurst parameters are seriously overestimated when $0 < H \leq 0.6$ and underestimated when $0.8 \leq H < 1$. So, the R/S method is not precise. The analysis results for LRD2 are presented in

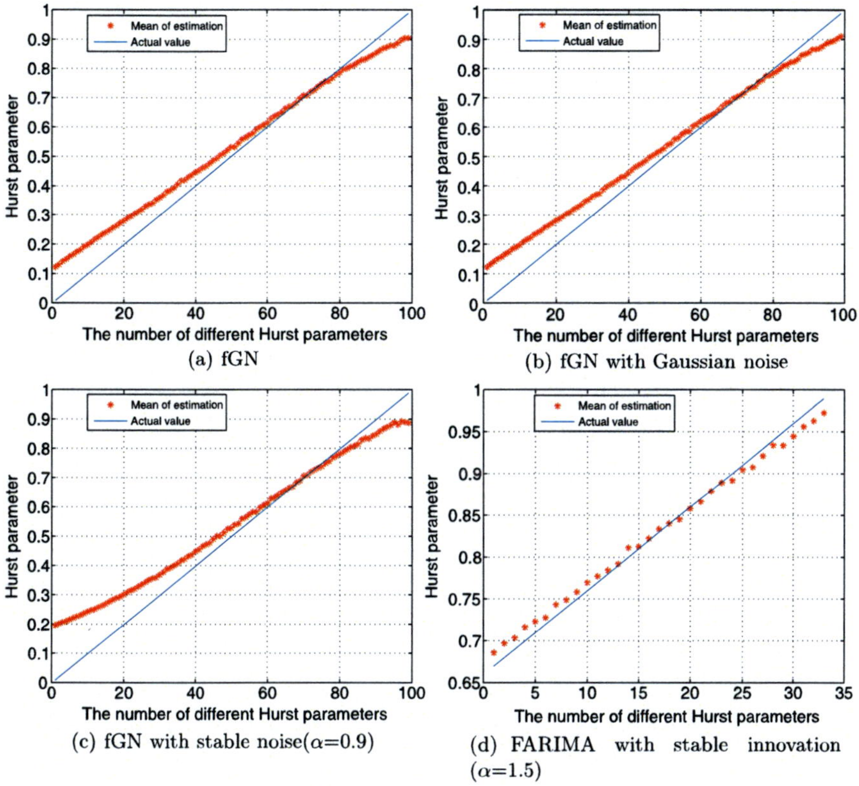

Fig. 3.5 R/S method: **(a)** fGN; **(b)** fGN with 30 dB SNR white Gaussian noise; **(c)** fGN with 30 dB SNR stable noise ($\alpha = 0.9$); **(d)** FARIMA with stable innovations ($\alpha = 1.5$)

Fig. 3.5(b). It is clear that the estimated values of the R/S method are still not satisfactory. Figure 3.5(c) indicates that the R/S method performs worse for LRD3. The Hurst parameter is more seriously overestimated when $0 < H \leq 0.6$ and underestimated when $0.8 \leq H < 1$. For LRD4, its Hurst parameters are slightly overestimated when $0.66 < H < 0.85$ and slightly underestimated when $0.85 \leq H < 1$. The standard deviations for these four LRD time series are all around 0.7. So the R/S method is short of accuracy and robustness for these four types of LRD time series, even though it is one of the best known Hurst estimators.

Results of Aggregated Variance Method

The robustness analysis results of the Aggregated Variance method for four types of LRD time series are presented in Fig. 3.6. The estimation results of this method for noise-free standard LRD time series are shown in Fig. 3.6(a). The Aggregated Variance method is almost unbiased when $0 < H \leq 0.7$, but it is underestimated when $0.7 < H < 1$. From the comparison we can see that, for the LRD process

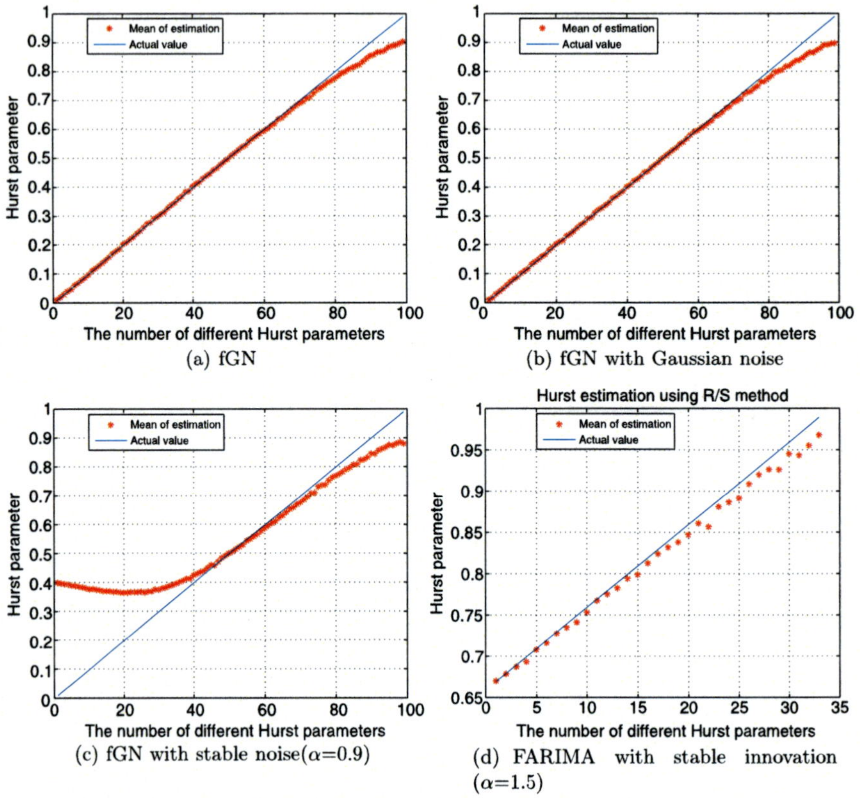

Fig. 3.6 Aggregated Variance method: **(a)** fGN; **(b)** fGN with 30 dB SNR white Gaussian noise; **(c)** fGN with 30 dB SNR stable noise ($\alpha = 0.9$); **(d)** FARIMA with stable innovations ($\alpha = 1.5$)

alone, the Aggregated Variance method is obviously better than the R/S method when $0 < H \leq 0.7$. Figure 3.6(b) presents the analysis results for LRD2. They show clearly that the Aggregated Variance method also performs better than the R/S method for LRD2. The estimation for LRD3 is presented in Fig. 3.6(c). The estimated value of the Aggregated Variance method is influenced obviously by 30 dB SNR stable noise with parameter $\alpha = 0.9$. The estimated Hurst parameters are biased through all the Hurst values. For LRD4, the estimation results are underestimated when $0.7 < H < 1$. The standard deviations for all four LRD time series are around 0.023.

Results of Absolute Value Method

The results of the Absolute Value method are presented in Fig. 3.7. The Absolute Value method has nearly the same estimation performance as the Aggregated Variance method for LRD1 and LRD2. Figures 3.7(a), (b) show that it exhibits almost

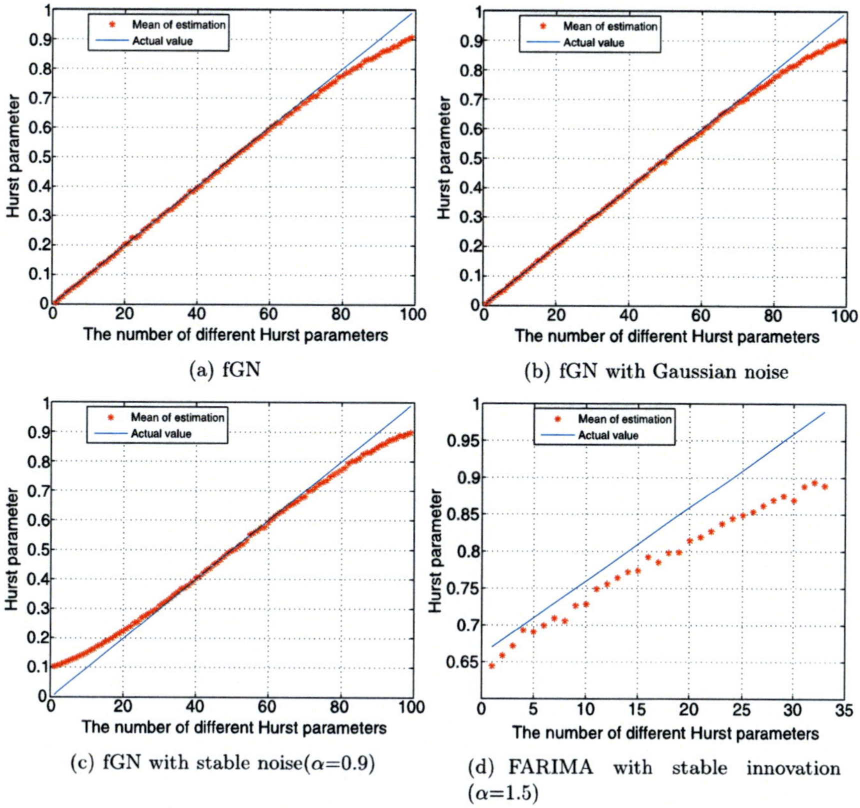

Fig. 3.7 Absolute Value method: **(a)** fGN; **(b)** fGN with 30 dB SNR white Gaussian noise; **(c)** fGN with 30 dB SNR stable noise ($\alpha = 0.9$); **(d)** FARIMA with stable innovations ($\alpha = 1.5$)

unbiased when $0 < H \leq 0.7$, but it is underestimated when $0.7 < H < 1$ for LRD1 and LRD2. Figure 3.7(c) presents the estimation results for LRD3. The analysis results show that the Absolute Value method has better robustness to 30 dB SNR stable noise than the Aggregated Variance method. The Hurst parameter is overestimated when $0 < H \leq 0.5$. The standard deviations for LRD1, LRD2 and LRD3 are around 0.024. For LRD4, it performs worse than the above two estimators. The estimated values are obviously underestimated when $0.66 < H < 1$. The standard deviation for LRD4 is about 0.05.

Results of Variance of Residuals Method

The results of the Variance of Residuals method are presented in Fig. 3.8. The estimated values of the Variance of Residuals method for LRD1 and LRD2 time series are presented in Figs. 3.8(a) and (b), respectively. This method is one of the most accurate among all the twelve estimators for LRD1 and LRD2, because it is unbiased

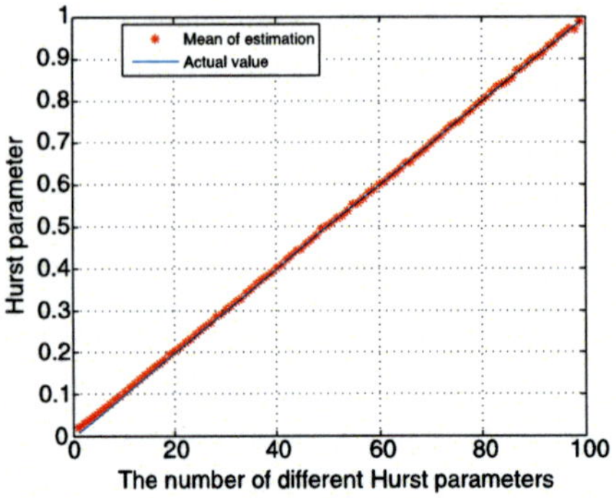

(a) fGN

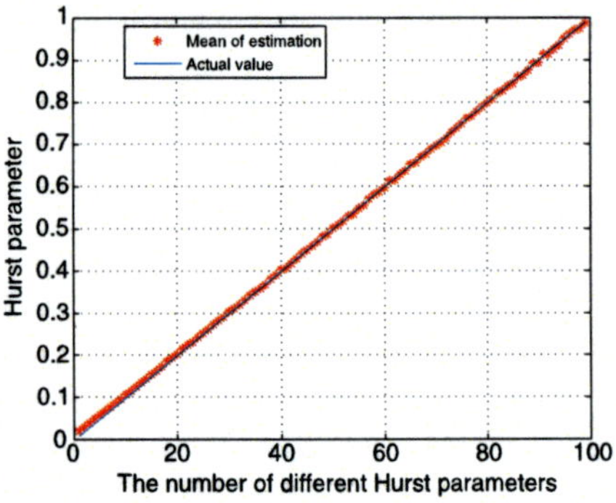

(b) fGN with Gaussian noise

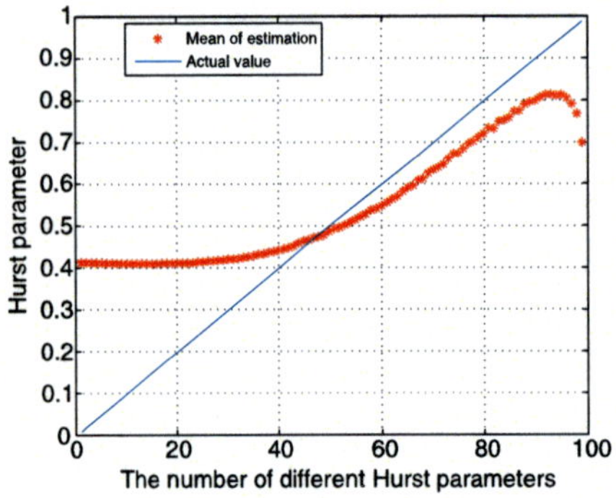

(c) fGN with stable noise(α=0.9)

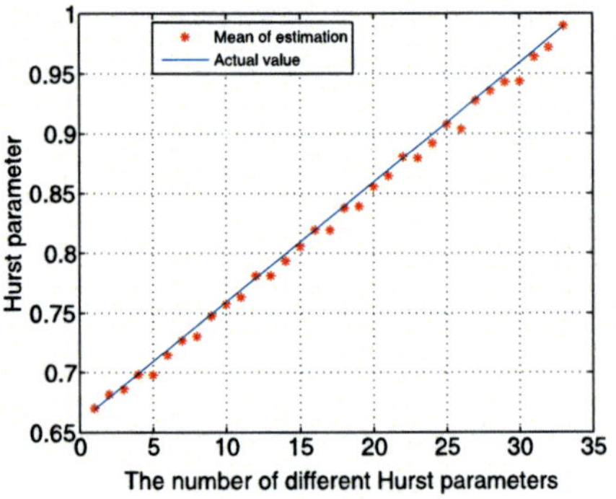

(d) FARIMA with stable innovation (α=1.5)

Fig. 3.8 Variance of Residuals method: **(a)** fGN; **(b)** fGN with 30 dB SNR white Gaussian noise; **(c)** fGN with 30 dB SNR stable noise ($\alpha = 0.9$); **(d)** FARIMA with stable innovations ($\alpha = 1.5$)

through almost all Hurst values ($0 < H < 1$). Figure 3.8(c) shows the estimation results for LRD3 time series. This method is influenced obviously by 30 dB SNR stable noise with parameter $\alpha = 0.9$ through all the Hurst values ($0 < H < 1$). So the robustness of the Variance of Residuals method to impulsive noise is very poor. For FARIMA with stable innovation series ($\alpha = 1.5$, $2/3 \le H < 1$), the estimated values are better than that of the above three methods. The standard deviations of the method for all four LRD time series are around 0.025.

Results of Periodogram Method

The results of the Periodogram method are presented in Fig. 3.9. The estimated values for the LRD1 time series are presented in Fig. 3.9(a). The estimated values of Periodogram method are almost unbiased when $0.25 < H < 1$, but obviously underestimated when $0 < H \le 0.25$. The analysis results for LRD2 are shown in Fig. 3.9(b), with a performance similar to that of the noise-free case. The estimation results when the noise is 30 dB stable noise with parameter $\alpha = 0.9$ are presented

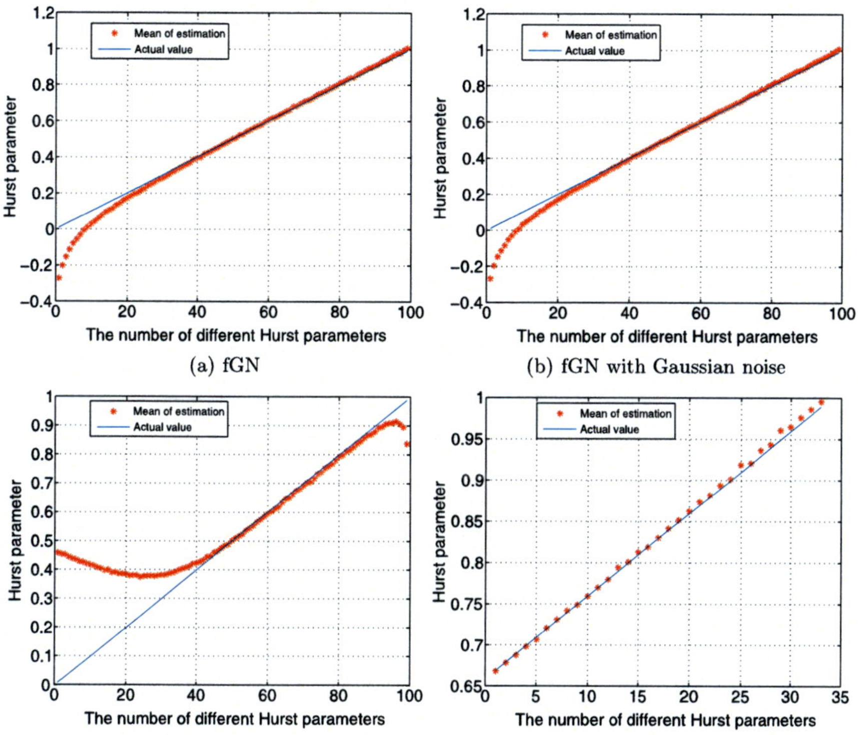

(a) fGN (b) fGN with Gaussian noise (c) fGN with stable noise(α=0.9) (d) FARIMA with stable innovation (α=1.5)

Fig. 3.9 Periodogram method: **(a)** fGN; **(b)** fGN with 30 dB SNR white Gaussian noise; **(c)** fGN with 30 dB SNR stable noise ($\alpha = 0.9$); **(d)** FARIMA with stable innovations ($\alpha = 1.5$)

in Fig. 3.9(c). It is clear that the Periodogram method is influenced by this kind of noise. The H is seriously over-estimated when $0 < H \leq 0.5$, and slightly underestimated when $0.5 \leq H < 1$. So the robustness of the Periodogram method to impulsive noise is poor. As for LRD4, the estimated values are a little bit better than those of the Variance of Residuals method, as seen in Fig. 3.9(d). The standard deviations for all four LRD time series are around 0.020.

Results of Modified Periodogram Method

The results of the Modified Periodogram method are presented in Fig. 3.10. The estimated values of the method for LRD1 and LRD2 time series are presented in Figs. 3.10(a) and (b), respectively. The Modified Periodogram method is almost unbiased when $0.1 < H < 1$, but H is underestimated when $0 < H \leq 0.1$ for LRD1 and LRD2. The estimation results when the noise is 30 dB SNR stable noise with parameter $\alpha = 0.9$ are presented in Fig. 3.10(c). The Modified Periodogram method is influenced slightly by this type of added noise. It accurately estimates the Hurst

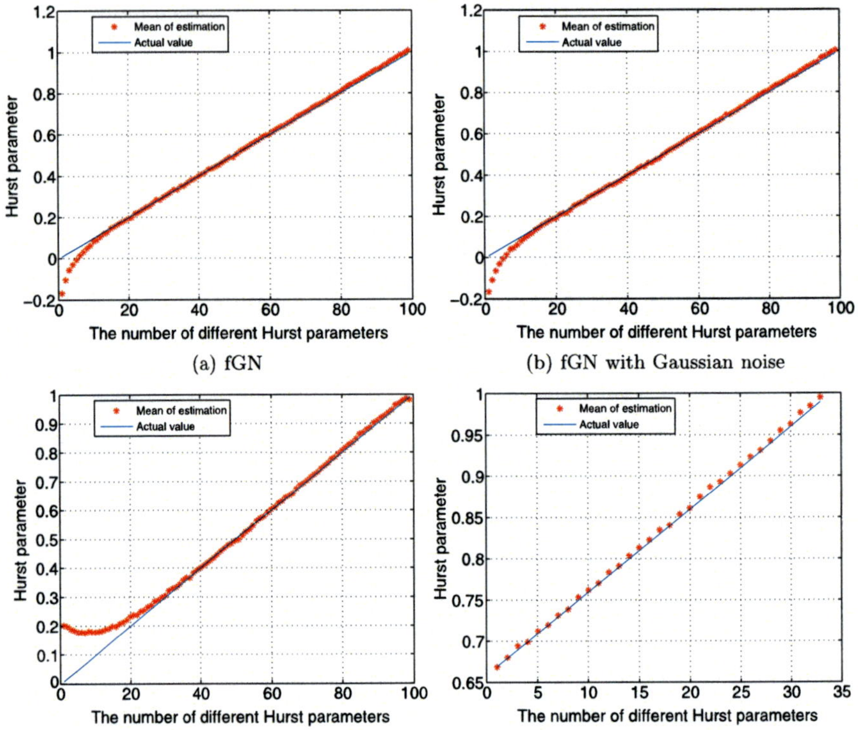

(a) fGN
(b) fGN with Gaussian noise
(c) fGN with stable noise(α=0.9)
(d) FARIMA with stable innovation (α=1.5)

Fig. 3.10 Modified Periodogram method: **(a)** fGN; **(b)** fGN with 30 dB SNR white Gaussian noise; **(c)** fGN with 30 dB SNR stable noise ($\alpha = 0.9$); **(d)** FARIMA with stable innovations ($\alpha = 1.5$)

value when $0.3 < H < 1$. The robustness of the Modified Periodogram method to impulsive noise is better than that of the Periodogram method. As for FARIMA with stable innovation series ($\alpha = 1.5$, $2/3 \le H < 1$), the method has similar performance to that of the Periodogram method. The standard deviations for all four LRD time series are around 0.020.

Results of Whittle Estimator

The results of the Whittle method are presented in Fig. 3.11. The estimated values of the method for LRD1 and LRD2 are presented in Figs. 3.11(a) and (b). With almost the same performance, the Whittle method is almost unbiased when $0.1 < H < 1$, and a little bit underestimated when $0 < H \le 0.1$. It can be seen in Fig. 3.11(c) that the Whittle method is influenced slightly by 30 dB SNR stable noise with parameter $\alpha = 0.9$ when $0.25 < H < 1$, but obviously overestimated

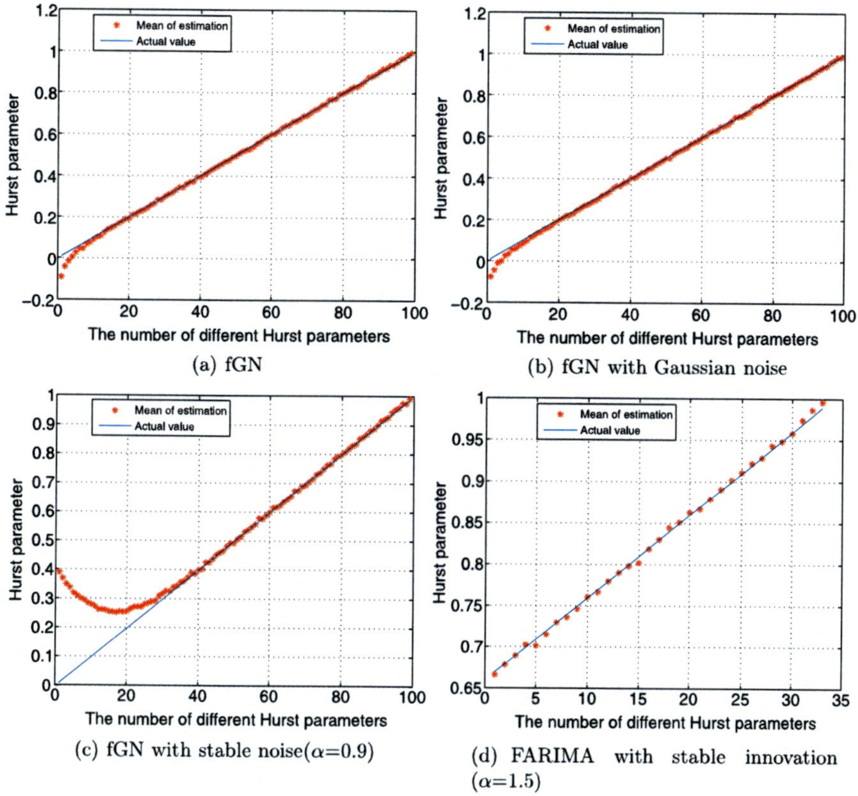

Fig. 3.11 Whittle method: **(a)** fGN; **(b)** fGN with 30 dB SNR white Gaussian noise; **(c)** fGN with 30 dB SNR stable noise ($\alpha = 0.9$); **(d)** FARIMA with stable innovations ($\alpha = 1.5$)

when $0 < H \leq 0.25$. Generally, the Whittle method has good robustness to impulsive noise. For the LRD4 time series, the method performs very well as shown in Fig. 3.11(d). The standard deviations for all four LRD time series are around 0.030.

Results of Diffusion Entropy Method

The results of the Diffusion Entropy method are presented in Fig. 3.12. The estimated values of the method for LRD1 and LRD2 are unbiased when $0.1 < H \leq 0.7$, and a little bit underestimated when $0.7 < H < 1$, as shown in Figs. 3.12(a) and (b). The results for LRD3 are presented in Fig. 3.12(c). It can be seen that the Diffusion Entropy method is influenced slightly by 30 dB SNR stable noise with parameter $\alpha = 0.9$. Compared with the results for LRD1, the H is only slightly over-estimated when $0 < H \leq 0.25$. So, the Diffusion Entropy method has good robustness to impulsive noise. For LRD4, it has a similar performance to that of the Absolute Value method. The standard deviations for all four LRD time series are around 0.015. In

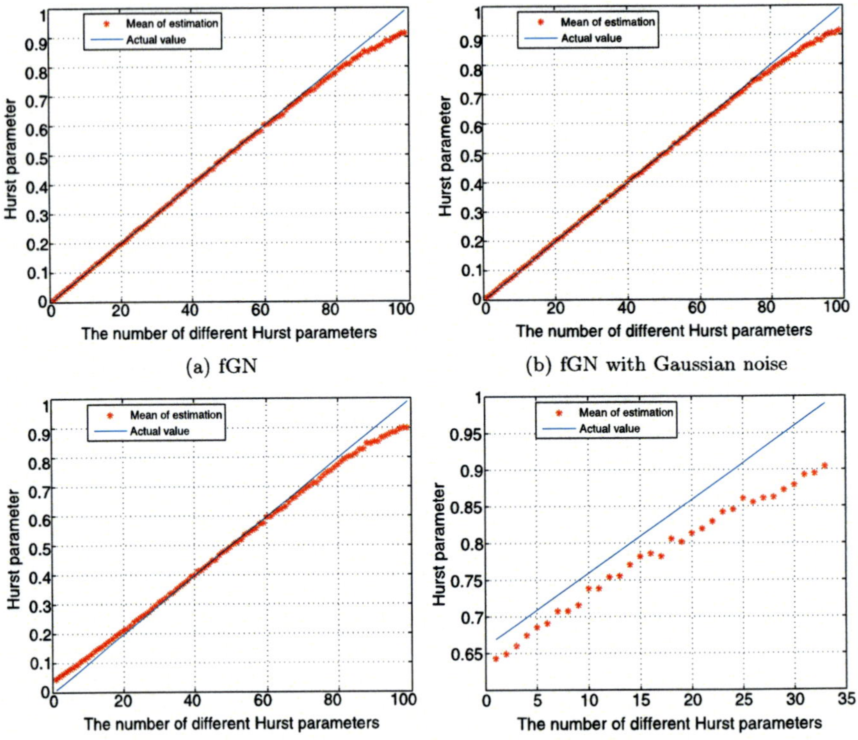

(a) fGN (b) fGN with Gaussian noise (c) fGN with stable noise(α=0.9) (d) FARIMA with stable innovation (α=1.5)

Fig. 3.12 Diffusion Entropy method: **(a)** fGN; **(b)** fGN with 30 dB SNR white Gaussian noise; **(c)** fGN with 30 dB SNR stable noise ($\alpha = 0.9$); **(d)** FARIMA with stable innovations ($\alpha = 1.5$)

practice, the Diffusion Entropy method can provide better estimation results when the heavy-tailed distribution parameter α of the time series is unknown.

Results of Kettani and Gubner's Method

The results of Kettani and Gubner's method are presented in Fig. 3.13. The estimated values of the method for LRD1 and LRD2 are quite similar to that of the Diffusion Entropy method. The estimated values are almost unbiased when $0.1 < H \leq 0.8$, and a little bit underestimated when $0.8 < H < 1$. The analysis results for LRD1 and LRD2 are presented in Figs. 3.13(a) and (b), respectively. The results for LRD3 are presented in Fig. 3.13(c). It can be seen that Kettani and Gubner's method is influenced slightly by 30 dB SNR stable noise with parameter $\alpha = 0.9$. The method has good robustness to impulsive noise. The standard deviations for LRD1, LRD2 and LRD3 are around 0.008. For LRD4, Kettani and Gubner's method performs as well as the Whittle method. The standard deviation for

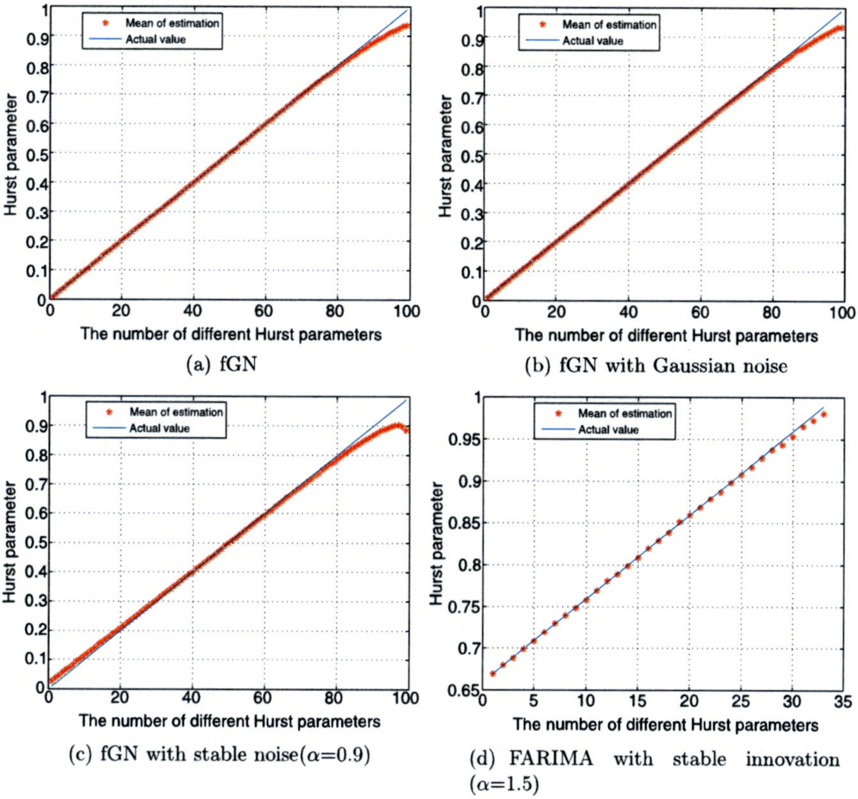

Fig. 3.13 Kettani and Gubner's method: **(a)** fGN; **(b)** fGN with 30 dB SNR white Gaussian noise; **(c)** fGN with 30 dB SNR stable noise ($\alpha = 0.9$); **(d)** FARIMA with stable innovations ($\alpha = 1.5$)

LRD4 is about 0.1766. From the above analysis results we can see that Kettani and Gubner's method has very good robustness to these four types of LRD time series.

Results of Abry and Veitch's Method

The results of Abry and Veitch's method are presented in Fig. 3.14. It can be seen from Figs. 3.14(a) and (b) that the estimated values of this method for LRD1 and LRD2 fluctuate around the true Hurst values when $0.2 < H < 1$. The estimated values are obviously underestimated when $0 < H \leq 0.2$. The results for LRD3 are presented in Fig. 3.14(c). Abry and Veitch's method is influenced slightly by 30 dB SNR stable noise with parameter $\alpha = 0.9$. The H is slightly over-estimated when $0 < H \leq 0.4$, and the estimated values fluctuate around the true values when $0.4 < H < 1$. So, Abry and Veitch's method has good robustness to impulsive noise. The standard deviations for LRD1, LRD2 and LRD3 are around 0.053. As for FARIMA with stable innovation series ($\alpha = 1.5$, $2/3 \leq H < 1$), the H is also fluctuating

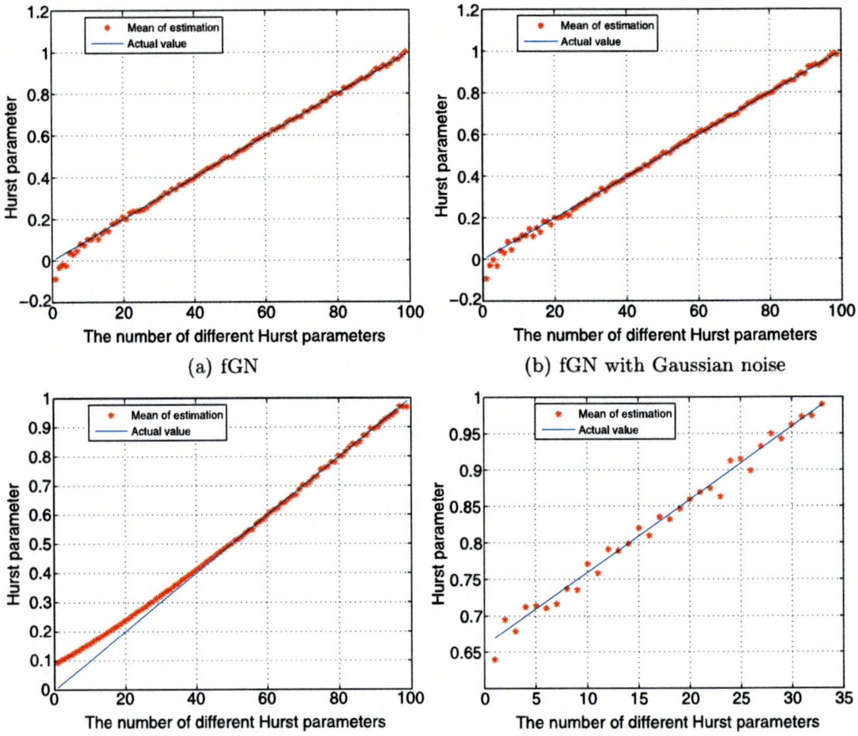

(a) fGN

(b) fGN with Gaussian noise

(c) fGN with stable noise(α=0.9)

(d) FARIMA with stable innovation (α=1.5)

Fig. 3.14 Abry and Veitch's method: **(a)** fGN; **(b)** fGN with 30 dB SNR white Gaussian noise; **(c)** fGN with 30 dB SNR stable noise ($\alpha = 0.9$); **(d)** FARIMA with stable innovations ($\alpha = 1.5$)

around the true Hurst value as seen in Fig. 3.14(d). The standard deviation for LRD4 is about 0.1766.

Results of Koutsoyiannis' Method

The results of Koutsoyiannis' method are presented in Fig. 3.15. The estimated values of the method for LRD1 and LRD2 time series are presented in Figs. 3.15(a) and (b), respectively. This method is the most accurate one among all the estimators for the standard LRD process and the 30 dB SNR white Gaussian noise corrupted LRD time series. It is unbiased almost through all Hurst values ($0 < H < 1$). The results for LRD3 are presented in Fig. 3.15(c). Koutsoyiannis' method shows almost no influence from the 30 dB SNR stable noise with parameter $\alpha = 0.9$. The H is only slightly underestimated when $0.97 < H < 1$. So, the Koutsoyiannis' method has very good robustness to impulsive noise. The standard deviations for LRD1, LRD2 and LRD3 are around 0.007. But for LRD4, the estimate results are underestimated

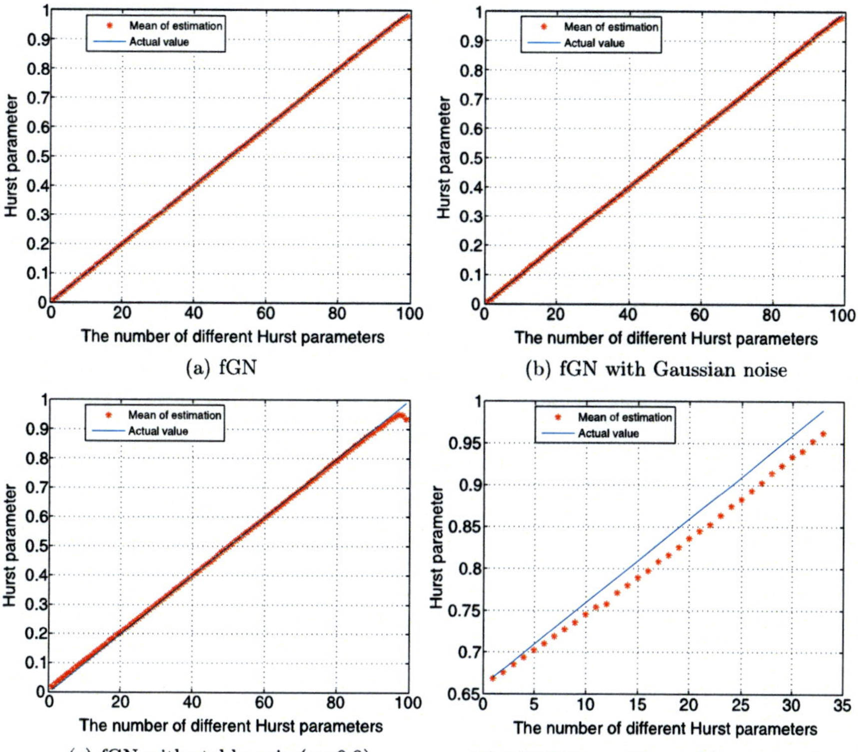

(a) fGN

(b) fGN with Gaussian noise

(c) fGN with stable noise(α=0.9)

(d) FARIMA with stable innovation (α=1.5)

Fig. 3.15 Koutsoyiannis' method: **(a)** fGN; **(b)** fGN with 30 dB SNR white Gaussian noise; **(c)** fGN with 30 dB SNR stable noise ($\alpha = 0.9$); **(d)** FARIMA with stable innovations ($\alpha = 1.5$)

when $0.66 < H < 1$. The standard deviation for LRD4 is about 0.1772. From the above analysis results we can conclude that Koutsoyiannis' method has very good robustness for LRD1, LRD2 and LRD3 time series, but has poor robustness for LRD4.

Results of Higuchi's Method

The results of Higuchi's method are presented in Fig. 3.16. The estimated values of Higuchi's method for LRD1 are almost unbiased when $0.1 < H \leq 0.8$, and a little bit underestimated when $0.8 < H < 1$. Figure 3.16(b) presents the analysis results for LRD2. Higuchi's method also has almost the same estimations for LRD1 and LRD2. The results for LRD3 are presented in Fig. 3.16(c). It can be seen that Higuchi's method is influenced slightly by 30 dB SNR stable noise with parameter $\alpha = 0.9$. The H is slightly over-estimated when $0 < H \leq 0.3$, and slightly underestimated when $0.8 < H < 1$. For LRD4, the method has similar performance to that

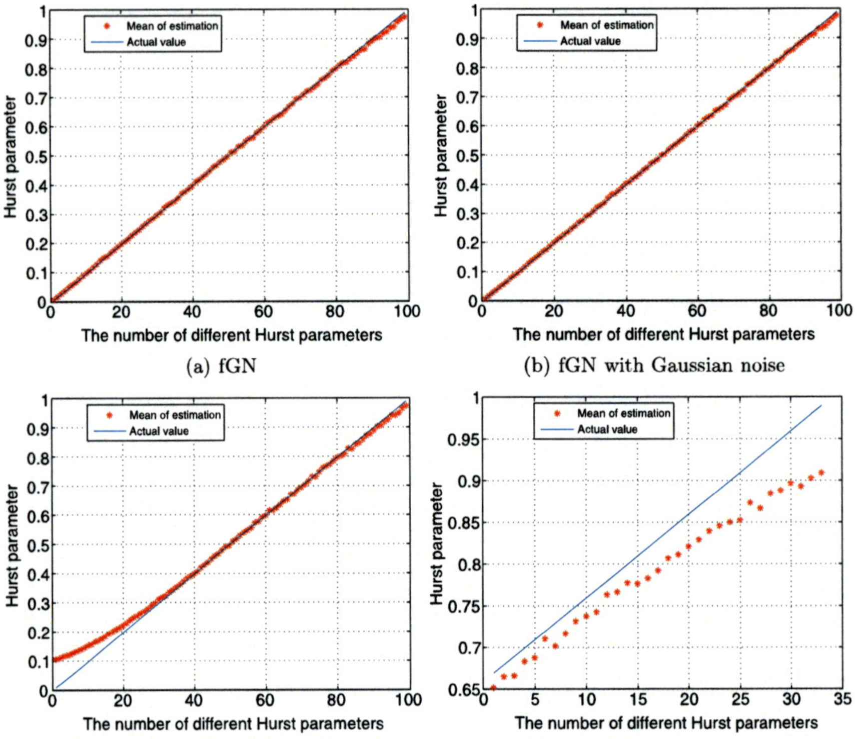

(a) fGN

(b) fGN with Gaussian noise

(c) fGN with stable noise(α=0.9)

(d) FARIMA with stable innovation (α=1.5)

Fig. 3.16 Higuchi's method: **(a)** fGN; **(b)** fGN with 30 dB SNR white Gaussian noise; **(c)** fGN with 30 dB SNR stable noise ($\alpha = 0.9$); **(d)** FARIMA with stable innovations ($\alpha = 1.5$)

of the Diffusion Entropy method. The standard deviations for all four LRD time series are around 0.03.

3.3.3 Quantitative Robustness Comparison and Guideline for Selection Estimator

From Figs. 3.5 to 3.16 we can roughly compare the robustness of the twelve Hurst parameter estimators. In order to quantify the robustness more accurately, we calculate the standard errors S of different estimators. S is defined as

$$S = \sqrt{\frac{\sum_{i=1}^{n}(x_i - u_i)^2}{n-1}}, \tag{3.39}$$

where u_i is the true value of the Hurst parameter and x_i is the estimated value of the Hurst parameter.

Table 3.1 Robustness comparison for noise-free LRD process; LRD process with 30 dB SNR white Gaussian noise added; LRD process with 30 dB SNR α-stable noise ($\alpha = 0.9$) added and FARIMA time series with α-stable innovations ($\alpha = 1.5$)

	fGN		fGN + Gaussian	
	$H \in (0, 1)$	$H \in (0.66, 1)$	$H \in (0, 1)$	$H \in (0.66, 1)$
R/S	0.0595	0.0228	0.0595	0.0224
Aggregated Variance	0.0251	0.0249	0.0255	0.0254
Absolute value	0.0249	0.0247	0.0253	0.0251
Variance of Residuals	0.0043	0.0019	0.0043	0.0020
Periodogram	0.0524	0.0063	0.0522	0.0063
Modified Periodogram	0.0271	0.0061	0.0271	0.0060
Whittle	0.0139	0.0025	0.0135	0.0021
Diffusion Entropy	0.0216	0.0214	0.0209	0.0207
Kettani and Gubner	0.0124	0.0123	0.0125	0.0125
Abry and Veitch	0.0152	0.0032	0.0161	0.0036
Koutsoyiannis	0.0014	0.0013	0.0015	0.0013
Higuchi	0.0052	0.0049	0.0049	0.0046

	fGN + stable		Stable FARIMA
	$H \in (0, 1)$	$H \in (0.66, 1)$	$H \in (0.66, 1)$
R/S	0.0787	0.0246	0.0107
Aggregated Variance	0.0934	0.0267	0.0123
Absolute value	0.0372	0.0256	0.0512
Variance of Residuals	0.1670	0.0636	0.0067
Periodogram	0.1544	0.0209	0.0042
Modified Periodogram	0.0469	0.0048	0.0040
Whittle	0.0965	0.0020	0.0035
Diffusion Entropy	0.0275	0.0245	0.0499
Kettani and Gubner	0.0214	0.0200	0.0032
Abry and Veitch	0.0303	0.0035	0.0112
Koutsoyiannis	0.0088	0.0075	0.0209
Higuchi	0.0279	0.0053	0.0431

Table 3.1 presents the standard errors of twelve estimators for four types of LRD time series. However, for FARIMA with stable innovations ($\alpha = 1.5, 2/3 \le H < 1$), we can only estimate the Hurst $H \in (0.66, 1)$. In order to compare the robustness of all twelve estimators for four types of LRD time series, we calculate the standard errors of $H \in (0, 1)$ and $H \in (0.66, 1)$ respectively. The comparison shows that 30 dB SNR white Gaussian noise almost has no influence on the accuracy of all estimators. Among all estimators, the Variance of Residuals method, Koutsoyiannis' method and Higuchi's method have the most accurate estimate values when

$H \in (0, 1)$ for LRD1 and LRD2. But the impulsive noise influences the accuracy obviously for most of estimators. Koutsoyiannis' method has the most accurate estimated value when $H \in (0, 1)$ for LRD3. As for the LRD time series with infinite variance, Kettani and Gubner's method, and the Whittle method have the best accurate estimate values. From the above robustness analysis we can say that, among all twelve estimators, Kettani and Gubner's method has the best robustness for noise corrupted LRD processes and FARIMA with stable innovations.

3.4 Chapter Summary

This chapter deals with the constant-order fractional processes and the Hurst parameter estimators evaluation. Typical constant-order fractional processes including fBm, fGn, fractional stable motion, and fractional stable noise were introduced in Sect. 3.1. Twelve Hurst parameter estimators were briefly introduced in Sect. 3.2. Section 3.3 aims at characterizing the robustness of twelve useful Hurst estimators for the LRD processes compared with noise corrupted LRD and LRD with infinite variance. From the analysis results we can conclude that, of all twelve estimators, the Kettani and Gubner's method performed the best on noise corrupted LRD time series and FARIMA time series with stable innovations. Although the Variance of Residuals method performed very well on noise-free LRD time series, it lacks robustness for time series corrupted with impulsive noise. The Koutsoyiannis' method has accurate estimate results for noise-free LRD, and has very good robustness to 30 dB SNR white Gaussian noise, and 30 dB SNR impulsive noise, but it has poor robustness for the LRD time series with infinite variance. In the next chapter, we will introduce the multifractional processes. Different from the LRD processes with constant long memory parameter, the multifractional processes exhibit a local memory property.

Chapter 4
Multifractional Processes

As shown in Chap. 3, constant-order fractional processes with a constant Hurst parameter H can be used to accurately characterize the long memory process and the short-range dependent stochastic processes [73, 123]. However, the results of recent empirical analyses of complex nonlinear dynamic systems have raised questions about the capabilities of constant-order fractional processes [39, 90]. At the root of these questions is the failure of the constant Hurst parameter to capture the multiscaling or multifractal characteristic of stochastic processes. In order to accurately analyze the complex signals which exhibit a local self-similarity property, multifractional processes with variable local Hölder exponent $H(t)$ were proposed in [232]. In recent years, it has been demonstrated that multifractional processes are able to describe complex or chaotic phenomena in several fields of sciences [134, 211, 232]. The multifractional processes are the extension of fractional processes by generalizing the constant Hurst parameter H to the case where H is indexed by a time-dependent local Hölder exponent $H(t)$ [232]. mGn and mBm are typical examples of multifractional processes. Multifractional stable noise and the multifractional stable motion are typical examples of multifractional processes with infinite second-order statistics.

To better take advantage of multifractional processes, efforts have been made to estimate the local Hölder exponent $H(t)$ [66, 134]. According to the local properties of the local Hölder exponent, the sliding window method is an effective way to estimate $H(t)$, but it is a difficult task. In practice, the signals measured are always corrupted by various types of noises, and sometimes even the signal itself may have infinite second-order statistics. In this chapter, after introducing some typical multifractional processes, we investigate both the tracking performance and the robustness issue of the twelve sliding-windowed Hurst estimators for noisy multifractional processes and for multifractional processes with infinite second-order statistics.

H. Sheng et al., *Fractional Processes and Fractional-Order Signal Processing*, Signals and Communication Technology,
DOI 10.1007/978-1-4471-2233-3_4, © Springer-Verlag London Limited 2012

4.1 Multifractional Processes

4.1.1 Multifractional Brownian Motion and Multifractional Gaussian Noise

The generalized fBm with time-varying local Hölder exponent is known as the multifractional Brownian motion (mBm) [232].

Definition 4.1 The mBm can be represented in an integral form [232],

$$B_{H(t)}(t) = \frac{\sigma}{\Gamma(H(t)+1/2)} \left\{ \int_{-\infty}^{0} [(t-s)^{H(t)-1/2} - (-s)^{H(t)-1/2}] \mathrm{d}B(s) + \int_{0}^{t} (t-s)^{H(t)-1/2} \mathrm{d}B(s) \right\}, \tag{4.1}$$

where $B(s)$ is the standard Brownian motion, $\sigma^2 = \mathrm{var}(B_{H(t)}(t))|_{t=1}$, and $\mathrm{var}(X)$ stands for the variance of X.

Based on this definition, the following statistical properties are summarized [134]:

- Mean value of mBm is

$$E[B_{H(t)}(t)] = 0. \tag{4.2}$$

- Variance function of mBm is

$$\mathrm{var}\left[B_{H(t)}(t)\right] = \frac{\sigma^2}{2}|t|^{2H(t)}. \tag{4.3}$$

From this mBm representation, it can be seen that the fBm is a special case of the mBm with a constant local Hölder exponent $H(t) = H$. Compared with the fBm which has stationary increments, the mBm is a non-stationary Gaussian process and in general does not possess independent stationary increments.

Definition 4.2 The harmonizable integral representation of mBm is given by [288]

$$B_{H(t)}(t) = \mathrm{Re} \int_{\mathbb{R}} \frac{e^{i\xi t} - 1}{|\xi|^{H(t)+1/2}} \tilde{W}(\mathrm{d}\xi), \tag{4.4}$$

where $\tilde{W} = W_1 + jW_2$, and W_1 and W_2 are two independent Wiener processes.

Based on the Riemann-Liouville fractional-order integral, mBm can also be described as [172]

$$B_{H(t)}(t) = {}_0D_t^{-1-\alpha(t)}\omega(t)$$
$$= \frac{1}{\Gamma(H(t)+1/2)}\int_0^t (t-\tau)^{H(t)-1/2}\omega(\tau)\mathrm{d}\tau, \quad 1/2 < H(t) < 1 \tag{4.5}$$

where $\omega(t)$ is wGn.

mGn is defined as the derivative of mBm [134]

$$G_{H(k),H(k-1)}(k) = B_{H(k)}(k) - B_{H(k-1)}(k-1), \quad k \in \mathbb{Z}. \tag{4.6}$$

Since the mGn, in general, does not have stationary increments, the mGn is a non-stationary process, but mGn provides a better model for non-stationary, nonlinear dynamic systems.

4.1.2 Linear Multifractional Stable Motion and Multifractional Stable Noise

The linear multifractional stable motion (LMFSM) is obtained by generalizing the constant Hurst parameter H to a time-dependent local Hölder exponent $H(t)$.

Definition 4.3 LMFSM is defined as [286]

$$L_{\alpha,H(t)} = \int_{-\infty}^{\infty}\Big\{a\Big[(t-s)_+{}^{H(t)-1/\alpha} - (-s)_+{}^{H(t)-1/\alpha}\Big]$$
$$+ b\Big[(t-s)_-{}^{H(t)-1/\alpha} - (-s)_-{}^{H(t)-1/\alpha}\Big]\Big\} M_{\alpha,\beta}(\mathrm{d}s), \tag{4.7}$$

where a, b are real constants, $M_{\alpha,\beta}$ is an α-stable random noise, $|a| + |b| > 0$, $0 < \alpha < 2$, $0 < H(t) < 1$ and for $x \in \mathbb{R}$

$$(x)_+^a = \begin{cases} x_+^a, & \text{when } x > 0, \\ 0, & \text{when } x \le 0, \end{cases} \tag{4.8}$$

and $x_-^a = (-x)_+^a$.

The increments of LMFSM is the multifractional stable noise. LMFSM and multifractional stable noise provide better characterizations for stochastic processes with local self-similarity character and heavy-tailed distribution.

4.2 Tracking Performance and Robustness of Local Hölder Exponent Estimator

In this section, we concentrate on the tracking performance and the robustness analysis of twelve sliding-windowed Hurst estimators, which were introduced in the previous chapter for multifractional processes. The tracking performance and the robustness analysis are based on some valuable efforts [65, 244, 263, 267, 297] made to evaluate the accuracy and validity of the constant Hurst estimators. It is known that most of the constant Hurst estimators are based on the characteristics of the power-law auto-covariance function and the strong dependence over large temporal lags, so these estimators can only be used to estimate the constant Hurst exponent of the whole time series. For multifractional Gaussian processes with a time varying local Hölder exponent, however, these methods cannot be directly used to estimate the time varying Hurst exponent $H(t)$. According to the local properties of the local Hölder exponent, it is easiest and most natural to combine these estimators with the sliding window method, and simply call them sliding-windowed Hurst estimators. In this study, an mGn and a multifractional α-stable noise with $H(t) = at + b$ $(0 < t < 1000)$, where the sampling interval is 0.01 second, are generated using the variable-order fractional calculus based synthetic method [271]. Furthermore, in our study, two types of noises are added to mGn when we analyze the accuracy and the robustness of sliding-windowed Hurst estimators: Gaussian and non-Gaussian. Most non-Gaussian noises are impulsive and can be well characterized by α-stable noise. Therefore, a 30 dB SNR Gaussian noise and 30 dB SNR α-stable noise are employed to test the robustness of the twelve local Hölder estimators.

4.2.1 Test Signal Generation and Estimation Procedures

As studied in Chap. 2, multifractional processes can be considered as the output of a variable-order fractional differential system represented by a variable-order fractional differential equation. The mGn can be generated by a variable-order fractional integrator with white noise excitation, and the multifractional stable noise can be simulated by a variable-order fractional integrator with white stable noise excitation. The detailed introduction of the synthesis method represented in Chap. 6. So, the algorithm in [271] and the related MATLAB® code [289] are used to generate the mGn and multifractional stable noise. Specifically, an mGn and a multifractional α-stable $(\alpha = 1.8)$ noise with $H(t) = at + b$, where $a = 4 \times 10^{-4}$ and $b = 0.5$, are generated using the above variable-order fractional calculus based synthesis method, where $0 < t < 1000$ (seconds), and the sampling frequency is 100 Hz. Figure 4.1 presents the synthesized mGn and multifractional stable noise. These two synthesized processes are used in the next section to evaluate the tracking performance and robustness of these twelve sliding-windowed Hurst estimators.

In order to accurately estimate the performance of these twelve Hurst parameters for a multifractional process and a multifractional stable process, the mGn and multifractional α-stable $(\alpha = 1.8)$ noise with $H(t) = at + b$ are synthesized. The reason we set the parameter of multifractional α-stable noise $\alpha = 1.8$, is that the estimation results of sliding-windowed Hurst estimators are greatly influence by the stable parameter α. The smaller the α is, the worse the estimation results are for multifractional stable process. Furthermore, in order to analyze the performance of these

Fig. 4.1 Synthesized mGn and multifractional stable noises

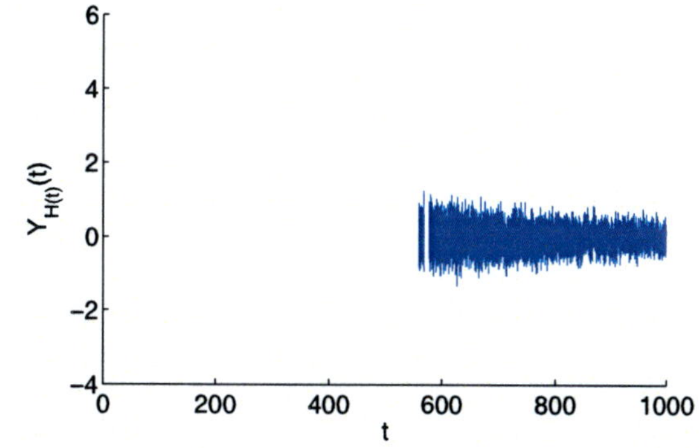

(a) mGn with $H(t) = at + b$, where $a = 4 \times 10^{-4}$, $b = 0.5$

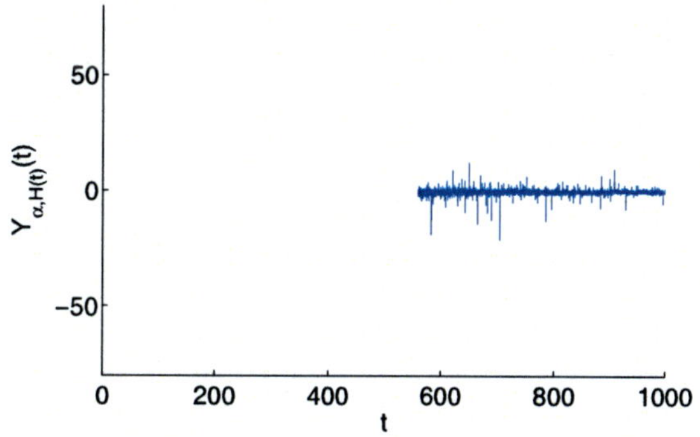

(b) multifractional α-stable noise with $H(t) = at + b$, where $a = 4 \times 10^{-4}$, $b = 0.5$

twelve estimators for additive noise corrupted multifractional processes, two types of noises, Gaussian and non-Gaussian are added. The α-stable noise is employed to simulate the impulsive non-Gaussian noise. Combined with sliding window-based method, these twelve sliding-windowed Hurst estimators are tested using the above synthesized processes.

The window size has a significant influence on the estimation results. A small window size may lead to instability, and a large window size cannot well capture the local property of the processes. So different window sizes are investigated. Figure 4.2 shows the estimation results of sliding-windowed Higuchi's method with window size $W_t = 1000$, $W_t = 2000$, $W_t = 4000$, and $W_t = 8000$, respectively. The blue lines in the figures are the true $H(t)$, and the red lines are estimated local Hölder exponents $\widehat{H}(t)$. It can be seen that the estimation results of the sliding windowed Higuchi's method become more smooth as the window size increases. However, to avoid loss of the local property, the window size cannot be too big. In order to obtain clear and accurate estimation results, the window size $W_t = 8000$ is chosen for all twelve sliding-windowed Hurst estimators.

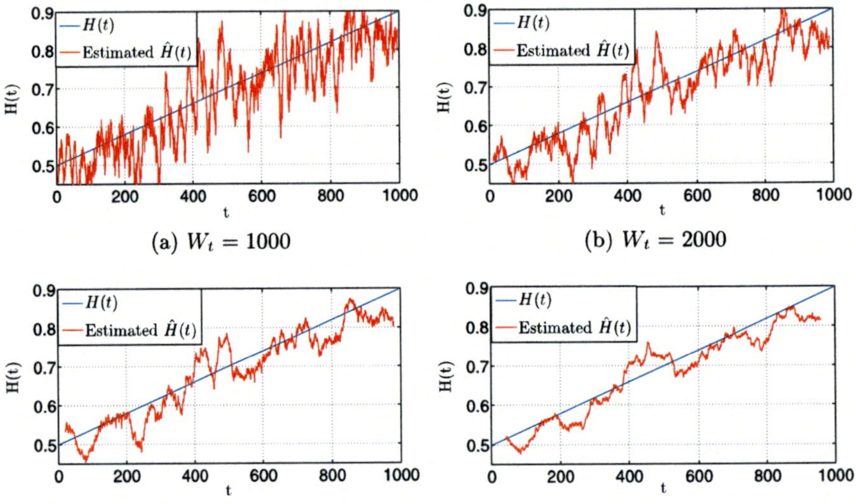

(a) $W_t = 1000$ (b) $W_t = 2000$

(c) $W_t = 4000$ (d) $W_t = 8000$

Fig. 4.2 Estimations results of sliding-windowed Higuchi's method with window size: **(a)** $W_t = 1000$; **(b)** $W_t = 2000$; **(c)** $W_t = 4000$; **(d)** $W_t = 8000$

4.2.2 Estimation Results

In this subsection, we analyze the tracking performance and robustness of twelve sliding-windowed Hurst estimators. As mentioned above, the twelve estimators are tested using four different types of multifractional processes:

1. (MGN1) mGn with $H(t) = at + b$ alone, where $a = 4 \times 10^{-4}$, $b = 0.5$, $0 < t < 1000$, and $H(t) \in (0.5, 1)$
2. (MGN2) mGn with $H(t) = at + b$, which is corrupted by 30 dB SNR white Gaussian noise, where $a = 4 \times 10^{-4}$, $b = 0.5$, $0 < t < 1000$, $H(t) \in (0.5, 1)$
3. (MGN3) mGn with $H(t) = at + b$, which is corrupted by 30 dB SNR α-stable noise, where $a = 4 \times 10^{-4}$, $b = 0.5$, $0 < t < 1000$, $H(t) \in (0.5, 1)$, and stable parameter $\alpha = 1.0$
4. (MGN4) multifractional α-stable process with $H(t) = at + b$, where $a = 4 \times 10^{-4}$, $b = 0.5$, $0 < t < 1000$, $H(t) \in (0.5, 1)$, and stable parameter $\alpha = 1.8$

The sampling interval of these multifractional processes is 0.01, and the window size $W_t = 8000$ are set for all twelve sliding-windowed Hurst parameter estimators. Figures 4.3 through 4.14 show the estimation results of all twelve sliding-windowed Hurst estimators for these four types of multifractional processes. In all these figures, the blue lines are true local Hölder exponents $H(t)$, and the red lines are estimated Hölder exponents $\widehat{H}(t)$.

The tracking performance and the robustness analysis results of the sliding-windowed R/S method for four types of multifractional processes are presented in Fig. 4.3. It can be seen that this method is almost biased through $0 < t < 1000$ for

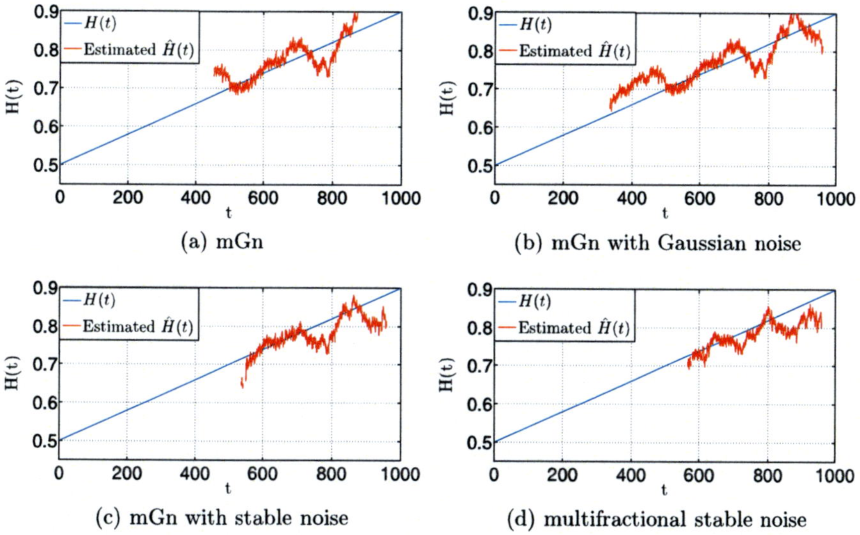

(a) mGn (b) mGn with Gaussian noise

(c) mGn with stable noise (d) multifractional stable noise

Fig. 4.3 Sliding-windowed R/S method: **(a)** mGn; **(b)** mGn with 30 dB SNR white Gaussian noise; **(c)** mGn with 30 dB SNR α-stable ($\alpha = 1.0$) noise; **(d)** multifractional stable noise

MGN1. The analysis result for MGN2 is presented in Fig. 4.3(b). It is clear that the estimated result for MGN2 is still not satisfactory. Figure 4.3(c) indicates that the 30 dB SNR α-stable ($\alpha = 1.0$) noise has little influence on the estimation result of the sliding-windowed R/S method, but the estimated result is biased. Figure 4.3(d) shows that the estimated result for MGN4 is a little bit better than that for noise-free mGn, but still has big errors when $300 < t < 400$ and $800 < t < 1000$.

The tracking performance and the robustness analysis results of the sliding-windowed Aggregated Variance method for four types of multifractional processes are presented in Fig. 4.4. The estimation results of this method for MGN1 and MGN2 are not accurate. It can be seen from Fig. 4.4(c) that this method is influenced obviously by 30 dB SNR α-stable ($\alpha = 1.0$) noise. The estimated local Hölder exponent $\widehat{H}(t)$ is obviously underestimated when $800 < t < 1000$. Similar to the R/S method, the estimated result for MGN4 is a little bit better than that for noise-free mGn. Therefore, the tracking performance of the sliding-windowed Aggregated Variance method for noise-free multifractional process is not satisfactory, and the robustness to 30 dB SNR α-stable ($\alpha = 1.0$) noise is also not good.

The tracking performance and the robustness analysis results of the sliding-windowed Absolute Value method for four types of multifractional processes are presented in Fig. 4.5. For MGN1 and MGN2, the estimation results are almost the same as that of the sliding-windowed Aggregated Variance method. Figure 4.5(c) presents the estimation result for MGN3. Compared with the sliding-windowed Aggregated Variance method, this method has better robustness to 30 dB SNR α-stable ($\alpha = 1.0$) noise. For MGN4, this method performs worse than the above two estimators.

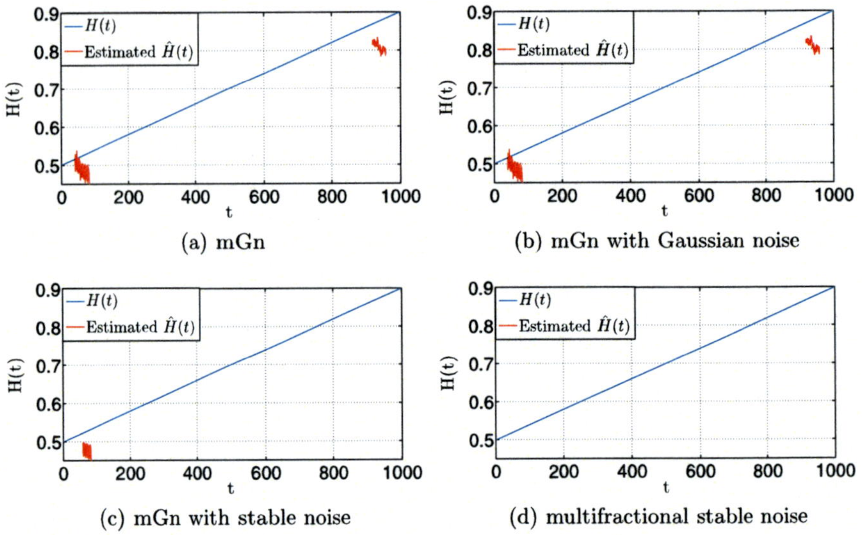

Fig. 4.4 Sliding-windowed Aggregated Variance method: **(a)** mGn; **(b)** mGn with 30 dB SNR white Gaussian noise; **(c)** mGn with 30 dB SNR α-stable ($\alpha = 1.0$) noise; **(d)** multifractional stable noise

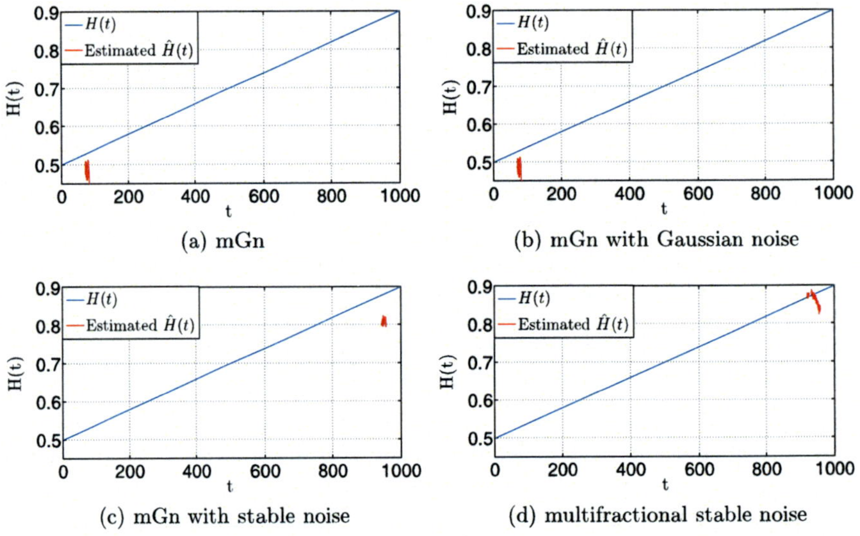

Fig. 4.5 Sliding-windowed Absolute Value method: **(a)** mGn; **(b)** mGn with 30 dB SNR white Gaussian noise; **(c)** mGn with 30 dB SNR α-stable ($\alpha = 1.0$) noise; **(d)** multifractional stable noise

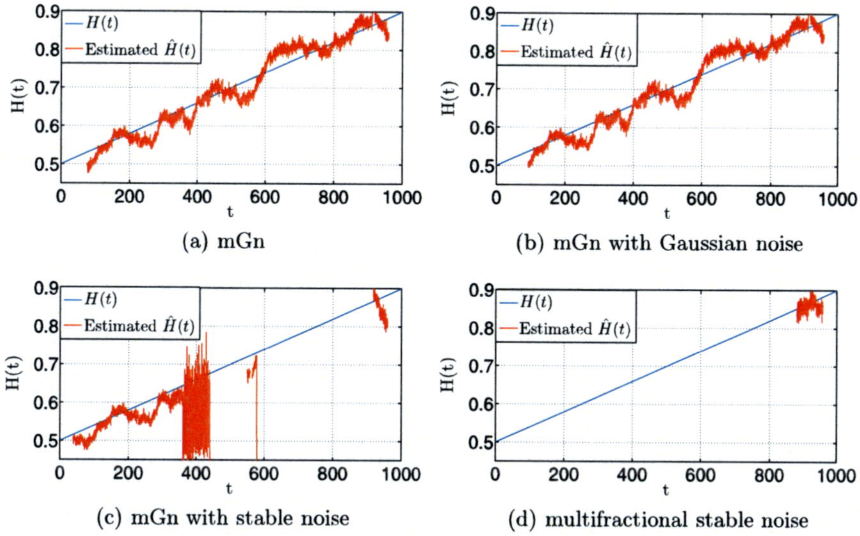

Fig. 4.6 Sliding-windowed Variance of Residuals method: **(a)** mGn; **(b)** mGn with 30 dB SNR white Gaussian noise; **(c)** mGn with 30 dB SNR α-stable ($\alpha = 1.0$) noise; **(d)** multifractional stable noise

The tracking performance and the robustness analysis results of the sliding-windowed Variance of Residuals method for four types of multifractional processes are presented in Fig. 4.6. The estimated results of this method for MGN1 and MGN2 are presented in Figs. 4.6(a) and (b), respectively. This method is a little bit better than the above three methods for MGN1 and MGN2. However, this method is obviously influenced by 30 dB SNR α-stable ($\alpha = 1.0$) noise. The estimated local Hölder exponent $\widehat{H}(t)$ displays large fluctuations when $350 < t < 450$ and $600 < t < 700$. Figure 4.6(d) shows the estimated result for MGN4, where the estimation result for the multifractional α-stable ($\alpha = 1.8$) process is worse than that for noise-free mGn.

The tracking performance and the robustness analysis results of the sliding-windowed Periodogram method for four types of multifractional processes are presented in Fig. 4.7. It can be seen that this method is better than the first three methods for both MGN1 and MGN2. The estimated local Hölder exponents $\widehat{H}(t)$ are close to the $H(t)$ in Figs. 4.7(a), (b). Figure 4.7(c) presents the estimated result of the method for MGN3. The estimated local Hölder exponent $\widehat{H}(t)$ is affected a little by 30 dB SNR α-stable ($\alpha = 1.0$) noise, when $500 < t < 800$. From Fig. 4.7(d) we can see that the sliding-windowed Periodogram method has almost the same estimation results for mGn and the multifractional α-stable ($\alpha = 1.8$) processes. Therefore, the sliding-windowed Periodogram method has good tracking performance for the noise-free multifractional process and the multifractional α-stable ($\alpha = 1.8$) processes, and has good robustness to 30 dB SNR white Gaussian noise and 30 dB SNR impulsive noise.

The tracking performance and the robustness analysis results of sliding-windowed Modified Periodogram method for four types of multifractional processes are presented in Fig. 4.8. It can be seen from Figs. 4.8(a), (b) that the estimation results

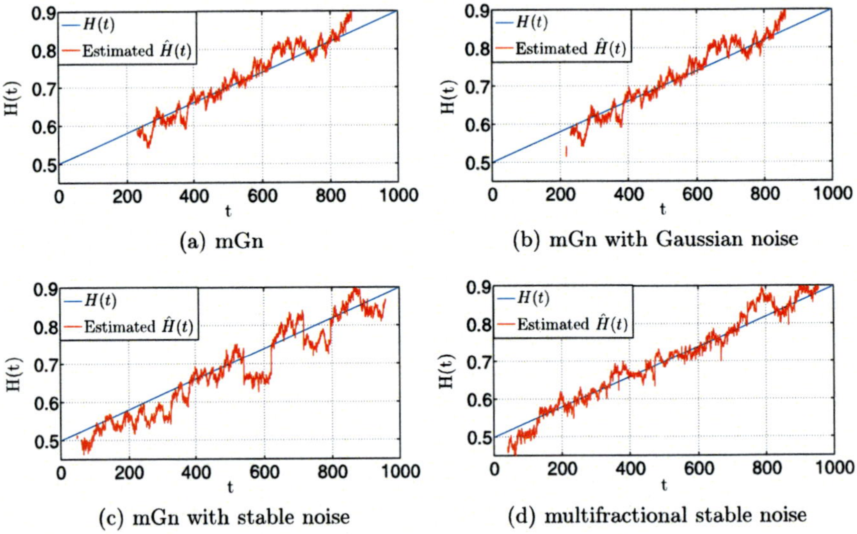

(a) mGn

(b) mGn with Gaussian noise

(c) mGn with stable noise

(d) multifractional stable noise

Fig. 4.7 Sliding-windowed Periodogram method: **(a)** mGn; **(b)** mGn with 30 dB SNR white Gaussian noise; **(c)** mGn with 30 dB SNR α-stable ($\alpha = 1.0$) noise; **(d)** multifractional stable noise

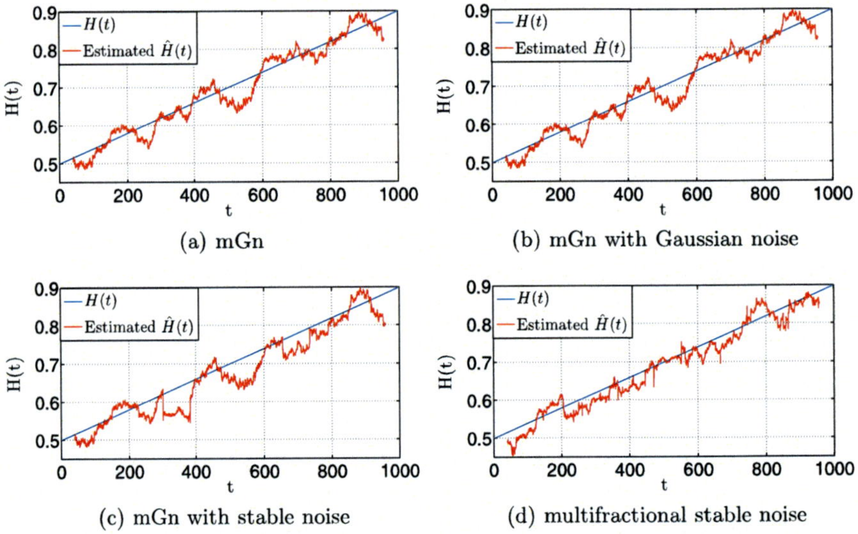

(a) mGn

(b) mGn with Gaussian noise

(c) mGn with stable noise

(d) multifractional stable noise

Fig. 4.8 Sliding-windowed Modified Periodogram method: **(a)** mGn; **(b)** mGn with 30 dB SNR white Gaussian noise; **(c)** mGn with 30 dB SNR α-stable ($\alpha = 1.0$) noise; **(d)** multifractional stable noise

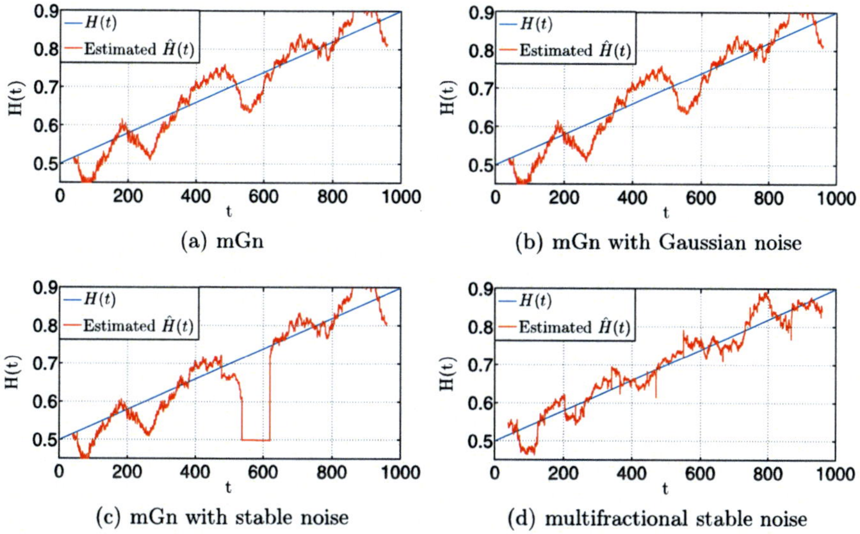

Fig. 4.9 Sliding-windowed Whittle method: **(a)** mGn; **(b)** mGn with 30 dB SNR white Gaussian noise; **(c)** mGn with 30 dB SNR α-stable ($\alpha = 1.0$) noise; **(d)** multifractional stable noise

for MGN1 and MGN2 are more smooth than those of the sliding-windowed Periodogram method, but the estimated local Hölder exponents $\widehat{H}(t)$ has larger fluctuations. Figure 4.8(c) shows that the estimated result is affected a little by 30 dB SNR α-stable ($\alpha = 1.0$) noise. It is interesting to note that the estimation result of sliding-windowed Modified Periodogram method for MGN4 is much better than that for noise-free mGn, which indicates that the sliding-windowed Modified Periodogram method has a better tracking performance for the multifractional α-stable ($\alpha = 1.8$) process than for the noise-free multifractional process.

The tracking performance and the robustness analysis results of the sliding-windowed Whittle method for four types of multifractional processes are presented in Fig. 4.9. The estimated results of this method for MGN1 and MGN2 are presented in Figs. 4.9(a) and (b), respectively. This method performs worse than the above five sliding-windowed Hurst estimators for MGN1 and MGN2. From Fig. 4.9(c) we can see that this method is obviously influenced by 30 dB SNR α-stable ($\alpha = 1.0$) noise when $t \approx 500$. Similar to the sliding-windowed Modified Periodogram method, the estimation result for MGN4 is better than that for a noise-free mGn. Overall, this method has bad tracking performance for a noise-free multifractional process, and lacks robustness to 30 dB SNR impulsive noise.

The tracking performance and the robustness analysis results of the sliding-windowed Diffusion Entropy method for four types of multifractional processes are presented in Fig. 4.10. The estimated results for MGN1 and MGN2 are presented in Figs. 4.10(a) and (b), respectively. This method is much better than all the above sliding-windowed Hurst estimators for MGN1 and MGN2. It has especially good robustness to 30 dB SNR α-stable ($\alpha = 1.0$) noise. The estimated local Hölder exponent $\widehat{H}(t)$ for MGN3 is almost the same as noise-free mGn. But for MGN4, the

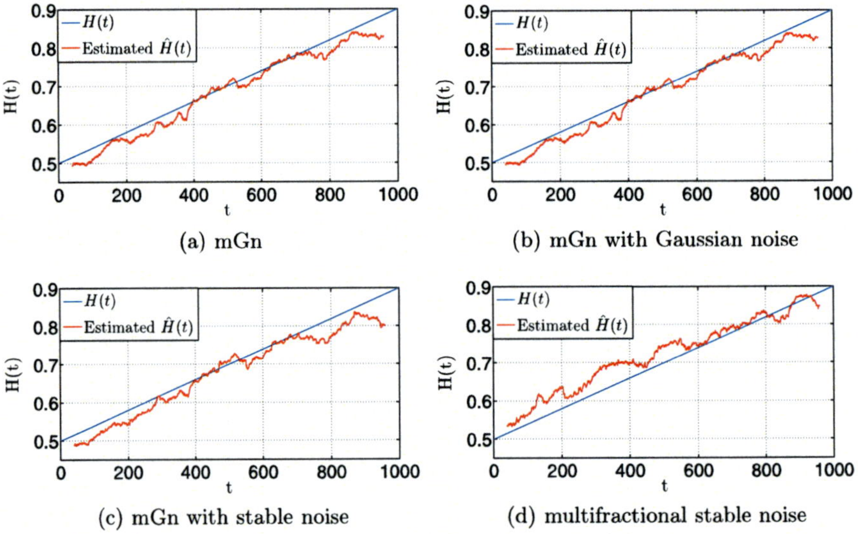

Fig. 4.10 Sliding-windowed Diffusion Entropy method: **(a)** mGn; **(b)** mGn with 30 dB SNR white Gaussian noise; **(c)** mGn with 30 dB SNR α-stable ($\alpha = 1.0$) noise; **(d)** multifractional stable noise

estimated local Hölder exponent $\widehat{H}(t)$ is slightly overestimated when $0 < t < 800$. So, the sliding-windowed Diffusion Entropy method has good robustness to 30 dB SNR white Gaussian and impulsive noise, and good tracking performance for the noise-free multifractional process, but it has bad tracking performance for the multifractional α-stable ($\alpha = 1.8$) process.

The tracking performance and the robustness analysis results of the sliding-windowed Kettani and Gubner's method for four types of multifractional processes are presented in Fig. 4.11. It can be seen from Figs. 4.11(a), (b), (d) that the estimation results of the sliding-windowed Kettani and Gubner's method for MGN1, MGN2 and MGN4 are much better than that of the sliding-windowed Hurst estimators presented above. The estimated local Hölder exponents $\widehat{H}(t)$ for MGN1, MGN2 and MGN4 are smooth and close to the $H(t)$. But the estimated result is slightly influenced by 30 dB SNR α-stable ($\alpha = 1.0$) noise. The estimated local Hölder exponent $\widehat{H}(t)$ is slightly underestimated when $500 < t < 1000$. Overall, the sliding-windowed Kettani and Gubner's method has very good tracking performance for the noise-free multifractional process and the multifractional α-stable ($\alpha = 1.8$) process, and acceptable robustness to 30 dB SNR with Gaussian noise and 30 dB SNR α-stable ($\alpha = 1.0$) noise.

The tracking performance and the robustness analysis results of the sliding-windowed Abry and Veitch's method for four types of multifractional processes are presented in Fig. 4.12. It can be seen that, the estimated result of this method for MGN1 is acceptable, but this method performs worse for the other three types of multifractional processes. The estimated local Hölder exponents $\widehat{H}(t)$ for MGN2, MGN3 and MGN4 all have large deviations. Therefore, the sliding-windowed Abry and Veitch's method has bad tracking performance for the multifractional α-stable

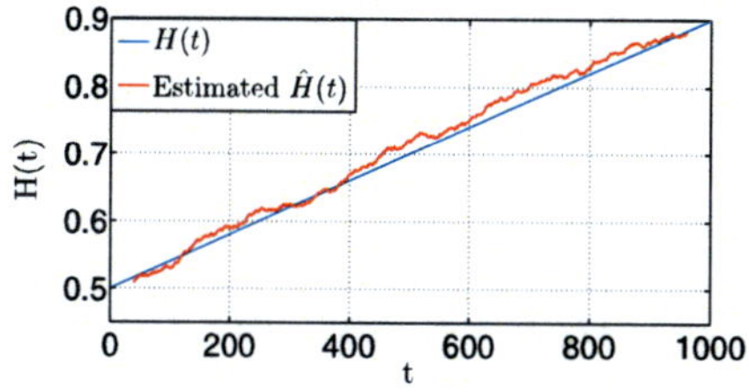

(a) mGn

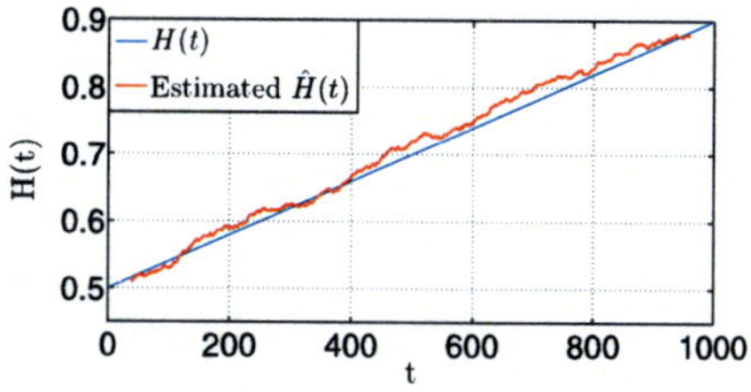

(b) mGn with Gaussian noise

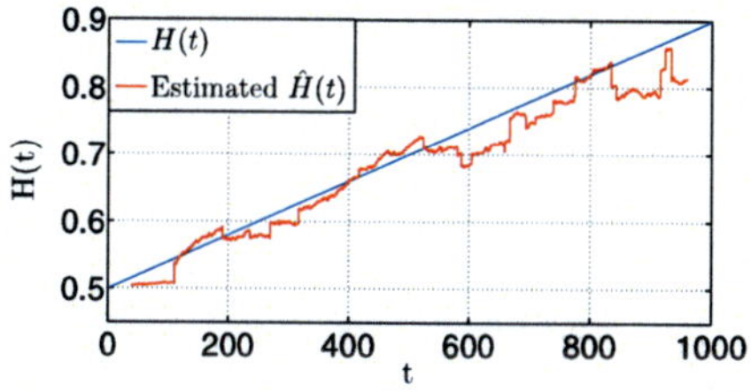

(c) mGn with stable noise

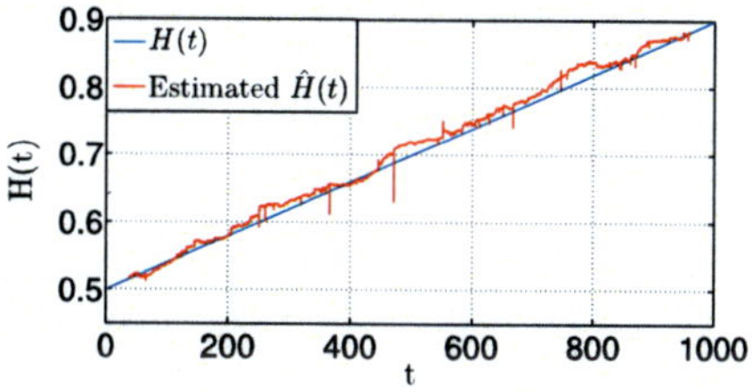

(d) multifractional stable noise

Fig. 4.11 Sliding-windowed Kettani and Gubner's method: **(a)** mGn; **(b)** mGn with 30 dB SNR white Gaussian noise; **(c)** mGn with 30 dB SNR α-stable ($\alpha = 1.0$) noise; **(d)** multifractional stable noise

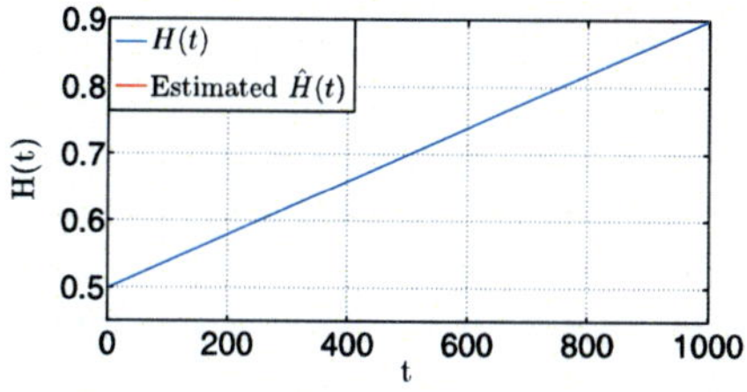

(a) mGn

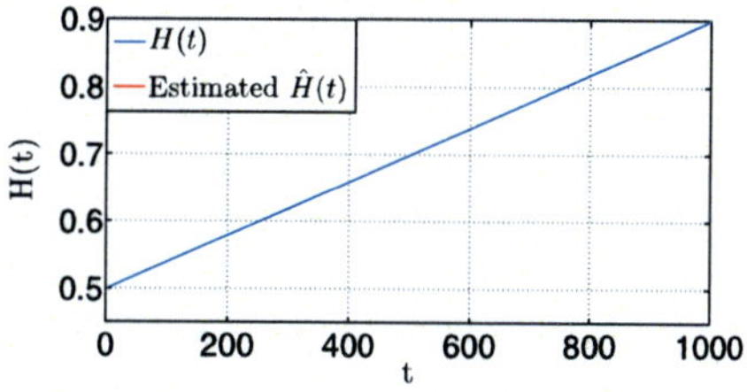

(b) mGn with Gaussian noise

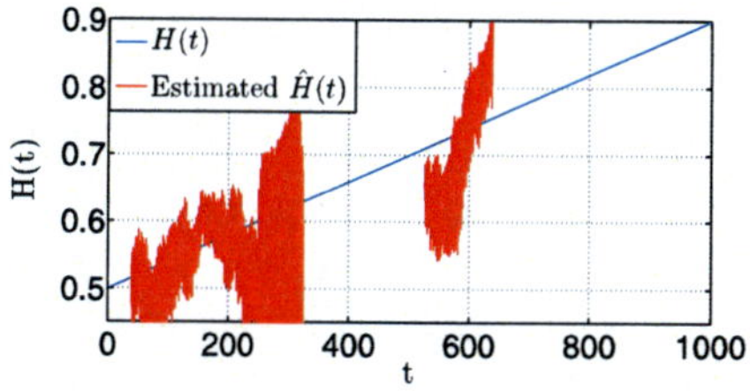

(c) mGn with stable noise

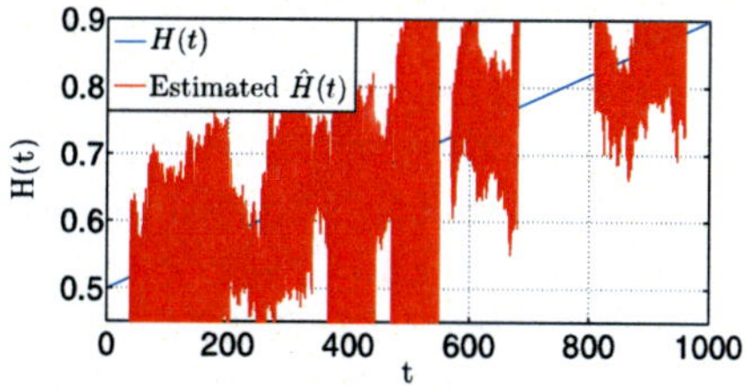

(d) multifractional stable noise

Fig. 4.12 Sliding-windowed Abry and Veitch's method: **(a)** mGn; **(b)** mGn with 30 dB SNR white Gaussian noise; **(c)** mGn with 30 dB SNR α-stable ($\alpha = 1.0$) noise; **(d)** multifractional stable noise

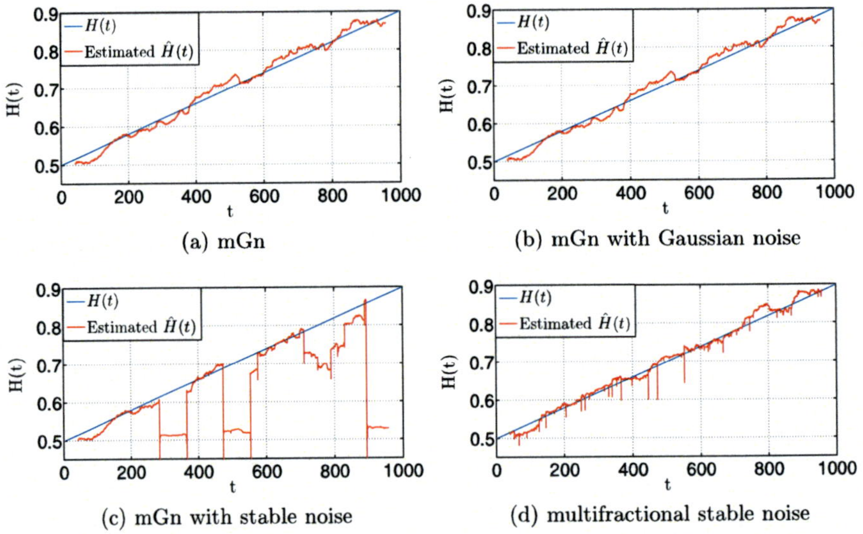

Fig. 4.13 Sliding-windowed Koutsoyiannis' method: **(a)** mGn; **(b)** mGn with 30 dB SNR white Gaussian noise; **(c)** mGn with 30 dB SNR α-stable ($\alpha = 1.0$) noise; **(d)** multifractional stable noise

($\alpha = 1.8$) process, and bad robustness to 30 dB SNR white Gaussian noise and 30 dB SNR α-stable ($\alpha = 1.0$) noise.

The tracking performance and the robustness analysis results of the sliding-windowed Koutsoyiannis' method for four types of multifractional processes are presented in Fig. 4.13. This method performs as well as the sliding-windowed Kettani and Gubner's method for MGN1, MGN2 and MGN4. The estimated local Hölder exponents $\widehat{H}(t)$ for MGN1, MGN2 and MGN4 are smooth and close to the ideal $H(t)$. But the sliding-windowed Koutsoyiannis' method is short of accuracy for 30 dB SNR α-stable ($\alpha = 1.0$) noise corrupted mGn. The estimated local Hölder exponents $H(t)$ is obviously underestimated when $300 < t < 4000$, $500 < t < 600$ and $t > 900$. Therefore, the sliding-windowed Koutsoyiannis' method has bad robustness to 30 dB SNR α-stable ($\alpha = 1.0$) noise, but has good performance for the other three types of multifractional processes.

The tracking performance and the robustness analysis results of the sliding-windowed Higuchi's method for four types of multifractional processes are presented in Fig. 4.14. It can be seen from Figs. 4.14(a), (b) that, the estimated results for MGN1 and MGN2 are worse than that of the sliding-windowed Abry and Veitch's method. The 30 dB SNR α-stable ($\alpha = 1.0$) noise almost has no influence on the sliding-windowed Higuchi's method, but the estimated result is biased. For MGN4, the sliding-windowed Higuchi's method does not perform very well. The estimated local Hölder exponent $\widehat{H}(t)$ is almost overestimated throughout the interval $0 < t < 1000$.

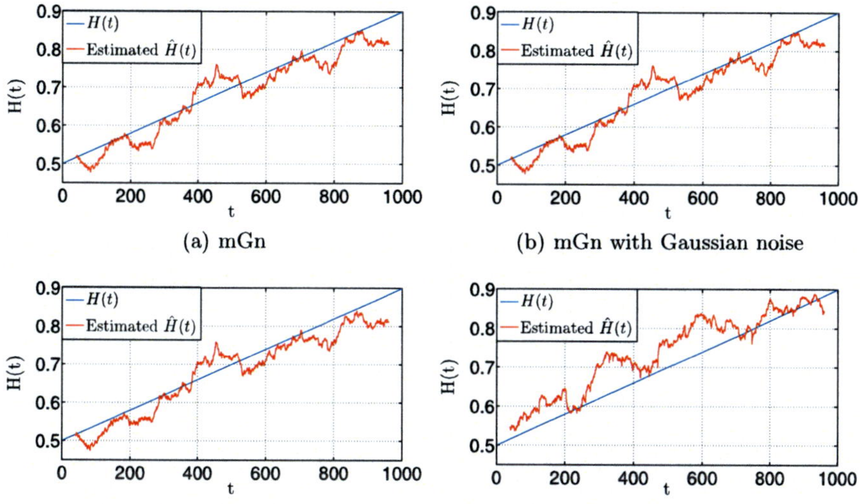

(a) mGn (b) mGn with Gaussian noise (c) mGn with stable noise (d) multifractional stable noise

Fig. 4.14 Sliding-windowed Higuchi's method: **(a)** mGn; **(b)** mGn with 30 dB SNR white Gaussian noise; **(c)** mGn with 30 dB SNR α-stable ($\alpha = 1.0$) noise; **(d)** multifractional stable noise

Table 4.1 Standard error comparison

No.	Method	mGN	mGN + Gaussian	mGN + Stable	Stable mGN
1	R/S	0.0324	0.0339	0.0336	0.0305
2	Aggregated variance	0.0349	0.0366	0.0520	0.0311
3	Absolute value	0.0340	0.0356	0.0348	0.0413
4	Variance of Residuals	0.0259	0.0271	0.1015	0.0246
5	Periodogram	0.0270	0.0283	0.0394	0.0240
6	Modified Periodogram	0.0269	0.0281	0.0401	0.0255
7	Whittle	0.0435	0.0456	0.0818	0.0310
8	Diffusion Entropy	0.0226	0.0236	0.0291	0.0333
9	Kettani and Gubner	0.0130	0.0137	0.0300	0.0120
10	Abry and Veitch	0.0205	0.0415	0.1442	0.0808
11	Koutsoyiannis	0.0142	0.0149	0.1215	0.0124
12	Higuchi	0.0336	0.0352	0.0360	0.0538

4.2.3 Guideline for Estimator Selection

In order to quantify the tracking performance and the robustness of these twelve sliding-windowed Hurst parameter estimators more accurately, we calculate and compare the standard errors S of different methods. Standard errors S is defined as (3.39). Table 4.1 summarizes the standard errors of these twelve sliding-windowed Hurst estimators for four types of multifractional processes. From Table 4.1 we can see that the sliding-windowed Kettani and Gubner's method, and the sliding-windowed Koutsoyiannis' method have the best tracking performance for the noise-free multifractional process and the multifractional α-stable ($\alpha = 1.8$) process. But the sliding windowed Koutsoyiannis' method is short of robustness to 30 dB SNR α-stable ($\alpha = 1.0$) noise. Among all twelve Hurst estimators, the sliding-windowed Diffusion Entropy method performs the best for a non-Gaussian noise corrupted multifractional process, and displays the best robustness to 30 dB SNR α-stable ($\alpha = 1.0$) noise. Besides, 30 dB SNR white Gaussian noise almost has no influence on all the sliding windowed Hurst parameter estimators except the sliding-windowed Abry and Veitch's method. From the above analyses we came to the conclusion that the sliding-windowed Kettani and Gubner's method has the best tracking performance for the multifractional process and the multifractional stable process, and has good robustness to 30 dB SNR white Gaussian noise and 30 dB SNR α-stable ($\alpha = 1.0$) noise.

4.3 Chapter Summary

This chapter deals with multifractional processes with a time varying local Hölder parameter, and the evaluation of various local Hölder estimators. Section 4.1 introduced the multifractional processes including mGn, mBm, multifractional stable motion, and the multifractional stable noise. Section 4.2 focused on the robustness analysis of twelve sliding-windowed Hurst estimators for noisy multifractional processes, and multifractional process with infinite second-order statistics. To evaluate the accuracy and robustness of all these sliding-windowed Hurst estimators, multifractional stable processes with $H(t) = at + b$ were synthesized using variable-order fractional calculus. All these twelve sliding-windowed Hurst estimators were tested using four different types of multifractional processes. According to the estimation results and the standard error analysis of these estimators, we conclude that most of the sliding-windowed Hurst estimators perform well to track local Hölder exponents $H(t)$ of multifractional processes. Except for the sliding-windowed Abry and Veitch method, all Hurst parameter estimators have very good robustness to 30 dB SNR white Gaussian noise, but few of them have good robustness to 30 dB SNR α-stable ($\alpha = 1.0$) noise. The sliding-windowed Kettani and Gubner's method, and the sliding-windowed Koutsoyiannis' method perform the best for multifractional process with infinite second-order statistics. Overall, Kettani and Gubner's method has the best accuracy and good robustness for multifractional processes among all these twelve sliding-windowed Hurst estimators.

In order to achieve more in-depth analysis of fractional processes, constant-order fractional processing techniques, variable-order fractional processing techniques, and distributed-order fractional filtering will be studied in the following three chapters, respectively.

Part III
Fractional-Order Signal Processing

Chapter 5
Constant-Order Fractional Signal Processing

5.1 Fractional-Order Differentiator/Integrator and Fractional Order Filters

Fractional-order differentiator and fractional-order integrator are fundamental building blocks for fractional-order signal processing. The transfer function of fractional-order integrator (FOI) is simply

$$G_{\mathrm{FOI}}(s) = \frac{1}{s^r},$$

where r is a positive real number. Without loss of generality, we only consider $0 < r < 1$. The transfer function of fractional-order differentiator (FOD) is simply

$$G_{\mathrm{FOD}}(s) = \frac{1}{G_{\mathrm{FOI}}(s)} = s^r.$$

In time domain, the impulse response of $G_{\mathrm{FOI}}(s)$ is

$$h(t) = L^{-1} G_{\mathrm{FOI}}(s) = \frac{t^{r-1}}{\Gamma(r)}, \quad t \geq 0.$$

Replacing r with $-r$ will give the impulse response of fractional differentiator s^r.

FOI or FOD is an infinite-dimensional system. So, when we implement it digitally, we must approximate it with a finite-dimensional discrete transfer function. This is called the "discretization" problem of FOI or FOD [59]. We refer to excellent reviews and tutorials on discretization issues [63, 154, 207].

In this section, we focus on continuous time approximation first and then on (direct) discrete time approximation. If continuous time approximation is done, we can discretize the obtained finite dimensional continuous transfer function by using MATLAB® command `c2d()`.

It should be noted that there is some work being done with 'passive' hardware devices for a fractional-order integrator, such as fractances (*e.g.*, RC transmission line circuit and Domino ladder network) [236] and Fractors [27]. However, there are

H. Sheng et al., *Fractional Processes and Fractional-Order Signal Processing*, Signals and Communication Technology,
DOI 10.1007/978-1-4471-2233-3_5,

some restrictions, since these devices are difficult to tune. Alternatively, it is feasible to implement fractional-order operators and controllers using finite-dimensional integer-order transfer functions.

As said, an integer-order transfer function representation to a fractional-order operator s^r is infinite-dimensional. However, it should be pointed out that a band-limit implementation of a fractional-order controller (FOC) is important in practice, *i.e.*, the finite-dimensional approximation of the FOC should be done in a proper range of frequencies of practical interest [227]. Moreover, the fractional-order r could be a complex number as discussed in [227]. In this book, we focus on the case where the fractional order is a real number.

This section describes different approximations or implementations of FOI or FOD and other fractional order filters. When fractional-order filters have to be implemented or simulations have to be performed, fractional-order transfer functions are usually replaced by integer-order transfer functions with a behavior close enough to the one desired, but much easier to handle.

There are many different ways of finding such approximations, but unfortunately it is hard to tell which one of them is the best. Even though some of them are better than others in regard to certain characteristics, the relative merits of each approximation depend on the differentiation order, on whether one is more interested in accurate frequency behavior or in accurate time responses, on how large admissible transfer functions may be, and on other factors like these.

5.1.1 Continuous-Time Implementations of Fractional-Order Operators

For the fractional-order operator, its Laplace representation is s^γ, which exhibits straight lines in both magnitude and phase Bode plots. Thus it is not possible to find a finite-order filter to fit the straight lines for all the frequencies. However, it is useful to fit the frequency responses over a frequency range of interest, $(\omega_\mathrm{b}, \omega_\mathrm{h})$.

Different continuous-time filters have been studied and some of the approximations can be constructed by relevant MATLAB functions in the N-Integer Toolbox.[1]

Continued Fraction Approximations

Continued fraction expansion (CFE) is often regarded as a useful type of rational-function approximation to a given function $f(s)$. It usually has better convergence than the power series functions, such as the Taylor series expansions. For the fractional-order operator $G(s) = s^r$, the continued fraction expansion can be written

[1] http://www.mathworks.com/matlabcentral/fileexchange/8312.

Table 5.1 Some rational function approximations to s^r

Approximation	nid() syntax	Comments
Carlson's method	$C=$ nid$(1, r, [\omega_b, \omega_h], N,$ 'carlson')	$\omega \in [\omega_b, \omega_h]$
Matsuda's method	$C=$ nid$(1, r, [\omega_b, \omega_h], N,$ 'matsuda')	$\omega \in [\omega_b, \omega_h]$
Low-frequency approx.	$C=$ nid$(1, r, \omega_b, N,$ 'cfelow')	$\omega \leq \omega_b$
High-frequency approx.	$C=$ nid$(1, r, \omega_h, N,$ 'cfehigh')	$\omega \geq \omega_h$

as

$$G(s)=\frac{b_0(s)}{a_0(s)+\frac{b_1(s)}{a_1(s)+\frac{b_2(s)}{a_2(s)+\cdots}}} \tag{5.1}$$

where $a_i(s)$ and $b_i(s)$ can be expressed by rational functions of s. One should first find the continued fraction expansion to the original fractional-order operator, then get the integer-order transfer function, *i.e.*, the rational function, representation.

There are several well-established continued fraction expansion based approximation methods to the fractional-order operator $G(s)=s^r$. The N-integer Toolbox provides a `nid()` function for finding the rational-function approximation. Some of the approximation can be obtained by direct calling of the `nid()` function, see Table 5.1, where N is the order of the transfer function $C(s)$, and r is the fractional order.

Example 5.1 Consider the fractional-order integrator with $r=0.5$. The rational function approximation using different continued fraction expansion based methods can be found in [313] as

$$\text{Low-frequency CFE:}\quad H_1(s)=\frac{0.351s^4+1.405s^3+0.843s^2+0.157s+0.009}{s^4+1.333s^3+0.478s^2+0.064s+0.002844},$$

$$\text{High-frequency CFE:}\quad H_2(s)=\frac{s^4+4s^3+2.4s^2+0.448s+0.0256}{9s^4+12s^3+4.32s^2+0.576s+0.0256},$$

$$\text{Carlson's method:}\quad H_3(s)=\frac{s^4+36s^3+126s^2+84s+9}{9s^4+84s^3+126s^2+36s+1},$$

$$\text{Matsuda's method:}\quad H_4(s)=\frac{0.08549s^4+4.877s^3+20.84s^2+12.99s+1}{s^4+13s^3+20.84s^2+4.876s+0.08551}.$$

The Bode plots with different approximations can be obtained as shown in Fig. 5.1. It can be seen that the fitting ranges are rather small and the quality of fitting is not satisfactory.

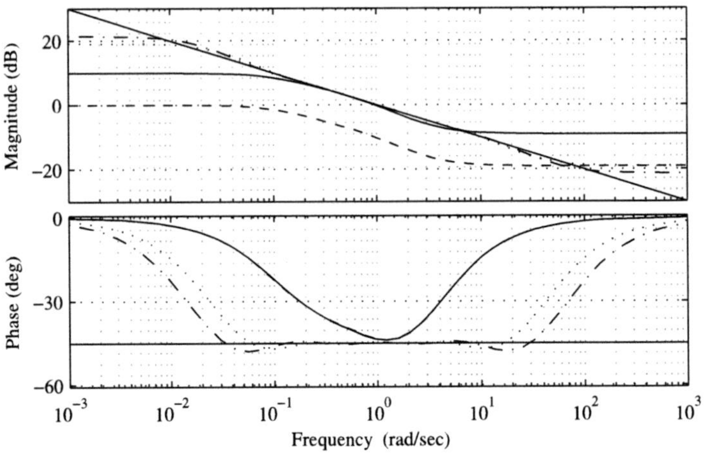

Fig. 5.1 Bode plots comparisons with different approximations, with *solid lines* for $H_1(s)$, *dashed lines* for $H_2(s)$, *dotted lines* for $H_3(s)$ and *dash-dotted lines* for $H_4(s)$. The *straight lines* are the theoretical results

Oustaloup Recursive Approximations

Oustaloup filter approximation to a fractional-order differentiator is widely used in fractional calculus [227]. A generalized Oustaloup filter can be designed as

$$G_f(s) = K \prod_{k=1}^{N} \frac{s + \omega_k'}{s + \omega_k}, \tag{5.2}$$

where, the poles, zeroes and gain are evaluated from

$$\omega_k' = \omega_\text{b}\omega_\text{u}^{(2k-1-\gamma)/N}, \quad \omega_k = \omega_\text{b}\omega_\text{u}^{(2k-1+\gamma)/N}, \quad K = w_\text{h}^\gamma, \tag{5.3}$$

where $\omega_\text{u} = \sqrt{\omega_\text{h}/\omega_\text{b}}$. We used the term "generalized" because N here can be either odd or even integers.

Based on the above algorithm, the following function can be written

```
function G=ousta_fod(gam,N,wb,wh)
k=1:N; wu=sqrt(wh/wb);
wkp=wb*wu.^((2*k-1-gam)/N); wk=wb*wu.^((2*k-1+gam)/N);
G=zpk(-wkp,-wk,wh^gam); G=tf(G);
```

and the Oustaloup filter can be designed with $G = \texttt{ousta\_fod}(\gamma, N, w_\text{b}, w_\text{h})$, where γ is the order of derivative, and N is the order of the filter.

Example 5.2 To illustrate the method, the approximation of the fractional-order integrator of order 0.45 can be obtained. In this particular case, the orders of the approximation are selected as 4 and 5, respectively, with $\omega_\text{h} = 1000$ rad/sec and

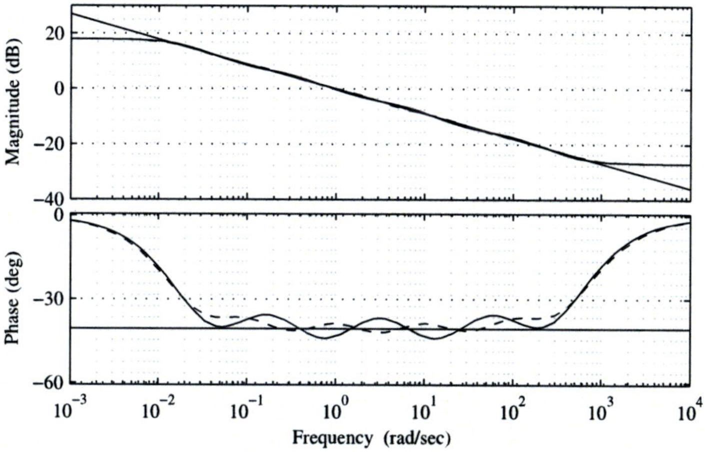

Fig. 5.2 Bode plots of $H_{\mathrm{Oust}}(s)$, corresponding to the approximation of a fractional-order integrator of order 0.45 with the Oustaloup method, with *solid lines* for $G_1(s)$, *dashed lines* for $G_2(s)$ and *dotted lines* for the theoretical Bode plot

$\omega_b = 0.01$ rad/sec. The filters can be designed with the following MATLAB commands

```
>> G1=ousta_fod(-0.45,4,1e-2,1e3);
   G2=ousta_fod(-0.45,5,1e-2,1e3);
   bode(G1,'-',G2,'--',{1e-3,1e4})
```

and the two filters are respectively obtained

$$G_1(s) = \frac{0.04467s^4 + 21.45s^3 + 548.2s^2 + 783.2s + 59.57}{s^4 + 131.5s^3 + 920.3s^2 + 360.1s + 7.499},$$

and

$$G_2(s) = \frac{0.04467s^5 + 26.35s^4 + 1413s^3 + 7500s^2 + 3942s + 188.4}{s^5 + 209.3s^4 + 3982s^3 + 7500s^2 + 1399s + 23.71}.$$

The Bode plots are shown in Fig. 5.2. It can be seen that the Bode plots of the two filters are relatively close to that of the theoretical one over the frequency range of interest. It can be seen that the fitting quality is much superior to those obtained with continued fraction based approaches.

Modified Oustaloup Filter

In practical applications, it is frequently found that the filter from using the `ousta_fod()` function cannot exactly fit the whole expected frequency range of interest. A new improved filter for a fractional-order derivative in the frequency range of interest (ω_b, ω_h), which is shown to perform better, is introduced in this subsection. The modified filter is

$$s^\gamma \approx \left(\frac{d\omega_h}{b}\right)^\gamma \left(\frac{ds^2 + b\omega_h s}{d(1-\gamma)s^2 + b\omega_h s + d\gamma}\right) \prod_{k=-N}^{N} \frac{s+\omega'_k}{s+\omega_k}, \tag{5.4}$$

and the filter is stable for $\gamma \in (0, 1)$, and

$$\omega'_k = \omega_b \omega_u^{(2k-1-\gamma)/N}, \qquad \omega_k = \omega_b \omega_u^{(2k-1+\gamma)/N}, \tag{5.5}$$

with $\omega_u = \sqrt{\omega_h/\omega_b}$.

Through a number of experimental confirmations and theoretical analyses, the modified filter achieves good approximation when $b = 10$ and $d = 9$. With the above algorithm, a MATLAB function `new_fod()` is written

```
function G=new_fod(r,N,wb,wh,b,d)
if nargin==4, b=10; d=9; end
k=1:N; wu=sqrt(wh/wb); K=(d*wh/b)^r;
wkp=wb*wu.^((2*k-1-r)/N); wk=wb*wu.^((2*k-1+r)/N);
G=zpk(-wkp',-wk',K)*tf([d,b*wh,0],[d*(1-r),b*wh,d*r]);
```

with the syntax $G_f = \texttt{new\_fod}(\gamma, N, \omega_b, \omega_h, b, d)$. Again here, the modified Oustaloup filter is also extended for use in handling odd and even order.

Example 5.3 Consider a fractional-order transfer function model

$$G(s) = \frac{s+1}{10s^{3.2} + 185s^{2.5} + 288s^{0.7} + 1}.$$

The approximations to the 0.2th-order derivative using the Oustaloup's filter and the modified Oustaloup's filter can be obtained as shown in Fig. 5.3(a). The frequency range of good fitting is larger with the improved filter. The exact Bode plot can be obtained with the `bode()` function. Also the two approximations to the $G(s)$ model is shown in Fig. 5.3(b). In the following commands, function `fotf()` is used to define the fractional-order transfer function (FOTF) object [207].

```
>> b=[1 1]; a=[10,185,288,1]; nb=[1 0]; na=[3.2,2.5,0.7,0];
   w=logspace(-4,4,200); G0=fotf(a,na,b,nb); H=bode(G0,w);
   s=zpk('s'); N=4; w1=1e-3; w2=1e3; b=10; d=9;
   g1=ousta_fod(0.2,N,w1,w2); g2=ousta_fod(0.5,N,w1,w2);
   a1=g1; g3=ousta_fod(0.7,N,w1,w2);
   G1=(s+1)/(10*s^3*g1+185*s^2*g2+288*g3+1);
   g1=new_fod(0.2,N,w1,w2,b,d);
   g2=new_fod(0.5,N,w1,w2,b,d);
   g3=new_fod(0.7,N,w1,w2,b,d); bode(g1,a1); figure
   G2=(s+1)/(10*s^3*g1+185*s^2*g2+288*g3+1); bode(H,G1,G2)
```

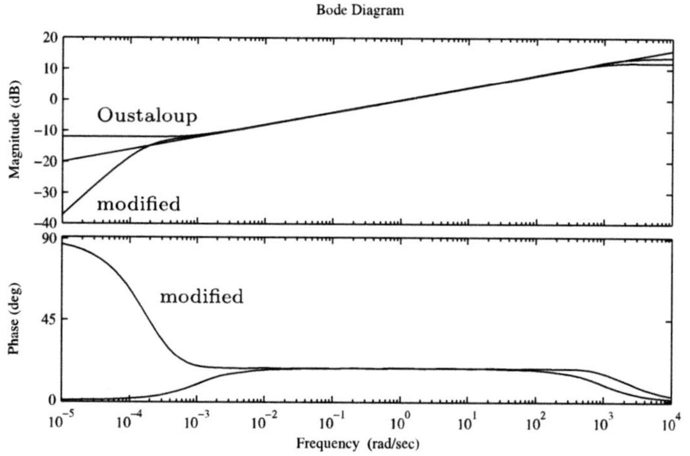

(a) $s^{0.2}$ fittings

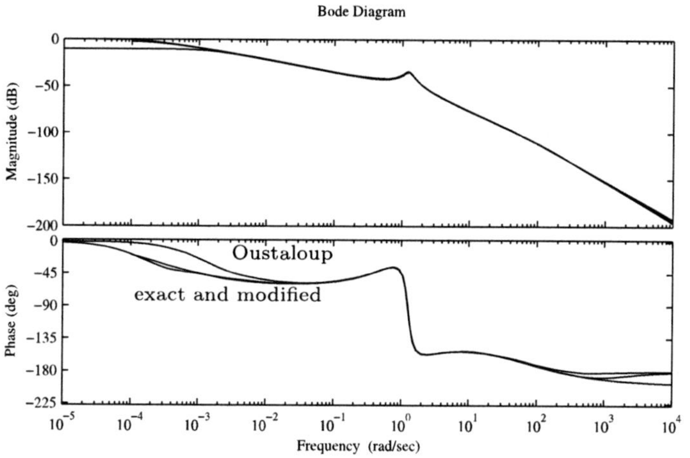

(b) Bode plot comparisons

Fig. 5.3 Bode plot comparisons, *straight lines* for exact ones, *dashed lines* for Oustaloup filters, and *dotted lines* for modified Oustaloup filters

It can be seen that the modified method provides a much better fit. Thus for certain fractional-order differentiators, the modified filter may be more appropriate.

5.1.2 Discrete-Time Implementation of Fractional-Order Operators

The key step in digital implementation of an FOC is the numerical evaluation or discretization of the fractional-order differentiator s^r. In general, there are two classes of discretization methods: *direct discretization* and *indirect discretization*. In *indirect discretization* methods [227], two steps are required, *i.e.*, frequency domain fitting in continuous time domain first, and then discretizing the fit s-transfer function. Other frequency-domain fitting methods can also be used but without guaranteeing the stable minimum-phase discretization. Existing *direct discretization* methods include the application of the direct power series expansion (PSE) of the Euler operator [185, 313], the continued fraction expansion (CFE) of the Tustin operator [59, 311, 313], and the numerical integration based method [59, 62, 185]. However, as pointed out in [3], the Tustin operator based discretization scheme exhibits large errors in a high frequency range. A new mixed scheme of Euler and Tustin operators is proposed in [59] which yields the so-called Al-Alaoui operator [3]. These discretization methods for s^r are in infinite impulse response (IIR) form. Recently, there are some reported methods to directly obtain the digital fractional-order differentiators in finite impulse response (FIR) form [304]. However, using an FIR filter to approximate s^r may be less efficient due to the very high order of the FIR filter. So, discretizing fractional-order differentiators in IIR forms is preferred.

In this section, FIR filter approximation and IIR filter discretization methods are presented. Then an introduction is made on finite-dimensional integer-order approximations retaining step and impulse response invariants of the fractional-order operators.

FIR Filter Approximation: Grünwald-Letnikov definition

Recall the approximate Grünwald-Letnikov definition given below, where the step size of h is assumed to be very small

$$ {}_aD_t^r f(t) \approx \frac{1}{h^r} \sum_{j=0}^{[(t-a)/h]} w_j^{(r)} f(t-jh), \tag{5.6}$$

where the binomial coefficients can be calculated recursively with the following formula

$$ w_0^{(r)} = 1, \qquad w_j^{(r)} = \left(1 - \frac{r+1}{j}\right) w_{j-1}^{(r)}, \quad j = 1, 2, \dots. \tag{5.7}$$

Based on the above algorithm, the γth-order derivative of a given function can be evaluated, and the syntax of $\boldsymbol{y}_1 = \mathtt{glfdiff}(\boldsymbol{y}, \boldsymbol{t}, \gamma)$, where $\boldsymbol{y}$ and $\boldsymbol{t}$ are signal and time vectors, respectively, and $\boldsymbol{y}_1$ is a vector of γth-order derivative of $f(t)$.

```
function dy=glfdiff(y,t,gam)
h=t(2)-t(1); dy(1)=0; y=y(:); t=t(:); w=1;
for j=2:length(t), w(j)=w(j-1)*(1-(gam+1)/(j-1)); end
for i=2:length(t), dy(i)=w(1:i)*[y(i:-1:1)]/h^gam; end
```

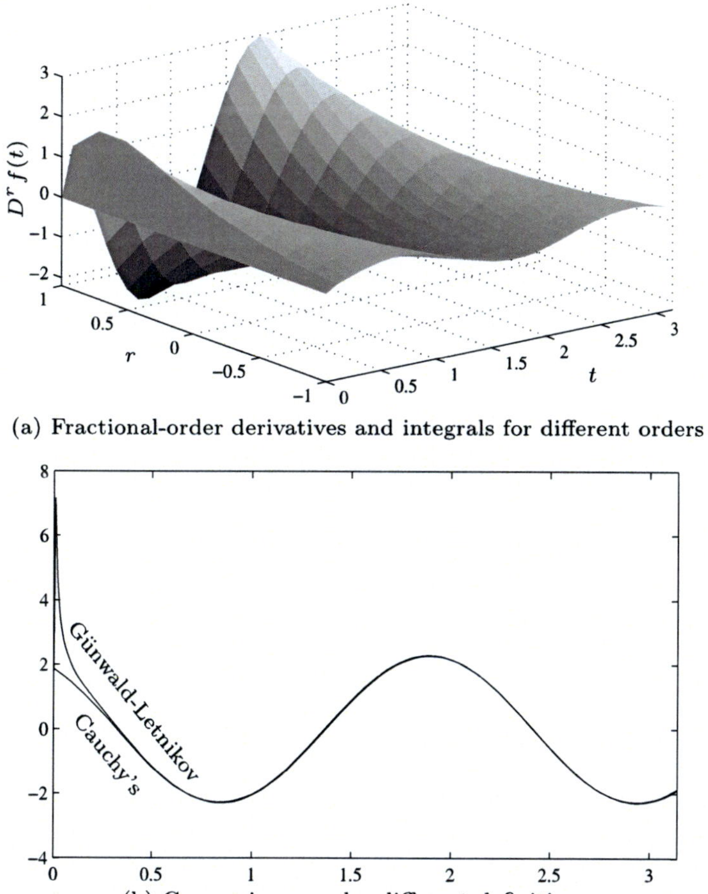

(a) Fractional-order derivatives and integrals for different orders

(b) Comparisons under different definitions

Fig. 5.4 Fractional-order derivatives and integrals

Example 5.4 Consider a sinusoidal function $f(t) = \sin(3t + 1)$. It is known from the Cauchy's formula that the kth-order derivative of the function is

$$f^{(k)}(t) = 3^k \sin(3t + 1 + k\pi/2),$$

and the formula also works for non-integer values of k. It is known from integer-order calculus that the integer-order derivatives can only be sinusoidal functions with a phase shift of multiples of $\pi/2$. The fractional-order derivatives may provide more intermediate information, since the phase shifts are no longer integer multiples of $\pi/2$. The 3D plot of the fractional-order integrals and derivatives is shown in Fig. 5.4(a), with the following MATLAB commands.

```
>> t=0:0.1:pi; y=sin(3*t+1); Y=[]; n_vec=[-1:0.2:1];
   for n=n_vec, Y=[Y; glfdiff(y,t,n)]; end
   surf(t,n_vec,Y), shading flat
```

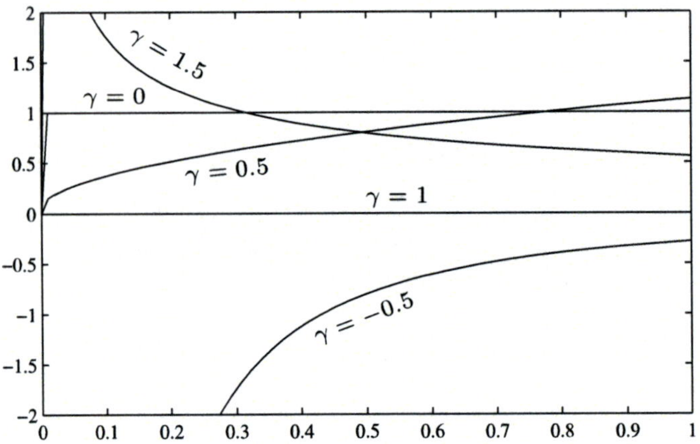

Fig. 5.5 Fractional-order differentiations of a unit step function

With the Grünwald-Letnikov definition, the 0.75th-order derivative of function $f(t)$ can be obtained as shown in Fig. 5.4(b), while the one with the Cauchy formula above can also be shown.

```
>> t=0:0.01:pi; y=sin(3*t+1);
   y1=3^0.75*sin(3*t+1+0.75*pi/2);
   y2=glfdiff(y,t,0.75); plot(t,y1,t,y2)
```

It can be seen that there exist some differences only at the initial time, since in the Grünwald-Letnikov definition, the initial values of function $f(t)$, for $t \leq 0$, are assumed to be zero, while in the Cauchy formula, the initial values of the function $f(t)$ is still assumed to be obtainable from $f(t) = \sin(3t + 1)$. Thus one must be careful with the differences in the definitions.

Example 5.5 It is well known in the field of integer-order calculus that the derivatives of a step function are a straight line. Now let us investigate the case for fractional-order derivatives and integrals. With the following MATLAB statements, the derivatives and integrals of selected orders can be obtained as shown in Fig. 5.5.

```
>> t=0:0.01:1; u=ones(size(t));
   n_vec=[-0.5,0,0.5,1,1.5]; Y=[];
   for n=n_vec, Y=[Y; glfdiff(u,t,n)]; end
   plot(t,Y), ylim([-2 2])
```

It can be seen that, when fractional calculus is involved, the fractional-order derivatives and integrals of a step function may not be straight lines depending on the definition used.

FIR Filter Approximation: Power Series Expansion

In general, the discretization of the fractional-order differentiator/integrator $s^{\pm r}$ ($r \in \mathbb{R}$) can be expressed by the so-called generating function $s = \omega(z^{-1})$. This generating function and its expansion determine both the form of the approximation and the coefficients. For example, when a backward difference rule is used, i.e., $\omega(z^{-1}) = (1 - z^{-1})/T$, performing the PSE (Power Series Expansion) of $(1 - z^{-1})^{\pm r}$ gives the discretization formula which happens to be the same as the Grünwald-Letnikov (GL) definition using the short memory principle. Then, the discrete equivalent of the fractional-order integro-differential operator, $(\omega(z^{-1}))^{\pm r}$, is given by

$$\left(\omega\left(z^{-1}\right)\right)^{\pm r} = T^{\mp r} z^{-[L/T]} \sum_{j=0}^{[L/T]} (-1)^j \binom{\pm r}{j} z^{[L/T]-j}, \tag{5.8}$$

where T is the sampling period, L is the memory length and $(-1)^j \binom{\pm r}{j}$ are binomial coefficients $c_j^{(r)}$ ($j = 0, 1, \ldots$), where

$$c_0^{(r)} = 1, \qquad c_j^{(r)} = \left(1 - \frac{1 + (\pm r)}{j}\right) c_{j-1}^{(r)}. \tag{5.9}$$

Petráš designed a MATLAB function `filt()`, which can be used in FIR filter approximation of fractional-order differentiators.[2] The key part of the function is simply

```
function H=dfod2(n,T,r)
if r>0
   bc=cumprod([1,1-((r+1)./[1:n])]); H=filt(bc,[T^r],T);
elseif r<0
   bc=cumprod([1,1-((-r+1)./[1:n])]);
   H=filt([T^(-r)],bc,T);
end
```

where n is the expected order or taps of the FIR filter, T is the sampling period, and r is the expected order of differentiation. Normally to achieve good approximation results, the order n must be assigned to a very high number, *i.e.*, $n = 100$.

Note that, for FIR approximation using PSE, $\omega(z^{-1})$ must be a polynomial of z^{-1}, that is, $\omega(z^{-1})$ must be FIR itself. Using $\omega(z^{-1}) = (1 - z^{-1})/T$ coincides

[2] http://www.mathworks.com/matlabcentral/fileexchange/3673.

with the Grünwald-Letnikov (GL) definition. Using the magic digital differentiator formula generator, as claimed in the Savitzky-Golay smoothing filter code,[3] we can expect various advanced generators $\omega(z^{-1})$ in FIR form.

IIR Filter Approximation: Tustin Method with Prewarping

As it is known, Tustin method relates the s and z domains with the following substitution formula

$$s = \frac{2}{T}\frac{z-1}{z+1}, \tag{5.10}$$

where T is the sampling period. In signal processing literature the Tustin method is frequently denoted the *bilinear transformation method*. The term bilinear is related to the fact that the imaginary axis in the complex s-plane for continuous-time systems is mapped or transformed onto the unity circle for the corresponding discrete-time system. In addition, the poles are transformed so that the stability property is preserved.

With the substitution formula in (5.10) the discrete version $H_{\mathrm{d}}(z)$ of a continuous transfer function $H_{\mathrm{c}}(s)$ is obtained. In general, the frequency responses of $H_{\mathrm{c}}(s)$ and $H_{\mathrm{disc}}(z)$ are not equal at the same frequencies. The Tustin method can be modified or enhanced so that a similar frequency response can be obtained for both $H_{\mathrm{c}}(s)$ and $H_{\mathrm{d}}(z)$ at one or more user-defined critical frequencies. This is done by modifying (*prewarping*) the critical frequencies of $H_{\mathrm{c}}(s)$ so that the frequency responses are equal after the discretization.

In our case, MATLAB function `c2d()` is used to obtain the discrete transfer function of a continuous system, whose syntax is $H_{\mathrm{d}} = \mathtt{c2d}(H_{\mathrm{c}}, T, \mathtt{METHOD})$, where H_{d} is the resulting discrete transfer function, H_{c} the continuous transfer function to discretize, and T the sampling period. The string `METHOD` selects the discretization method among the following:

- `'zoh'`: Zero-order hold on the inputs.
- `'foh'`: Linear interpolation of inputs.
- `'tustin'`: Bilinear approximation.
- `'prewarp'`: Tustin approximation with frequency prewarping. The critical frequency ω_{c} (in rad/sec) is specified as fourth input by $H_{\mathrm{d}} = \mathtt{c2d}(H_{\mathrm{c}}, T, \mathtt{'prewarp'}, \omega_{\mathrm{c}})$. In our case, the critical frequency will be the gain crossover frequency, that is, $\omega_{\mathrm{c}} = \omega_{\mathrm{cg}}$.
- `'matched'`: Matched pole-zero method (for SISO systems only).
- The default option is `'zoh'` when `METHOD` is omitted.

Example 5.6 To illustrate this method, the discrete-time transfer function $H_{\mathrm{invf}}(z)$ corresponding to the continuous approximation $H_1(s)$ from the previous section is obtained with the following statements,

[3] http://www.mathworks.com/matlabcentral/fileexchange/3514.

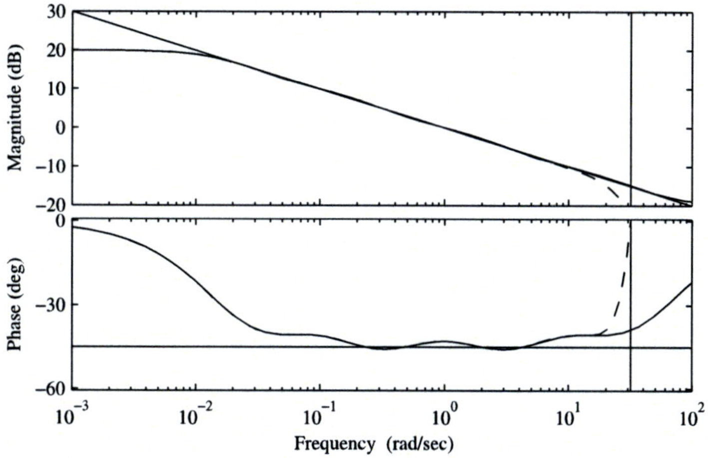

Fig. 5.6 Bode plots of the transfer functions $H_1(s)$ and $H_2(z)$, with *solid lines* for $H_1(s)$ and *dashed lines* for $H_2(z)$. The *straight lines* are for the theoretical results

```
>> H1=ousta_fod(-0.5,4,1e-2,1e2);
   H2=c2d(H1,0.1,'prewarp',1), bode(H1,'-',H2,'--')
```

resulting respectively

$$H_1(s) = \frac{0.1s^4 + 6.248s^3 + 35.45s^2 + 19.76s + 1}{s^4 + 19.76s^3 + 35.45s^2 + 6.248s + 0.1},$$

and

$$H_2(z) = \frac{0.2425z^4 - 0.491z^3 + 0.2033z^2 + 0.106z - 0.06079}{z^4 - 2.875z^3 + 2.802z^2 - 0.974z + 0.0478},$$

with $\omega_c = 1$ rad/sec and $T = 0.001$ sec. From the Bode plots of Fig. 5.6, the similarity between the frequency responses of $H_2(z)$ and $H_1(s)$ can be observed in Fig. 5.17.

Direct Discretization: First-Order IIR Generating Functions

In general, discretization of the fractional-order differentiator s^r (r is a real number) can be expressed by the so-called generating function $s = \omega(z^{-1})$. This generating function and its expansion determine both the form of the approximation and the coefficients. For example, as shown in the last section, when a backward difference rule is used, i.e., $\omega(z^{-1}) = (1 - z^{-1})/T$ with T the sampling period, performing the power series expansion (PSE) of $(1 - z^{-1})^{\pm r}$ gives the discretization formula which is actually in FIR filter form. In other words, when $\omega(z^{-1})$ is a 2 tap FIR, it is equivalent to the Grünwald-Letnikov formula.

In this section, we consider several options for the generating function $\omega(z^{-1})$ in IIR form. Let us first consider the first order IIR forms of $\omega(z^{-1})$. The first option is the trapezoidal (Tustin) formula used as the generating function

$$\left(\omega\left(z^{-1}\right)\right)^{\pm r}=\left(\frac{2}{T}\frac{1-z^{-1}}{1+z^{-1}}\right)^{\pm r}. \tag{5.11}$$

The above integer-order digital fractional-order differentiator can then be obtained by using the CFE. The second option is the so-called Al-Alaoui operator which is a mixed scheme of Euler and Tustin operators [3, 59]. Correspondingly, the generating function for discretization is

$$\left(\omega\left(z^{-1}\right)\right)^{\pm r}=\left(\frac{8}{7T}\frac{1-z^{-1}}{1+z^{-1}/7}\right)^{\pm r}. \tag{5.12}$$

Clearly, both (5.11) and (5.12) are rational discrete-time transfer functions of infinite orders. To approximate it with a finite order rational one, continued fraction expansion (CFE) is an efficient way. In general, any well-behaved function $G(z)$ can be represented by continued fractions in the form of

$$G(z)\simeq a_0(z)+\cfrac{b_1(z)}{a_1(z)+\cfrac{b_2(z)}{a_2(z)+\cfrac{b_3(z)}{a_3(z)+\cdots}}}, \tag{5.13}$$

where the coefficients a_i and b_i are either rational functions of the variable z or constants. By truncation, an approximate rational function, $\widehat{G}(z)$, can be obtained.

CFE Tustin Operator Let the resulting discrete transfer function, approximating fractional-order operators, be expressed by

$$\begin{aligned}D^{\pm r}(z)&=\frac{Y(z)}{F(z)}=\left(\frac{2}{T}\right)^{\pm r}\mathrm{CFE}\left\{\left(\frac{1-z^{-1}}{1+z^{-1}}\right)^{\pm r}\right\}_{p,q}\\&=\left(\frac{2}{T}\right)^{\pm r}\frac{P_p(z^{-1})}{Q_q(z^{-1})},\end{aligned} \tag{5.14}$$

where T is the sample period, CFE$\{u\}$ denotes the function resulting from applying the continued fraction expansion to the function u, $Y(z)$ is the Z transform of the output sequence $y(nT)$, $F(z)$ is the Z transform of the input sequence $f(nT)$, p and q are the orders of the approximation, and P and Q are polynomials of degrees p and q, correspondingly, in the variable z^{-1}.

Table 5.2 Expressions of $D^r(z)$ in (5.14) for different orders

$p=q$	$P_p(z^{-1})$ $(k=1)$, and $Q_q(z^{-1})$ $(k=0)$
1	$(-1)^k z^{-1} r + 1$
3	$(-1)^k(r^3-4r)z^{-3}+(6r^2-9)z^{-2}+(-1)^k 15z^{-1}r+15$
5	$(-1)^k(r^5-20r^3+64r)z^{-5}+(-195r^2+15r^4+225)z^{-4}+(-1)^k(105r^3-735r)z^{-3}$ $+(420r^2-1050)z^{-2}+(-1)^k 945z^{-1}r+945$
7	$(-1)^k(784r^3+r^7-56r^5-2304r)z^{-7}+(10612r^2-1190r^4-11025+28r^6)z^{-6}$ $+(-1)^k(53487r+378r^5-11340r^3)z^{-5}+(99225-59850r^2+3150r^4)z^{-4}$ $+(-1)^k(17325r^3-173250r)z^{-3}+(-218295+62370r^2)z^{-2}+(-1)^k 135135z^{-1}r$ $+135135$
9	$(-1)^k(-52480r^3+147456r+r^9-120r^7+4368r^5)z^{-9}+(45r^8+120330r^4$ $-909765r^2-4410r^6+893025)z^{-8}+(-1)^k(-5742495r-76230r^5+1451835r^3$ $+990r^7)z^{-7}+(-13097700+9514890r^2-796950r^4+13860r^6)z^{-6}$ $+(-1)^k(33648615r-5405400r^3+135135r^5)z^{-5}+(-23648625r^2+51081030$ $+945945r^4)z^{-4}+(-1)^k(-61486425r+4729725r^3)z^{-3}+(16216200r^2$ $-72972900)z^{-2}+(-1)^k 34459425z^{-1}r+34459425$

By using the MAPLE call

```
Drp:=cfrac(((1-x)/(1+x))^r,x,p)
```

where $x=z^{-1}$, the obtained symbolic approximation has the following form:

$$D^r(z)=1+\cfrac{z^{-1}}{-\cfrac{1}{2}\cfrac{1}{r}+\cfrac{z^{-1}}{-2+\cfrac{z^{-1}}{\cfrac{3}{2}\cfrac{r}{r^2-1}+\cfrac{z^{-1}}{2+\cfrac{z^{-1}}{-\cfrac{5}{2}\cfrac{r^2-1}{r(-4+r^2)}+\cfrac{z^{-1}}{-2+\cdots}}}}}}. \tag{5.15}$$

In MATLAB Symbolic Math Toolbox, we can get the same result by the following script:

```
syms x r;
maple('with(numtheory)');
f = ((1-x)/(1+x))^r; %
maple(['cf:=cfrac(' char(f) ',x,10);']) %
maple('nd5 :=nthconver','cf',10)
maple('num5 := nthnumer','cf',10)
maple('den5 := nthdenom','cf',10)
```

In Table 5.2, the general expressions for numerator and denominator of $D^r(z)$ in (5.14) are listed for $p=q=1,3,5,7,9$.

With $r=0.5$ and $T=0.001$ sec. the approximate models for $p=q=1,3,7,9$ are:

$$G_1(z)=44.72\frac{z-0.5}{z+0.5},\qquad G_3(z)=44.72\frac{z^3-0.5z^2-0.5z+0.125}{z^3+0.5z^2-0.5z-0.125},$$

$$G_7(z)=44.72\frac{\begin{array}{c}z^7-0.5z^6-1.5z^5+0.625z^4\\+0.625z^3-0.1875z^2-0.0625z+0.007813\end{array}}{\begin{array}{c}z^7+0.5z^6-1.5z^5-0.625z^4+0.625z^3\\+0.1875z^2-0.0625z-0.007813\end{array}},$$

$$G_9(z)=44.72\frac{\begin{array}{c}z^9-0.5z^8-2z^7+0.875z^6+1.313z^5-0.4688z^4\\-0.3125z^3+0.07813z^2+0.01953z-0.001953\end{array}}{\begin{array}{c}z^9+0.5z^8-2z^7-0.875z^6+1.313z^5+0.4688z^4\\-0.3125z^3-0.07813z^2+0.01953z+0.001953\end{array}}.$$

In Fig. 5.7, the Bode plots and the distributions of zeros and poles of the approximations are presented. In Fig. 5.7, the effectiveness of the approximations fitting the ideal responses in a wide range of frequencies, in both magnitude and phase, can be observed. In Fig. 5.8, it can be observed that the approximations fulfill the two desired properties: (i) all the poles and zeros lie inside the unit circle, and (ii) the poles and zeros are interlaced along the segment of the real axis corresponding to $z\in(-1,1)$.

Al-Alaoui Operator Now, let us show how to perform CFE of Al-Alaoui operator (5.12). The resulting discrete transfer function, approximating fractional-order operators, can be expressed as:

$$\begin{aligned}D^{\pm r}(z)&\approx\left(\frac{8}{7T}\right)^{\pm r}\mathrm{CFE}\left\{\left(\frac{1-z^{-1}}{1+z^{-1}/7}\right)^{\pm r}\right\}_{p,q}\\&=\left(\frac{8}{7T}\right)^{\pm r}\frac{P_p(z^{-1})}{Q_q(z^{-1})}.\end{aligned}\tag{5.16}$$

Normally, we can set $p=q=n$. In MATLAB Symbolic Math Toolbox, we can easily get the approximate direct discretization of the fractional order derivative by the following script, for a given n (replace `14` by $2n$):

```
clear all;close all; syms x z r
%Al-Alouoi's scheme
x=((1-z)/(1+z/7) )^r;
[RESULT,STATUS] = maple('with(numtheory)')
%7-th order; put 2*7 here.
h7=maple('cfrac',x,z,14);
h7n=maple('nthnumer(%%,14)');
```

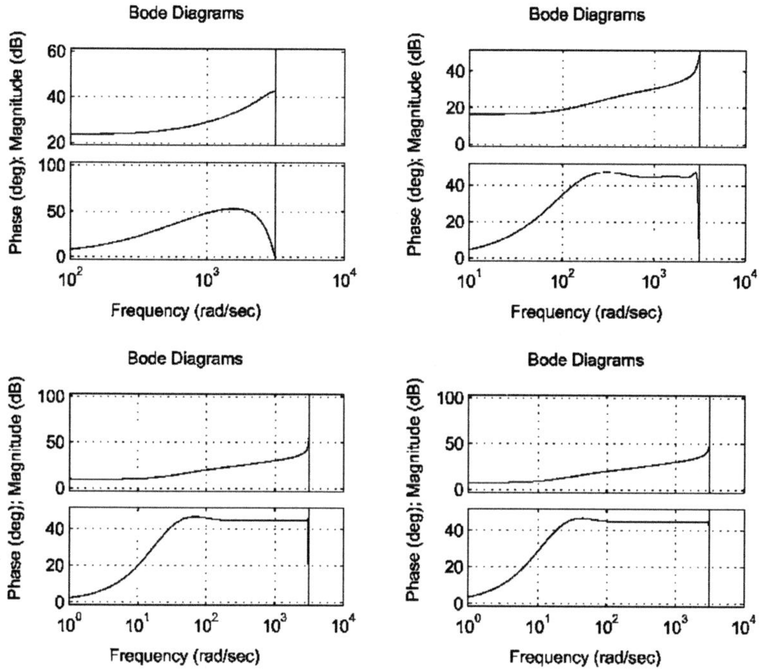

Fig. 5.7 Bode plots (approximation orders 1, 3, 7, 9) by Tustin CFE approximate discretization of $s^{0.5}$ at $T = 0.001$ sec

```
h7d=maple('nthdenom(%%%,14)');
h7ns=sym(h7n);h7ds=sym(h7d);
num7=collect(h7ns,z);den7=collect(h7ds,z);
fn7=subs(num7,z,1/z),fd7=subs(den7,z,1/z)
```

The CFE scheme presented in the above (Tustin and Al-Alaoui) contains two tuning parameters, namely p and q. The optimal choice of these two parameters is possible based on a quantitative measure. One possibility is the use of the least squares (LS) error between the continuous frequency response and discretized frequency response. Note that in practice, p and q can usually be set to be equal.

The discretization of the half-differentiator $s^{0.5}$ sampled at 0.001 sec. is studied numerically, and the approximate models are

$$G_1(z) = \frac{236.6z - 169}{7z - 1}, \qquad G_3(z) = \frac{1657z^3 - 2603z^2 + 1048z - 62.78}{49z^3 - 49z^2 + 7z + 1},$$

$$G_5(z) = \frac{2.47e04z^5 - 5.999e04z^4 + 4.941e04z^3 - 1.512e04z^2 + 956.9z + 98.48}{730.7z^5 - 1357z^4 + 745.7z^3 - 89.48z^2 - 15.52z + 1},$$

$$G_7(z) = \frac{\begin{array}{l}3.128e05z^7 - 1.028e06z^6 + 1.283e06z^5 \\ \quad - 7.433e05z^4 + 1.87e05z^3 - 9772z^2 - 2140z + 104.5\end{array}}{\begin{array}{l}9253z^7 - 2.512e004z^6 + 2.436e004z^5 \\ \quad - 9577z^4 + 905.7z^3 + 219.7z^2 - 23.67z - 1\end{array}}.$$

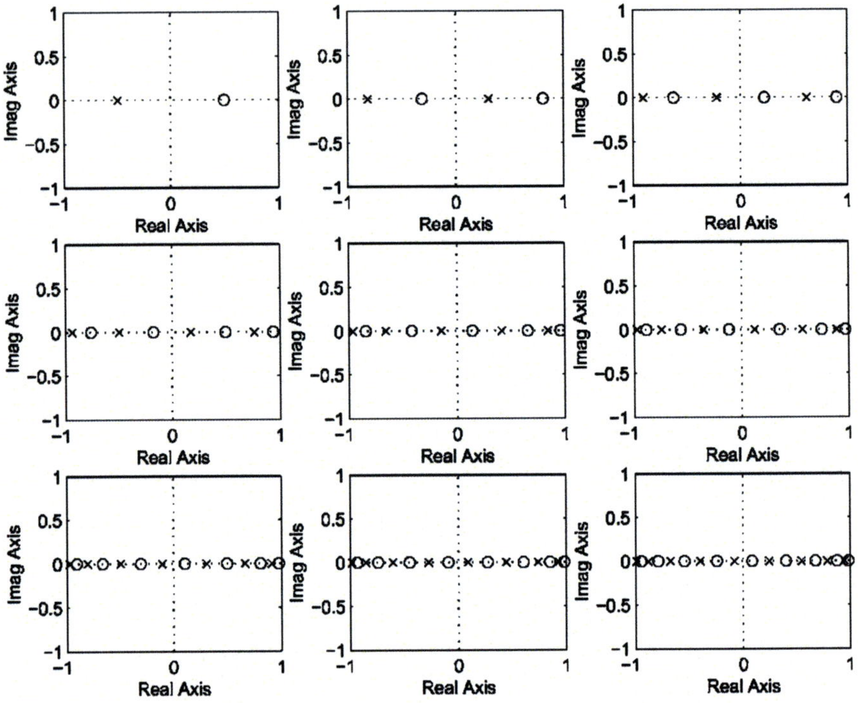

Fig. 5.8 Zero-pole distribution (approximation orders $1, 2, \ldots, 9$) by Tustin CFE approximate discretization of $s^{0.5}$ at $T = 0.001$ sec

We present four plots, shown in Fig. 5.9, to demonstrate the effectiveness of the approximate discretization. We can observe from Fig. 5.9 that this scheme is much better than the Tustin scheme in the magnitude fit to the original s^r. After the linear phase compensation, the maximum phase error of the Al-Alaoui operator based discretization scheme is around $r \times 8.25°$ at 55% of the Nyquist frequency (around 275 Hz in this example) as shown in Fig. 5.9. To compensate for the linear phase drop, a half sample phase advance is used which means that we should cascade $z^{0.5r}$ to the obtained approximately discretized transfer function $G(z)$. However, in this example, the phase compensator is $z^{0.25}$ which is noncausal. In implementation, we can simply use $z^{-0.75}/z^{-1}$ instead.

Direct Discretization: Second-Order IIR Generating Function Method

Now, let us consider the second order IIR type generating function $\omega(z^{-1})$. To start with, we first look into the integer order IIR-type digital integrator by a weighted Simpson and Tustin scheme in order to derive the proper second order IIR type generating function $\omega(z^{-1})$.

It was pointed out in [3] that the magnitude of the frequency response of the ideal integrator $1/s$ lies between that of the Simpson and trapezoidal digital integrators. It is reasonable to "interpolate" the Simpson and trapezoidal digital integrators to

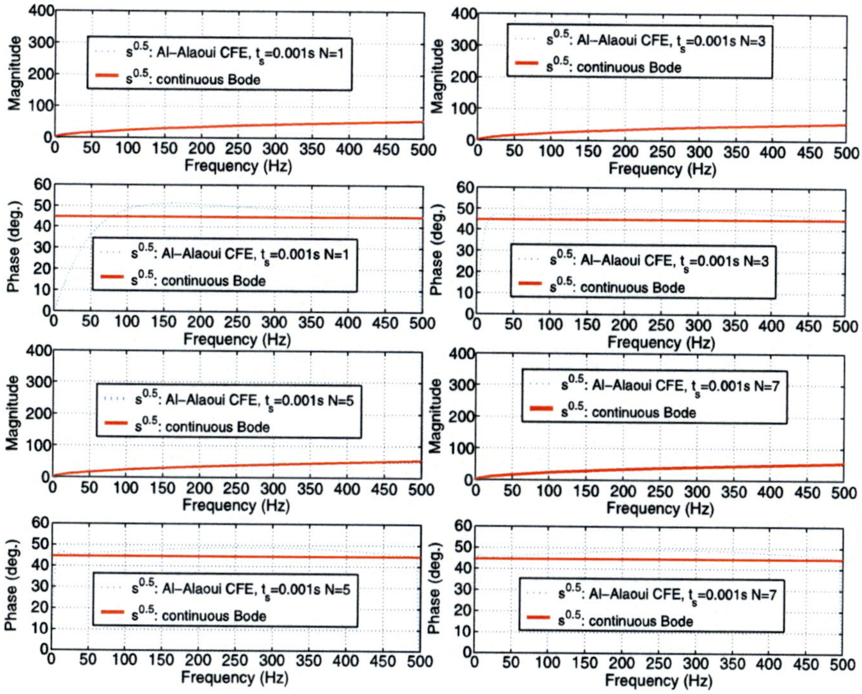

Fig. 5.9 CFE (Al-Alaoui) discretization of $s^{0.5}$ at $T = 0.001$ sec. (Bode plots of *top left*: $G_1(z)$; *top right*: $G_3(z)$; *bottom left*: $G_5(z)$; *bottom right*: $G_7(z)$)

compromise the high frequency accuracy in frequency response. This leads to the following hybrid digital integrator

$$H(z) = aH_S(z) + (1-a)H_T(z), \quad a \in [0, 1] \tag{5.17}$$

where a is actually a weighting factor or tuning knob. $H_S(z)$ and $H_T(z)$ are the z-transfer functions of the Simpson's and the trapezoidal integrators given respectively as follows:

$$H_S(z) = \frac{T(z^2+4z+1)}{3(z^2-1)} \tag{5.18}$$

and

$$H_T(z) = \frac{T(z+1)}{2(z-1)}. \tag{5.19}$$

The overall weighted digital integrator with the tuning parameter a is hence given by

$$\begin{aligned} H(z) &= \frac{T(3-a)\{z^2+[2(3+a)/(3-a)]z+1\}}{6(z^2-1)} \\ &= \frac{T(3-a)(z+r_1)(z+r_2)}{6(z^2-1)}, \end{aligned} \tag{5.20}$$

where

$$r_1 = \frac{3+a+2\sqrt{3a}}{3-a}, \qquad r_2 = \frac{3+a-2\sqrt{3a}}{3-a}.$$

It is interesting to note that $r_1 = \frac{1}{r_2}$ and $r_1 = r_2 = 1$ only when $a = 0$ (trapezoidal). For $a \neq 0$, $H(z)$ must have one non-minimum phase (NMP) zero.

Now, we can obtain a family of new integer-order digital differentiators from the digital integrators introduced in the above. Direct inversion of $H(z)$ will give an unstable filter since $H(z)$ has an non-minimum phase (NMP) zero r_1. By reflecting the NMP r_1 to $1/r_1$, i.e. r_2, we have

$$\tilde{H}(z) = K\frac{T(3-a)(z+r_2)^2}{6(z^2-1)}.$$

To determine K, let the final values of the impulse responses of $H(z)$ and $\tilde{H}(z)$ be the same, i.e., $\lim_{z\to 1}(z-1)H(z) = \lim_{z\to 1}(z-1)\tilde{H}(z)$, which gives $K = r_1$. Therefore, the new family of first-order digital differentiators are given by

$$\omega(z) = \frac{1}{\tilde{H}(z)} = \frac{6(z^2-1)}{r_1 T(3-a)(z+r_2)^2} = \frac{6r_2(z^2-1)}{T(3-a)(z+r_2)^2}. \tag{5.21}$$

We can regard $\omega(z)$ in (5.21) as the generating function. Finally, we can obtain the expression for a family of digital fractional order differentiator as

$$G(z^{-1}) = (\omega(z^{-1}))^r = k_0\left(\frac{1-z^{-2}}{(1+bz^{-1})^2}\right)^r, \tag{5.22}$$

where $r \in [0,1]$, $k_0 = (\frac{6r_2}{T(3-a)})^r$ and $b = r_2$.

Using CFE, an approximation for an irrational function $G(z^{-1})$ can be expressed in the form of (5.13). Similar to (5.12), here, the irrational transfer function $G(z^{-1})$ in (5.22) can be expressed by an infinite order of rational discrete-time transfer function by CFE method as shown in (5.13).

The CFE expansion can be automated by using a symbolic computation tool such as the MATLAB Symbolic Math Toolbox. For illustrations, let us denote $x = z^{-1}$. Referring to (5.22), the task is to perform the following expansion:

$$\mathrm{CFE}\left(\frac{1-x^2}{(1+bx)^2}\right)^r$$

to the desired order n. The following MATLAB script will generate the above CFE with `p1` and `q1` containing, respectively, the numerator and denominator polynomials in `x` or z^{-1} with their coefficients being functions of `b` and `r`.

```
clear all;close all;syms x r b;maple('with(numtheory)');
aas = ((1-x*x)/(1+b*x)^2 )^r; n=3; n2=2*n;
maple(['cfe := cfrac('char(aas) ',x,n2);']);
```

```
pq=maple('P_over_Q := nthconver','cfe',n2);
p0=maple('P := nthnumer','cfe',n2);
q0=maple('Q := nthdenom','cfe',n2);
p=(p0(5:length(p0))); q=(q0(5:length(q0)));
p1=collect(sym(p),x); q1=collect(sym(q),x);
```

Here we present some results for $r = 0.5$. The values of the truncation order n and the weighting factor a are denoted as subscripts of $G_{(n,a)}(z)$. Let $T = 0.001$ sec. We have the following:

$$\begin{aligned}
G_{(2,0.00)}(z^{-1}) &= \frac{178.9 - 89.44z^{-1} - 44.72z^{-2}}{4 + 2z^{-1} - z^{-2}},\\
G_{(2,0.25)}(z^{-1}) &= \frac{138.8 + 98.07z^{-1} - 158.2z^{-2}}{4 + 5.034z^{-1} - z^{-2}},\\
G_{(2,0.50)}(z^{-1}) &= \frac{127 + 41.26z^{-1} - 112.6z^{-2}}{4 + 2.98z^{-1} - z^{-2}},\\
G_{(2,0.75)}(z^{-1}) &= \frac{119.3 + 25.56z^{-1} - 97.96z^{-2}}{4 + 2.19z^{-1} - z^{-2}},\\
G_{(2,1.00)}(z^{-1}) &= \frac{113.4 + 17.74z^{-1} - 89.81z^{-2}}{4 + 1.698z^{-1} - z^{-2}};
\end{aligned} \tag{5.23}$$

$$\begin{aligned}
G_{(3,0.00)}(z^{-1}) &= \frac{357.8 - 178.9z^{-1} - 178.9z^{-2} + 44.72z^{-3}}{8 + 4z^{-1} - 4z^{-2} - z^{-3}},\\
G_{(3,0.25)}(z^{-1}) &= \frac{392.9 - 78.04z^{-1} - 349.8z^{-2} + 88.97z^{-3}}{11.32 + 4z^{-1} - 5.66z^{-2} - z^{-3}},\\
G_{(3,0.50)}(z^{-1}) &= \frac{1501 - 503.6z^{-1} - 1289z^{-2} + 446.5z^{-3}}{47.26 + 4z^{-1} - 23.63z^{-2} - z^{-3}},\\
G_{(3,0.75)}(z^{-1}) &= \frac{968.1 - 442z^{-1} - 820.8z^{-2} + 363z^{-3}}{32.47 - 4z^{-1} - 16.24z^{-2} + z^{-3}},\\
G_{(3,1.00)}(z^{-1}) &= \frac{353.1 - 208z^{-1} - 297.4z^{-2} + 164.7z^{-3}}{12.46 - 4z^{-1} - 6.228z^{-2} + z^{-3}};
\end{aligned} \tag{5.24}$$

$$\begin{aligned}
G_{(4,0.00)}(z^{-1}) &= \frac{715.5 - 357.8z^{-1} - 536.7z^{-2} + 178.9z^{-3} + 44.72z^{-4}}{16 + 8z^{-1} - 12z^{-2} - 4z^{-3} + z^{-4}},\\
G_{(4,0.25)}(z^{-1}) &= \frac{555.3 - 392.9z^{-1} - 477.2z^{-2} + 349.8z^{-3} - 19.56z^{-4}}{16 - 2.489z^{-1} - 12z^{-2} + 1.245z^{-3} + z^{-4}},\\
G_{(4,0.50)}(z^{-1}) &= \frac{508.1 - 1501z^{-1} - 4.478z^{-2} + 1289z^{-3} - 382.9z^{-4}}{16 - 40.54z^{-1} - 12z^{-2} + 20.27z^{-3} + z^{-4}},
\end{aligned} \tag{5.25}$$

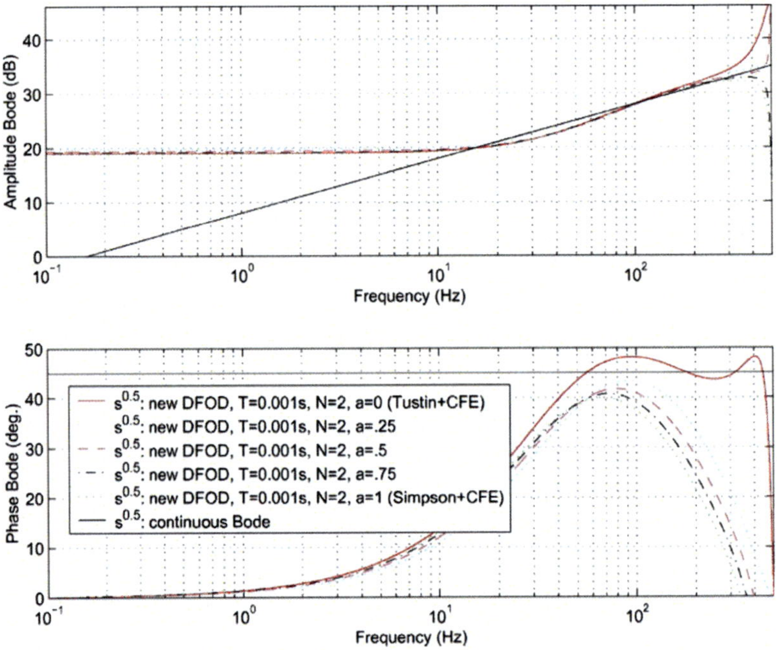

Fig. 5.10 Bode plot comparison for $r = 0.5$, $n = 2$ and $a = 0, .25, .5, .75, 1$

$$G_{(4,0.75)}\left(z^{-1}\right) = \frac{477 + 968.1z^{-1} - 919z^{-2} - 820.8z^{-3} + 422.7z^{-4}}{16 + 37.8z^{-1} - 12z^{-2} - 18.9z^{-3} + z^{-4}},$$

$$G_{(4,1.00)}\left(z^{-1}\right) = \frac{453.6 + 353.1z^{-1} - 661.7z^{-2} - 297.4z^{-3} + 221.5z^{-4}}{16 + 16.74z^{-1} - 12z^{-2} - 8.371z^{-3} + z^{-4}}.$$

The Bode plot comparisons for the above three groups of approximate fractional order digital differentiators are summarized in Figs. 5.10, 5.11 and 5.12 respectively. We can observe the improvement in high frequency magnitude response. If a trapezoidal scheme is used, the high frequency magnitude response is far from the ideal one. The role of the tuning knob a is obviously useful in some applications. MATLAB code for this new digital fractional-order differentiator is available upon request.

Remark 5.7 The phase approximations in Figs. 5.10, 5.11 and 5.12 did not consider the linear phase lag compensation as is done in [59]. For a given a and r, a pure linear phase lead compensation can be added without affecting the magnitude approximation. For example, when $a = r = 0.5$, a pure phase lead $z^{0.5}$ can be cascaded to $G_{(4,0.50)}(z^{-1})$ and the phase approximation can be improved as shown in Fig. 5.13. Note that $z^{0.5}$ can be realized by $z^{-0.5}/z^{-1}$ which is causally realizable.

For $n = 3$ and $n = 4$, the pole-zero maps are shown respectively in Figs. 5.14 and 5.15 for some different values of a. First of all, we observe that there are no

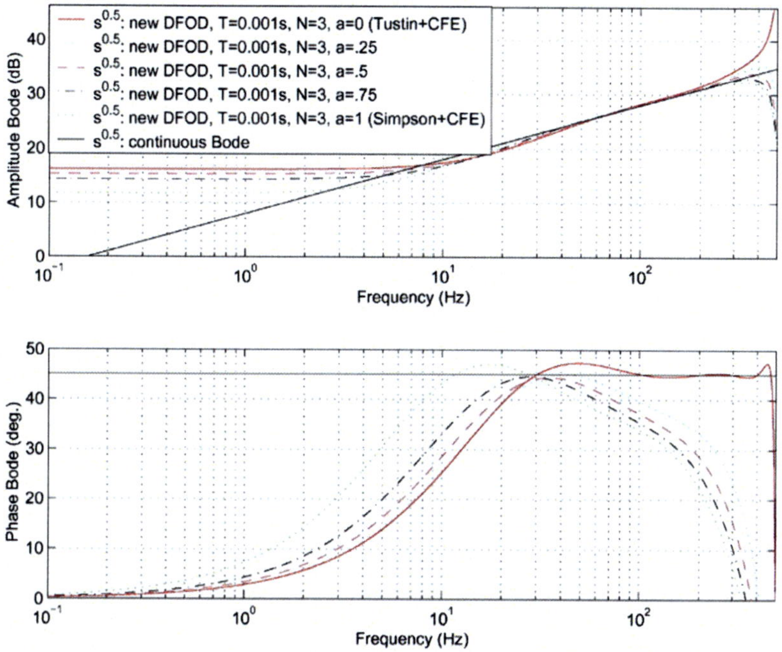

Fig. 5.11 Bode plot comparison for $r = 0.5$, $n = 3$ and $a = 0, .25, .5, .75, 1$

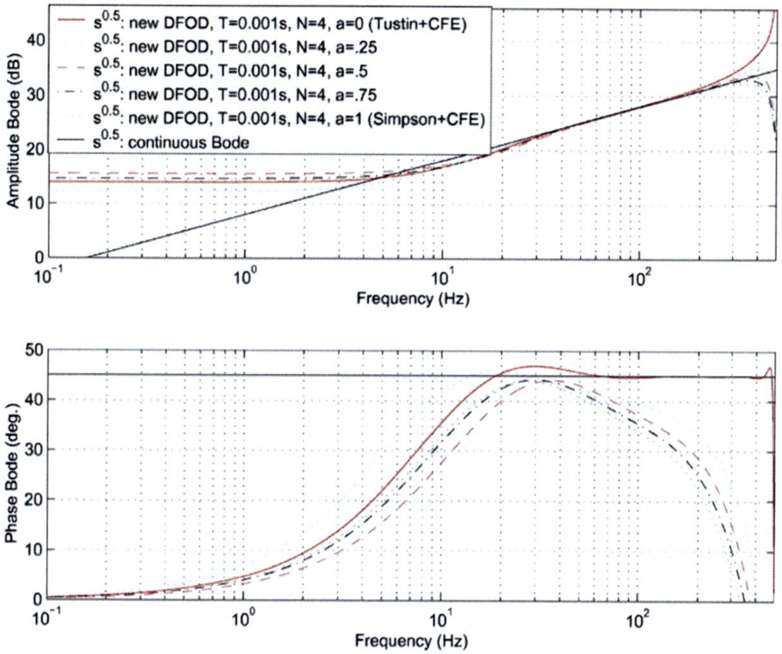

Fig. 5.12 Bode plot comparison for $r = 0.5$, $n = 4$ and $a = 0, .25, .5, .75, 1$

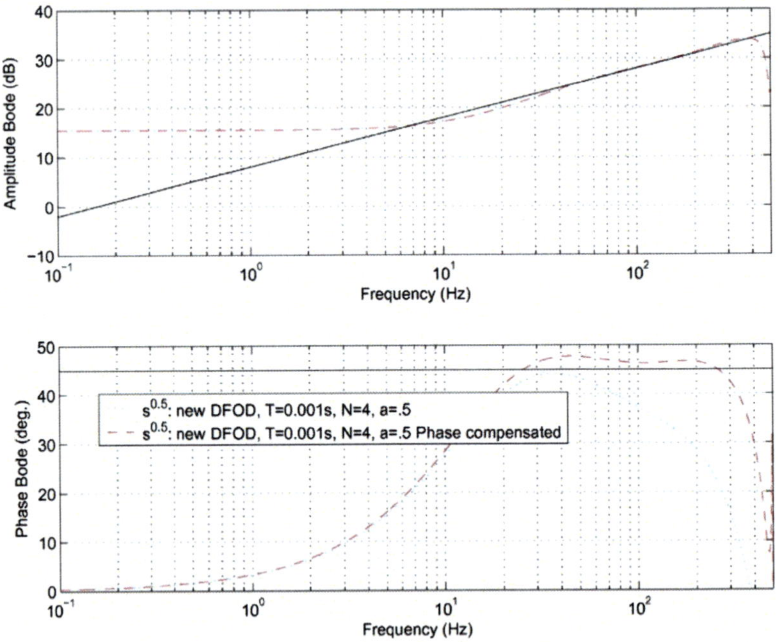

Fig. 5.13 Effect of linear phase compensation for $r = 0.5$, $n = 4$ and $a = .5$

complex conjugate poles or zeros. We can further observe that for odd order of CFE ($n = 3$), the pole-zero maps are nicely behaved, that is, all the poles and zeros lie inside the unit circle and the poles and zeros are interlaced along the segment of the real axis corresponding to $z \in (-1, 1)$. However, when n is even, and when a is near 1, there may have one canceling pole-zero pair as seen in Fig. 5.15 which may not be desirable. We suggest the use an odd n when applying this discretization scheme.

Direct Discretization: Step or Impulse Response Invariant Method

Table 5.3 shows a set of MATLAB functions for discrete-time implementation of the fractional-order differentiator and integrator, as well as the complicated transfer functions with non-integer powers developed based on the fitting of step response (SR) invariants and impulse response (IR) invariants. These functions can be downloaded for free and used directly in establishing the discrete-time implementations.[4] With the use of the functions, a discrete-time implementation to the fractional-order terms can easily be constructed.

[4] http://www.mathworks.com/matlabcentral/fileexchange/authors/9097.

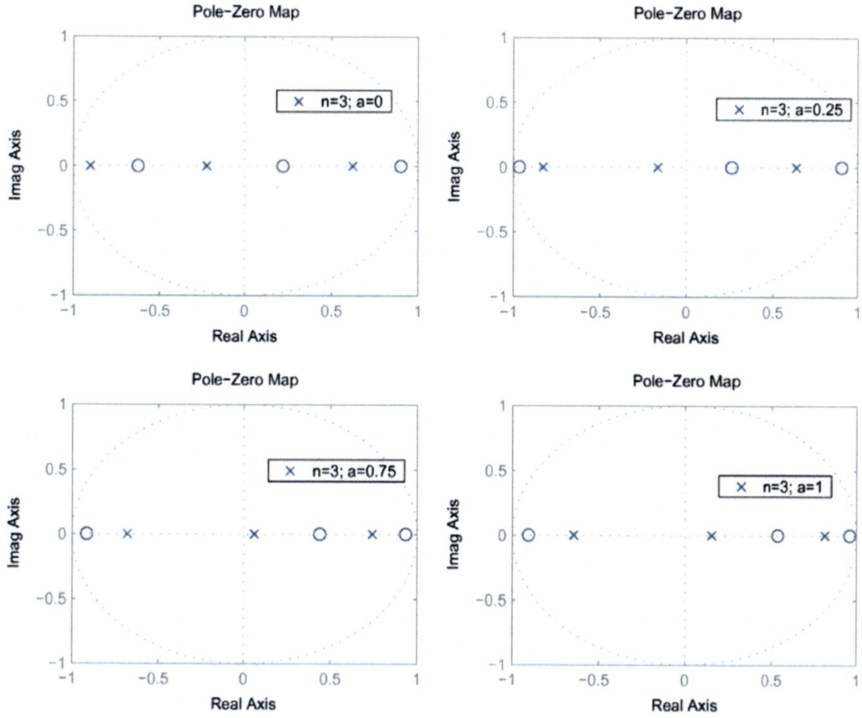

Fig. 5.14 Pole-zero maps for $r = 0.5$, $n = 3$ and $a = 0, .25, .75, 1$

Table 5.3 MATLAB functions for discrete-time implementations

Function	Syntax	Descriptions
`irid_fod()`	$G =$ `irid_fod`(r, T, N)	s^r fitting with IR invariants
`srid_fod()`	$G =$ `srid_fod`(r, T, N)	s^r fitting with SR invariants
`irid_folpf()`	$G =$ `irid_folpf`(τ, r, T, N)	$(\tau s + 1)^{-r}$ fitting with IR invariants

Example 5.8 Selecting a sampling period of $T = 0.1$ sec, and the order of 5, the 0.5th-order integrator can be implemented with the step response invariants and impulse response invariants using the following statements

```
>> G1=irid_fod(-0.5,0.1,5); G2=srid_fod(-0.5,0.1,5);
   bode(G1,'--',G2,':')
```

with the results

$$G_1(z) = \frac{0.09354z^5 - 0.2395z^4 + 0.2094z^3 - 0.06764z^2 + 0.003523z + 0.0008224}{z^5 - 3.163z^4 + 3.72z^3 - 1.966z^2 + 0.4369z - 0.02738}, \tag{5.26}$$

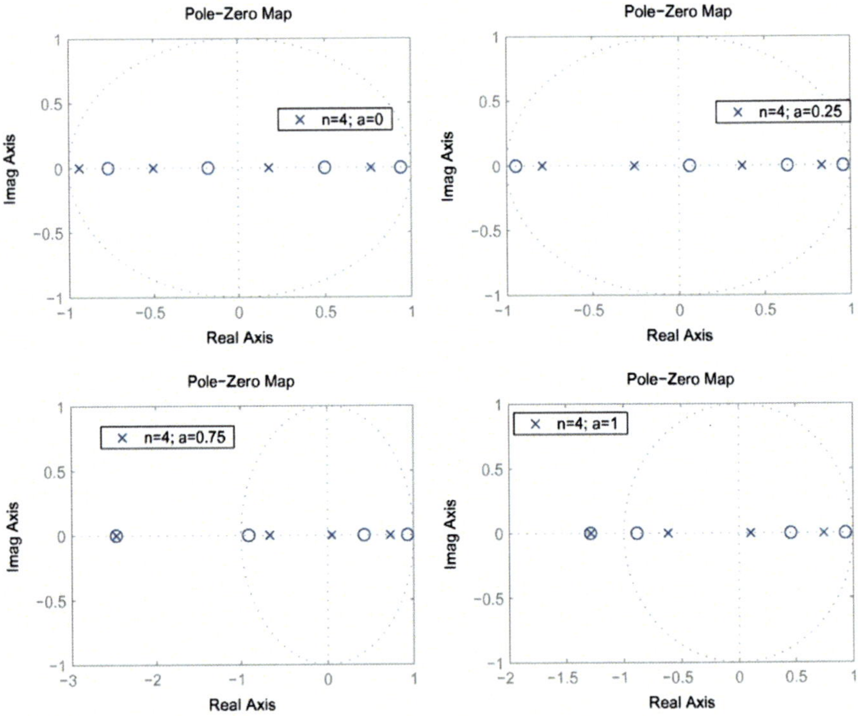

Fig. 5.15 Pole-zero maps for $r = 0.5$, $n = 4$ and $a = 0, .25, .75, 1$

and

$$G_2(z) = \frac{2.377 \times 10^{-6} z^5 + 0.1128 z^4 - 0.367 z^3 + 0.4387 z^2 - 0.2269 z + 0.04241}{z^5 - 3.671 z^4 + 5.107 z^3 - 3.259 z^2 + 0.882 z - 0.05885}, \tag{5.27}$$

and the Bode plot comparisons given in Fig. 5.16. It can be seen that the fittings are satisfactory.

5.1.3 Frequency Response Fitting of Fractional-Order Filters

Continuous-Time Approximation

In general, any available method for frequency domain identification can be applied in order to obtain a rational function, whose frequency response fits the one corresponding to the filter's original transfer function. For example, a minimization of the cost function of the ISE form is generally aimed, *i.e.*,

$$J = \int W(\omega) \left| G(\omega) - \hat{G}(\omega) \right|^2 \mathrm{d}\omega, \tag{5.28}$$

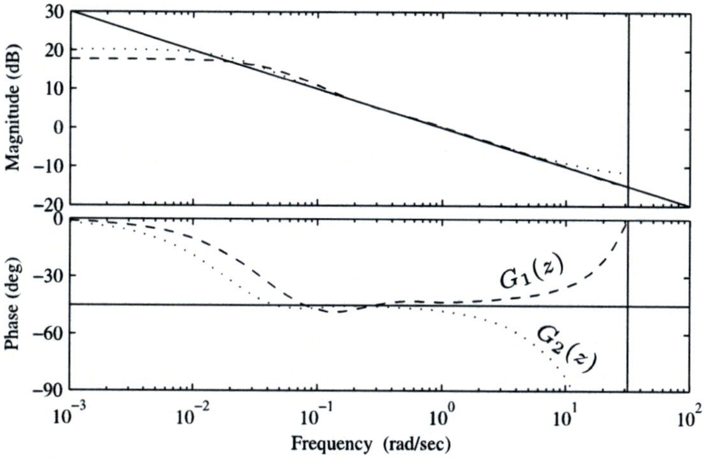

Fig. 5.16 Bode plots comparisons with discrete-time implementations

where $W(\omega)$ is a weighting function, $G(\omega)$ is the original frequency response, and $\hat{G}(\omega)$ is the frequency response of the approximated rational function.

MATLAB function `invfreqs()` follows this criterion, with the next syntax: $[\boldsymbol{B}, \boldsymbol{A}] = \texttt{invfreqs}(H, \boldsymbol{w}, n_\text{b}, n_\text{a})$. This function gives real numerator and denominator coefficients $\boldsymbol{B}$ and $\boldsymbol{A}$ of orders n_b and n_a, respectively. H is the desired complex frequency response of the system at frequency points $\boldsymbol{w}$, and $\boldsymbol{w}$ contains the frequency values in rad/sec. Function `invfreqs()` yields a filter with real coefficients. This means that it is sufficient to specify positive frequencies only.

The approximation of the fractional-order integrator of order 0.5 has been obtained using this method. The order of the approximation is 4, that is $n_\text{b} = n_\text{a} = 4$, and the frequency range $\boldsymbol{w}$ goes from 0.01 rad/sec to 100 rad/sec. The identified model can be obtained with the following statements

```
>> w=logspace(-2,2,100); H=1./(sqrt(-1)*w).^0.5;
   [n,d]=invfreqs(H,w,4,4); G=tf(n,d), bode(G)
```

The resulting transfer function is

$$G(s) = \frac{B(s)}{A(s)} = \frac{0.02889s^4 + 17.08s^3 + 1102s^2 + 10270s + 4567}{s^4 + 172.1s^3 + 4378s^2 + 11480s + 459.8},$$

and the Bode plots are shown in Fig. 5.17.

Discrete-Time Approximation

If the frequency response of a fractional-order filter is given, discrete-time implementation can also be obtained. There are several ways for finding the discrete-time transfer function approximations to the fractional-order filters. One may use

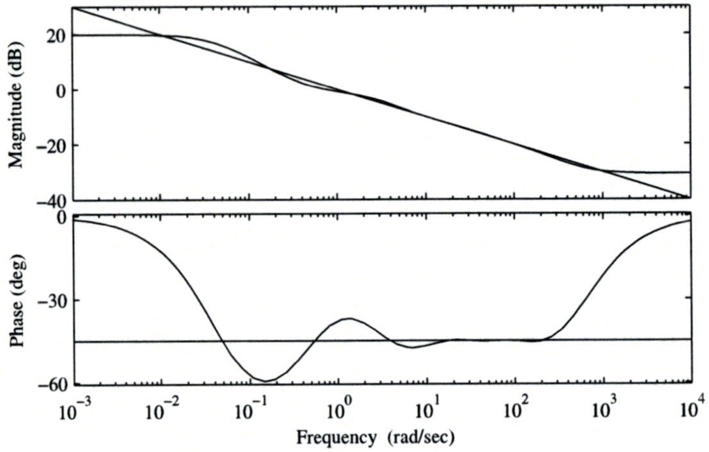

Fig. 5.17 Bode plots of $G(s)$, corresponding to the approximation of a fractional-order integrator of order 0.5 with MATLAB function `invfreqs()`

MATLAB function `invfreqz()` for a direct approximation to the given frequency response data. On the other hand, the continuous-time approximation can be obtained first, and then, with the use of `c2d()` function, the discrete-time implementation can be obtained. One may also use special algorithms for specific types of fractional-order filters. For instance, the impulse response invariant function $G = \texttt{irid\_folpf}(\tau, r, T, N)$ given in Table 5.3 can be used for fitting a fractional-order low-pass filter of the form $(\tau s + 1)^{-r}$.

Example 5.9 Consider a fractional-order low-pass filter given by $G(s) = (3s + 2)^{-0.4}$. One may simple rewrite the model by $G(s) = 2^{-0.4}(1.5s + 1)^{-0.4}$. It can be seen that $\tau = 1.5$ and $r = 0.4$. Selecting sampling periods as $T = 0.1$ sec, with order $N = 4$, the discrete-time implementation using impulse response invariants can be obtained as

```
>> tau=1.5; a=0.4; T=0.1; N=4;
   G1=2^(-0.4)*irid_folpf(tau,a,T,N);
```

The approximation model is

$$G_1(z) = \frac{0.2377z^4 - 0.4202z^3 + 0.2216z^2 - 0.02977z - 0.00138}{z^4 - 2.222z^3 + 1.663z^2 - 0.4636z + 0.03388}.$$

The Bode plot comparisons of the fitting model and the original model are shown in Fig. 5.18. It can be seen that the fitting results are good for this example.

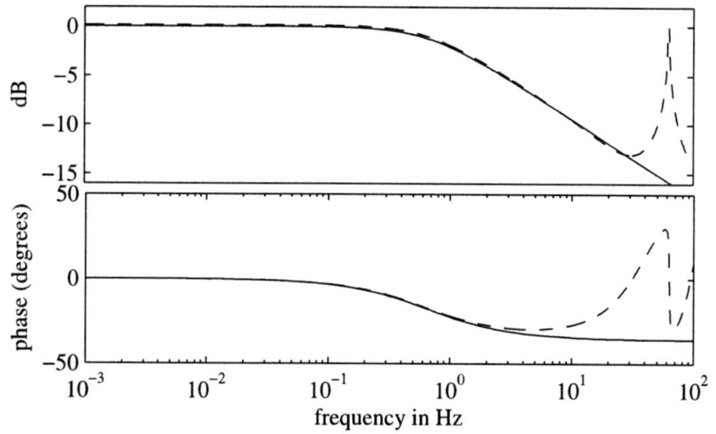

Fig. 5.18 Bode plots comparisons, with *solid lines* for exact filter and *dashed lines* for the discrete-time implementation model

5.1.4 Transfer Function Approximations to Complicated Fractional-Order Filters

In signal processing or control applications, sometimes the fractional-order filter designed may be rather complicated. To implement the controllers in continuous-time form, the following procedures should be taken:

1. Get the exact frequency response of the fractional-order controller.
2. Select appropriate orders for the numerator and denominator of the integer-order filters.
3. Identify the continuous-time integer-order controllers with the use of `invfreqs()` function.
4. Verify frequency response fitting. If the fitting is not satisfactory, go back to Step 2 to select another set of orders, or another frequency range of interest, until satisfactory approximations can be obtained.

Example 5.10 Consider a QFT controller studied in [207] given as

$$G_c(s) = 1.8393\left(\frac{s+0.011}{s}\right)^{0.96}\left(\frac{8.8\times 10^{-5}s+1}{8.096\times 10^{-5}s+1}\right)^{1.76}\frac{1}{(1+s/0.29)^2}.$$

It should be noted that the filter is too complicated to implement with the impulse response invariant fitting method given earlier. With MATLAB, the function `frd()` can be used to get the frequency response of an integer-order block, and the `RespnseData` membership of the frequency response object can be used to extract the frequency response data. Then dot multiplications and dot powers in MATLAB can be used to evaluate the exact frequency response data. Selecting the orders of numerator and denominator as 4 for continuous-time fitting, and the fitting frequency range of $\omega \in (10^{-4}, 10^{0})$ rad/sec, the following commands can be used

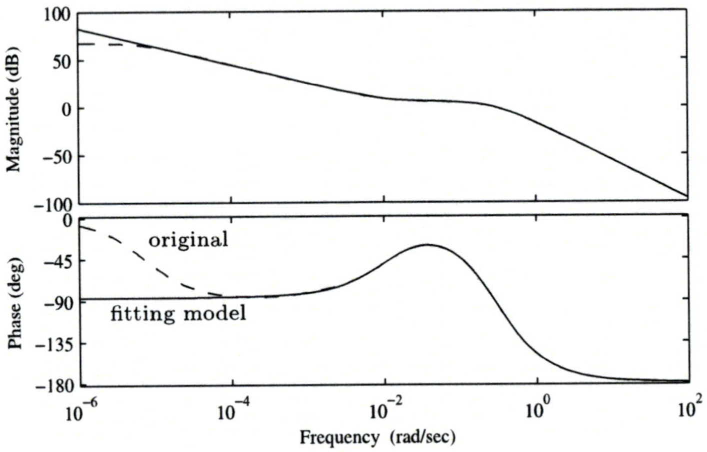

Fig. 5.19 Bode plot comparisons for a fractional-order QFT controller

```
>> w=logspace(-4,0); G1=tf([1 0.011],[1 0]); F1=frd(G1,w);
   G2=tf([8.8e-5 1],[8.096e-5 1]); F2=frd(G2,w);
   s=tf('s'); G3=1/(1+s/0.29)^2; F3=frd(G3,w); F=F1;
   h1=F1.ResponseData; h2=F2.ResponseData;
   h3=F3.ResponseData;
   h=1.8393*h1.^0.96.*h2.^1.76.*h3; F.ResponseData=h;
   %exact
   [n,d]=invfreqs(h(:),w,4,4); G=tf(n,d);
```

The continuous-time approximate integer-order controller can be obtained as

$$G(s)=\frac{2.213\times 10^{-7}s^4+1.732\times 10^{-6}s^3+0.1547s^2+0.001903s+2.548\times 10^{-6}}{s^4+0.5817s^3+0.08511s^2+0.000147s+1.075\times 10^{-9}}.$$

To verify the controller from the view point of frequency response fitting, we should compare the original and fitted controller over a larger frequency interval. The following commands can be used to compare the two controller in the frequency range of $(10^{-6}, 10^2)$ rad/sec.

```
>> w=logspace(-6,2,200); F1=frd(G1,w); F2=frd(G2,w); F=F1;
F3=frd(G3,w); h1=F1.ResponseData; h2=F2.ResponseData;
h3=F3.ResponseData; h=1.8393*h1.^0.96.*h2.^1.76.*h3;
F.ResponseData=h; bode(F,'-',G,'--',w)
```

The Bode plots of both controllers over the new frequency range are shown in Fig. 5.19. It can be seen that the frequency response of the controller is satisfactory, albeit there is small discrepancy at very-low frequency range. If such extremely low-frequency range is to be fitted, we should go to Step 2 to generate more frequency response points in the range.

5.1.5 Sub-optimal Approximation of Fractional-Order Transfer Functions

In this section, we consider the general fractional-order FO-LTI systems with non-commensurate fractional orders as follows:

$$G(s) = \frac{b_m s^{\gamma_m} + b_{m-1} s^{\gamma_{m-1}} + \cdots + b_1 s^{\gamma_1} + b_0}{a_n s^{\eta_n} + a_{n-1} s^{\eta_{n-1}} + \cdots + a_1 s^{\eta_1} + a_0}. \tag{5.29}$$

Using the aforementioned approximation schemes for a single s^r and then again for the general FO-LTI system (5.29) could be very tedious, leading to a very high order model. In this section, we propose to use a numerical algorithm to achieve a good approximation of the overall transfer function (5.29) using the finite-dimensional integer-order rational transfer function with a possible time delay term, then we illustrate how to use the approximated integer-order model for integer-order controller design.

Our target now is to find an approximate integer-order model with a relatively low order, possibly with a time delay in the following form [326]:

$$G_{r/m,\tau}(s) = \frac{\beta_1 s^r + \cdots + \beta_r s + \beta_{r+1}}{s^m + r_1 s^{m-1} + \cdots + r_{m-1} s + r_m} e^{-\tau s}. \tag{5.30}$$

An objective function for minimizing the $\mathcal{H}_2$-norm of the reduction error signal $e(t)$ can be defined as

$$J = \min_{\boldsymbol{\theta}} \|\widehat{G}(s) - G_{r/m,\tau}(s)\|_2, \tag{5.31}$$

where $\boldsymbol{\theta}$ is the set of parameters to be optimized such that

$$\boldsymbol{\theta} = [\beta_1, \ldots, \beta_r, r_1, \ldots, r_m, \tau]. \tag{5.32}$$

For an easy evaluation of the criterion J, the delayed term in the reduced order model $G_{r/m,\tau}(s)$ can be further approximated by a rational function $\widehat{G}_{r/m}(s)$ using the Padé approximation technique. Thus, the revised criterion can then be defined by

$$J = \min_{\boldsymbol{\theta}} \|\widehat{G}(s) - \widehat{G}_{r/m}(s)\|_2. \tag{5.33}$$

Suppose that for a stable transfer function type $E(s) = \widehat{G}(s) - \widehat{G}_{r/m}(s) = B(s)/A(s)$, the polynomials $A_k(s)$ and $B_k(s)$ can be defined such that,

$$A_k(s) = a_0^k + a_1^k s + \cdots + a_k^k s^k, \qquad B_k(s) = b_0^k + b_1^k s + \cdots + b_{k-1}^k s^{k-1}. \tag{5.34}$$

The values of a_i^{k-1} and b_i^{k-1} can be evaluated recursively from

$$a_i^{k-1} = \begin{cases} a_{i+1}^k, & i \text{ even} \\ a_{i+1}^k - r_k a_{i+2}^k, & i \text{ odd,} \end{cases} \qquad i = 0, \ldots, k-1 \tag{5.35}$$

and

$$b_i^{k-1} = \begin{cases} b_{i+1}^k, & i \text{ even} \\ b_{i+1}^k - \beta_k a_{i+2}^k, & i \text{ odd}, \end{cases} \quad i = 1, \ldots, k-1 \tag{5.36}$$

where, $r_k = a_0^k / a_1^k$, and $\beta_k = b_1^k / a_1^k$.

The $\mathcal{H}_2$-norm of the approximate reduction error signal $\hat{e}(t)$ can be evaluated from

$$J = \sum_{k=1}^{n} \frac{\beta_k^2}{2r_k} = \sum_{k=1}^{n} \frac{(b_1^k)^2}{2a_0^k a_1^k}. \tag{5.37}$$

The sub-optimal $\mathcal{H}_2$-norm reduced order model for the original high-order fractional-order model can be obtained using the following procedure [326]:

1. Select an initial reduced model $\widehat{G}^0_{r/m}(s)$.
2. Evaluate an error $\|\widehat{G}(s) - \widehat{G}^0_{r/m}(s)\|_2$ from (5.37).
3. Use an optimization algorithm to iterate one step for a better estimated model $\widehat{G}^1_{r/m}(s)$.
4. Set $\widehat{G}^0_{r/m}(s) \leftarrow \widehat{G}^1_{r/m}(s)$, go to Step 2 until an optimal reduced model $\widehat{G}^*_{r/m}(s)$ is obtained.
5. Extract the delay from $\widehat{G}^*_{r/m}(s)$, if any.

Based on the above approach, a MATLAB function `opt_app()` can be designed with the syntax $G_r = \texttt{opt\_app}(G, r, d, \texttt{key}, G_0)$, where `key` is the indicator of whether omit a delay is required in the reduced order model. G_0 is the initial reduced order model, optional. The listings of the function is

```
function G_r=opt_app(G_Sys,r,k,key,G0)
GS=tf(G_Sys); num=GS.num{1}; den=GS.den{1};
Td=totaldelay(GS); GS.ioDelay=0;
GS.InputDelay=0;GS.OutputDelay=0;
if nargin<5,
   n0=[1,1]; for i=1:k-2, n0=conv(n0,[1,1]); end
   G0=tf(n0,conv([1,1],n0));
end
beta=G0.num{1}(k+1-r:k+1); alph=G0.den{1}; Tau=1.5*Td;
x=[beta(1:r),alph(2:k+1)]; if abs(Tau)<1e-5, Tau=0.5; end
dc=dcgain(GS); if key==1, x=[x,Tau]; end
y=opt_fun(x,GS,key,r,k,dc);
x=fminsearch('opt_fun',x,[],GS,key,r,k,dc);
alph=[1,x(r+1:r+k)]; beta=x(1:r+1); if key==0, Td=0; end
beta(r+1)=alph(end)*dc;
if key==1, Tau=x(end)+Td; else, Tau=0; end
G_r=tf(beta,alph,'ioDelay',Tau);
```

Two lower-level MATLAB functions should also be designed as

```
function y=opt_fun(x,G,key,nn,nd,dc)
ff0=1e10; alph=[1,x(nn+1:nn+nd)];
beta=x(1:nn+1); beta(end)=alph(end)*dc; g=tf(beta,alph);
if key==1,
   tau=x(end); if tau<=0, tau=eps; end
   [nP,dP]=pade(tau,3); gP=tf(nP,dP);
else, gP=1; end
G_e=G-g*gP;
G_e.num{1}=[0,G_e.num{1}(1:end-1)];
[y,ierr]=geth2(G_e);
if ierr==1, y=10*ff0; else, ff0=y; end
%---sub function geth2
function [v,ierr]=geth2(G)
G=tf(G); num=G.num{1}; den=G.den{1}; ierr=0;
n=length(den); v=0;
if abs(num(1))>eps
   disp('System not strictly proper'); ierr=1; return
else, a1=den; b1=num(2:end); end
for k=1:n-1
  if (a1(k+1)<=eps), ierr=1; v=0; return
  else,
     aa=a1(k)/a1(k+1); bb=b1(k)/a1(k+1);
     v=v+bb*bb/aa; k1=k+2;
     for i=k1:2:n-1
        a1(i)=a1(i)-aa*a1(i+1); b1(i)=b1(i)-bb*a1(i+1);
end, end, end
v=sqrt(0.5*v);
```

We consider the above procedure sub-optimal since the Oustaloup's method is used for each single term s^γ in (5.29), and the Padé approximation is used for pure delay terms.

Example 5.11 Consider the following non-commensurate FO-LTI system:

$$G(s) = \frac{5s^{0.6}+2}{s^{3.3}+3.1s^{2.6}+2.89s^{1.9}+2.5s^{1.4}+1.2}.$$

Using the following MATLAB scripts,

```
>> N=5; w1=1e-3; w2=1e3;
   g1=ousta_fod(0.3,N,w1,w2); g2=ousta_fod(0.6,N,w1,w2);
   g3=ousta_fod(0.9,N,w1,w2); g4=ousta_fod(0.4,N,w1,w2);
   s=tf('s');
   G=(5*g2+2)/(s^3*g1+3.1*s^2*g2+2.89*s*g3+2.5*s*g4+1.2);
   G=minreal(G)
```

an extremely high-order model can be obtained with the Oustaloup's filter, such that

$$G(s)=\frac{\begin{array}{l}39.97s^{20}+7.68\times 10^4s^{19}+5.16\times 10^7s^{18}+1.53\times 10^{10}s^{17}\\ \quad +2.06\times 10^{12}s^{16}+1.339\times 10^{14}s^{15}+4.388\times 10^{15}s^{14}+7.32\times 10^{16}s^{13}\\ \quad +6.053\times 10^{17}s^{12}+2.515\times 10^{18}s^{11}+5.422\times 10^{18}s^{10}+6.149\times 10^{18}s^9\\ \quad +3.597\times 10^{18}s^8+1.067\times 10^{18}s^7+1.671\times 10^{17}s^6+1.41\times 10^{16}s^5\\ \quad +6.229\times 10^{14}s^4+1.344\times 10^{13}s^3+1.459\times 10^{11}s^2+7.703\times 10^8s\\ \quad +1.577\times 10^6\end{array}}{\begin{array}{l}s^{23}+2211s^{22}+1.782\times 10^6s^{21}+6.524\times 10^8s^{20}+1.122\times 10^{11}s^{19}\\ \quad +9.336\times 10^{12}s^{18}+4.013\times 10^{14}s^{17}+9.167\times 10^{15}s^{16}+1.114\times 10^{17}s^{15}\\ \quad +7.328\times 10^{17}s^{14}+2.739\times 10^{18}s^{13}+6.03\times 10^{18}s^{12}+8.058\times 10^{18}s^{11}\\ \quad +6.695\times 10^{18}s^{10}+3.651\times 10^{18}s^9+1.462\times 10^{18}s^8+3.952\times 10^{17}s^7\\ \quad +6.472\times 10^{16}s^6+6.007\times 10^{15}s^5+2.934\times 10^{14}s^4+6.775\times 10^{12}s^3\\ \quad +7.745\times 10^{10}s^2+4.275\times 10^8s+9.103\times 10^5\end{array}}$$

and the order of rational approximation to the original order model is the 23th, for $N=5$. For larger values of N, the order of rational approximation may be even much higher. For instance, the order of the approximation may reach the 30th- and 40th-order respectively for the selections $N=7$ and $N=9$, with extremely large coefficients. Thus the model reduction algorithm should be used with the following MATLAB statements

```
>> G2=opt_app(G,2,3,0); G3=opt_app(G,3,4,0);
   G4=opt_app(G,4,5,0); step(G,G2,G3,G4,60).
```

The step responses can be compared in Fig. 5.20 where it can be seen that the seventh-order approximation is satisfactory and the fourth order fitting gives a better approximation. The obtained optimum approximated results are listed in the following:

$$G_2(s)=\frac{0.41056s^2+0.75579s+0.037971}{s^3+0.24604s^2+0.22176s+0.021915},$$

$$G_3(s)=\frac{-4.4627s^3+5.6139s^2+4.3354s+0.15330}{s^4+7.4462s^3+1.7171s^2+1.5083s+0.088476},$$

$$G_4(s)=\frac{1.7768s^4+2.2291s^3+10.911s^2+1.2169s+0.010249}{s^5+11.347s^4+4.8219s^3+2.8448s^2+0.59199s+0.0059152}.$$

It can be seen that with the lower-order models obtained, the system response of the system may not change much. The sub-optimum fitting algorithm presented may be useful in a class of linear fractional-order system approximation.

Example 5.12 Let us consider the following FO-LTI plant model:

$$G(s)=\frac{1}{s^{2.3}+3.2s^{1.4}+2.4s^{0.9}+1}.$$

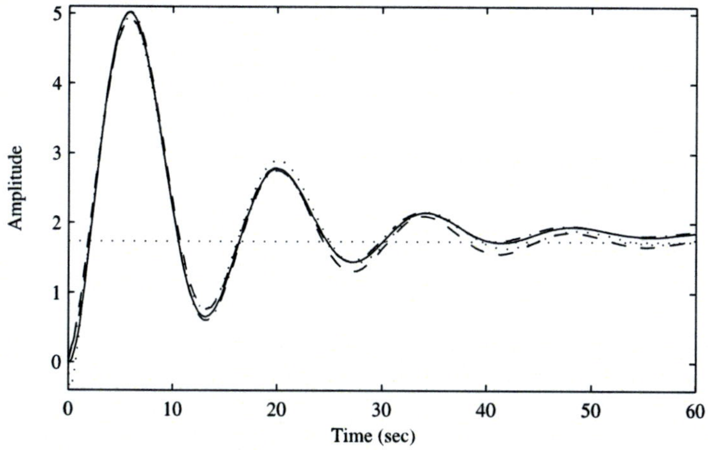

Fig. 5.20 Step responses comparisons: *solid line* for the original system, and the *rest lines* are respectively for $G_2(s)$, $G_3(s)$ and $G_4(s)$

Let us first approximate it with Oustaloup's method and then fit it with a fixed model structure known as first-order lag plus deadtime (FOLPD) model, where $G_r(s) = \frac{K}{Ts+1}e^{-Ls}$. The following MATLAB scripts

```
>> N=5; w1=1e-3; w2=1e3; g1=ousta_fod(0.3,N,w1,w2);
   g2=ousta_fod(0.4,N,w1,w2);
   g3=ousta_fod(0.9,N,w1,w2);
   s=tf('s'); G=1/(s^2*g1+3.2*s*g2+2.4*g3+1);
   G2=opt_app(G,0,1,1); step(G,'-',G2,'--')
```

can perform this task and the obtained optimal FOLPD model is given as follows:

$$G_r(s) = \frac{0.9951}{3.5014s+1}e^{-1.634s}.$$

The comparison of the open-loop step response is shown in Fig. 5.21. It can be observed that the approximation is fairly effective.

5.2 Synthesis of Constant-Order Fractional Processes

5.2.1 Synthesis of Fractional Gaussian Noise

It was suggested in [60, 221] that fGn can be considered as the output of a fractional integrator with wGn as the input, and the fGn is the derivative of the fBm. In this

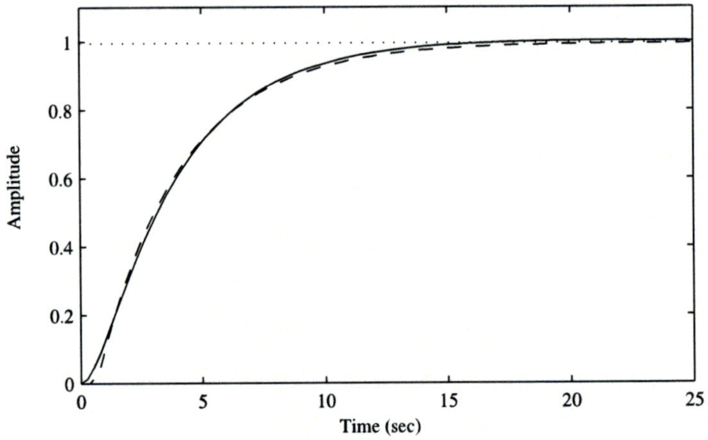

Fig. 5.21 Step response comparison of the optimum FOLPD and the original model

section, we analyze the relationship of wGn, fGn and fBm. The self-similar process $Y(n)$ can be modeled by a power-law decay of the autocorrelation:

$$R_Y(\tau) = E\big[Y(n)Y(n-\tau)\big] \sim C_Y|\tau|^{-\gamma}, \quad \tau \to \infty,\ 0 < \gamma < 1, \tag{5.38}$$

where C_Y is a positive constant, and '$\sim$' means the ratio of the left and the right sides converges to 1. Imposing the condition (5.38) on the spectral density S_Y of Y, as $\xi \to 0$ we get

$$S_Y(\xi) \sim C_s|\xi|^{-\beta}, \quad 0 < \beta < 1, \tag{5.39}$$

where the constant $C_s > 0$. Let $\omega(t)$ be a continuous-time white noise with variance σ^2, then the αth order integration of $\omega(t)$ can be expressed as

$$\begin{aligned} Y_H(t) &= {}_0D_t^{-\alpha}\omega(t) \\ &= \frac{1}{\Gamma(\alpha)}\int_0^t (t-\tau)^{\alpha-1}\omega(\tau)\mathrm{d}\tau, \quad 0 < \alpha < 1/2. \end{aligned} \tag{5.40}$$

Its power spectrum

$$G(\xi) = \sigma^2|\xi|^{-2\alpha}, \tag{5.41}$$

has the same form as (5.39). So the fGn can be considered as the output of a fractional integrator with wGn as the input. The Hurst parameter of fGn is related to α by $H = 1/2 + \alpha$. According to the definition of fGn, fBm can be considered as the integration of fGn, that is $B_H(t) = {}_0D_t^{-1}Y_H(t)$. Therefore, the fBm is the $(\alpha+1)$th integration of wGn. fBm can thus be described as

$$\begin{aligned} B_H(t) &= {}_0D_t^{-1-\alpha}\omega(t) \\ &= \frac{1}{\Gamma(H+1/2)}\int_0^t (t-\tau)^{H-1/2}\omega(\tau)\mathrm{d}\tau, \quad 1/2 < H < 1, \end{aligned} \tag{5.42}$$

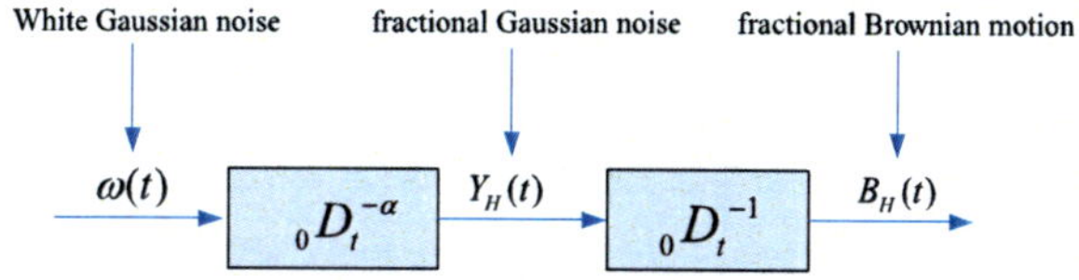

Fig. 5.22 Relationship of wGn, fGn and fBm

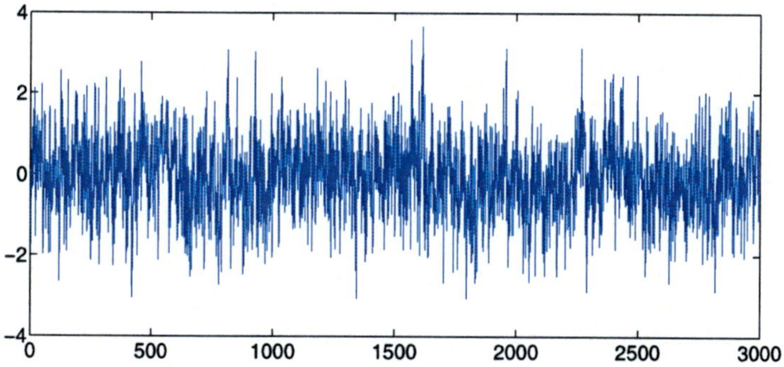

Fig. 5.23 Fractional Gaussian noise with $H = 0.75$

where $\omega(t)$ is wGn. (5.42) is the definition of 'one-sided' fBm introduced in [20] based on the Riemann-Liouville fractional integral. The relationship of wGn, fGn and fBm is presented in Fig. 5.22, where $\omega(t)$ is a white Gaussian noise, $Y_H(t)$ is the fGn process, and $B_H(t)$ is the fBm. Figure 5.23 shows an example of synthetic fGn with $H = 0.75$ using this fractional integration method.

5.2.2 Synthesis of Fractional Stable Noise

It was introduced in [235] that the fractional α-stable processes can be viewed as the output of a fractional-order integrator driven by a white α-stable noise [253]. Similar to the synthesis of fGn, multifractional stable noise can be generated by αth order integration of the white stable noise

$$\begin{aligned} Y_H(t) &= {}_0D_t^{-\alpha}\omega_\beta(t) \\ &= \frac{1}{\Gamma(\alpha)}\int_0^t (t-\tau)^{\alpha-1}\omega_\beta(\tau)\mathrm{d}\tau, \quad 0 < \alpha < 1/2, \end{aligned} \tag{5.43}$$

where $\omega_\beta(t)$ is the white stable noise. Figure 5.24 shows an example of synthetic fractional stable noise with $H = 0.75$ and $\beta = 1.6$.

5.3 Constant-Order Fractional System Modeling

In this section, three well-known models: FARIMA, FIGARCH and FARIMA with stable innovations models are introduced.

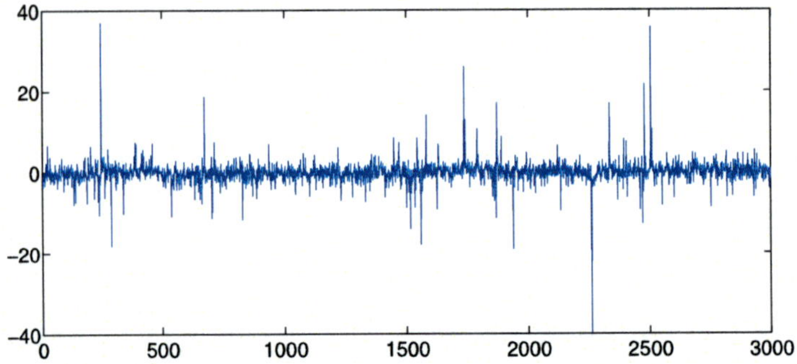

Fig. 5.24 Fractional stable noise with $H = 0.75$ and $\beta = 1.6$

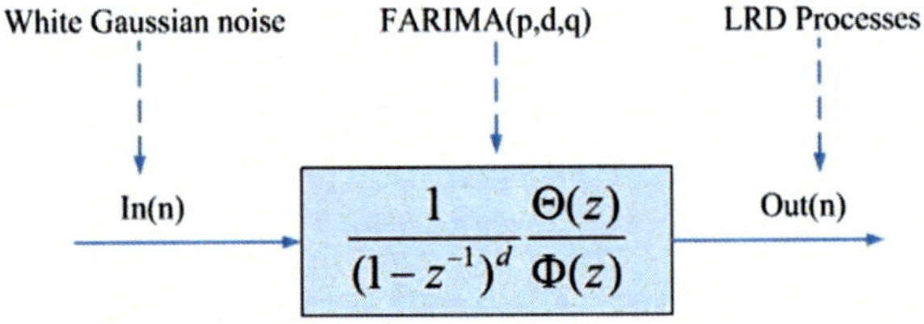

Fig. 5.25 Fractional autoregressive integrated moving average model

5.3.1 Fractional Autoregressive Integrated Moving Average Model

FARIMA(p, d, q) processes are widely used in modeling LRD time series, where p is the autoregression order, d is the level of differencing, and q is the moving average order [37]. Both p and q have non-negative integer values while d has non-integer value. Figure 5.25 presents the discrete FARIMA process which can be described as the output of the fractional-order system driven by a discrete wGn, for $d \in (-0.5, 0.5)$. FARIMA processes are the natural generalization of the standard ARIMA(p, d, q) processes.

A FARIMA(p, d, q) process $X_t : t = \ldots, -1, 0, 1, \ldots$ is defined as

$$\Phi(B)(1 - B)^d X_t = \Theta(B)\epsilon_t, \tag{5.44}$$

where ϵ_t is the wGn and $d \in (-0.5, 0.5)$, B is the backshift operator, defined by $BX_t = X_{t-1}$,

$$\Phi(B) = 1 - \Phi_1 B - \Phi_2 B^2 - \cdots - \Phi_p B^p, \tag{5.45}$$

$$\Theta(B) = 1 + \Theta_1 B + \Theta_2 B^2 + \cdots + \Theta_p B^q, \tag{5.46}$$

and $(1 - B)^d$ is the fractional differencing operator defined by

$$(1 - B)^d = \sum_{k=0}^{\infty} \binom{d}{k} (-B)^k, \tag{5.47}$$

where

$$\binom{d}{k} = \frac{\Gamma(d+1)}{\Gamma(k+1)\Gamma(d-k+1)}, \tag{5.48}$$

and Γ denotes the Gamma function. The parameter d is allowed to assume any real value. Clearly, if $d = 0$, FARIMA(p, d, q) processes are the usual ARMA(p, q) processes. FARIMA$(0, d, 0)$ process is the simplest and most fundamental form of FARIMA processes. The property of FARIMA$(0, d, 0)$ process is similar to that of the fGn process. The parameter d in FARIMA$(0, d, 0)$ process is the indicator for the strength of LRD, just like the Hurst parameter H in fGn process, and $H = d + 0.5$.

5.3.2 Gegenbauer Autoregressive Moving Average Model

A generalized FARIMA model named the Gegenbauer autoregressive moving average (GARMA) model was introduced in [104]. It is used in modeling LRD time series which exhibit a strong cyclic or seasonal behavior. GARMA model is defined as

$$\Phi(B)\big(1 - 2uB + B^2\big)^d X_t = \Theta(B)\epsilon_t, \tag{5.49}$$

where ϵ_t is the white noise, $\Phi(B)$ and $\Theta(B)$ are defined as (5.45) and (5.46), respectively. $u \in [-1, 1]$ is a parameter which governs the frequency at which the long memory occurs, and d controls the rate of decay of the autocovariance function. The GARMA model involves the Gegenbauer polynomial

$$\big(1 - 2uB + B^2\big)^{-d} = \sum_{n=0}^{\infty} C_n B^n, \tag{5.50}$$

where

$$C_n = \sum_{k=0}^{\lfloor n/2 \rfloor} \frac{(-1)^k (2u)^{n-2k} \Gamma(d-k+n)}{k!(n-2k)!\Gamma(d)}. \tag{5.51}$$

When $u = 1$, the GARMA model reduces to an ARFIMA model. When $|u| < 1$ and $0 < d < 1/2$ or $|u| = 1$ and $0 < d < 1/4$, the stationary GARMA process is a long memory process. The spectral density of a GARMA series is

$$f(\lambda) = c\frac{|\Theta(e^{i\lambda})|^2}{|\Phi(e^{i\lambda})|^2}\big[\cos(\lambda) - u\big]^{-2d}, \quad \lambda \in [-\pi, \pi), \tag{5.52}$$

where $c = \sigma^2/(\pi 2^{2d} + 1)$ is a constant.

The GARMA model was extended to the so-called "k-factor GARMA model" that allows for long-memory behavior to be associated with each of k frequencies

(Gegenbauer frequencies) in [0, 0.5] [322]. The k-factor GARMA model is defined as

$$\Phi(B)\prod_{j=1}^{k}\left(1-2u_jB+B^2\right)^{d_j}X_t=\Theta(B)\epsilon_t. \tag{5.53}$$

The spectral density of a k-factor GARMA process is given by

$$f(\lambda)=c\frac{|\Theta(e^{i\lambda})|^2}{|\Phi(e^{i\lambda})|^2}\prod_{j=1}^{k}|\cos(\lambda)-u_j|^{-d_j}, \tag{5.54}$$

where $c>0$ is a constant and the u_js are in $[-1,1]$. The k-factor GARMA model is long memory if u_j are distinct and $0<d_j<1/2$ whenever $u_j\neq 1$, and $0<d_j<1/4$ when $u_j=1$ [322].

5.3.3 Fractional Autoregressive Conditional Heteroscedasticity Model

Bollerslev and Mikkelsen constructed and evaluated the FIGARCH model [29]. Their results provide evidence against short memory specifications where $d=0$, and reject the integrated process where $d=1$. In their research, it shows that the effects of a shock on the conditional variance decrease at a hyperbolic rate when d is between 0 and 1. This is different from the FARIMA model where $0<d<0.5$. An FIGARCH model is defined as:

$$x_t=\mu_t+\epsilon_t, \tag{5.55}$$

$$\epsilon_t=\sigma_te_t, \tag{5.56}$$

$$\sigma_t^2=\frac{\omega}{1-\beta(L)}+\lambda(L)\epsilon_t^2, \tag{5.57}$$

$$\lambda(L)=1-\frac{\phi(L)(1-L)^d}{1-\beta(L)}, \tag{5.58}$$

where μ_t represents the regression function for the conditional mean, e_t is a white noise with mean 0 and variance 1, and L is the lag operator similar to B. When $d=1$, the FIGARCH model will reduce to an integrated GARCH model. When $d=0$, the FIGARCH model will reduce to GARCH model.

5.3.4 Fractional Autoregressive Integrated Moving Average with Stable Innovations Model

In classical time series literature, the innovations of the FARIMA process are white noise with finite variance. Although those processes can capture both short and long

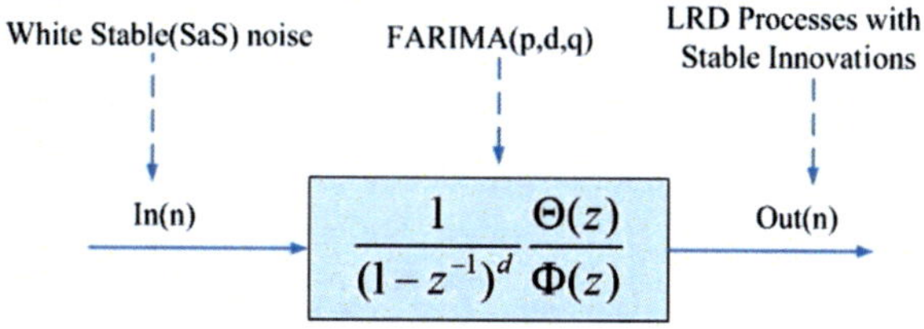

Fig. 5.26 FARIMA with stable innovations

memories, they concentrate their mass around the mean. α-stable distributions with $0 < \alpha < 2$, on the other hand, allow for much greater variability [253]. By assuming that the innovations of the FARIMA process follow the α-stable distribution, we are, in fact, dealing with powerful models that can exhibit both short/long-range dependence and heavy-tailed. Infinite variance α-stable distributions are a rich class of distributions with numerous applications in telecommunications, engineering, finance, insurance, physics etc. Figure 5.26 presents a discrete FARIMA process with stable innovations which can be described as the output of a fractional-order system driven by a discrete white $S\alpha S$ noise.

Let X_t be a time series, and consider the model

$$\Phi(B)(1-B)^d X_t = \Theta(B)\xi_t, \tag{5.59}$$

where $\Phi(B) = 1 - \Phi_1 B - \Phi_2 B^2 - \cdots - \Phi_p B^p$; $\Theta(B) = 1 + \Theta_1 B + \Theta_2 B^2 + \cdots + \Theta_p B^q$ are stationary autoregressive and invertible moving average operators respectively, and ξ_t is a sequence of i.i.d. $S\alpha S$ random variables. X_t is causal when $\alpha(d-1) < -1$ and $0 < \alpha \le 2$, and can be written as

$$X_t = \sum_{j=0}^{\infty} c_j \xi_{t-j}, \tag{5.60}$$

where c_js are the coefficients in

$$C_d(z) = \frac{\Theta(B)}{\Phi(B)}(1-B)^{-d} = \sum_{j=0}^{\infty} c_j B^j. \tag{5.61}$$

In the case of FARIMA$(0,d,0)$, the moving average coefficients $c(j) := b(j)$, $j \in \mathbb{Z}$ are given by

$$b(0) = 1, \qquad b(j) = \frac{\Gamma(j+d)}{\Gamma(d)\Gamma(j+1)}, \quad j = 1, 2, \ldots. \tag{5.62}$$

The parameter d determines the long-range behavior. So, the FARIMA time series with innovations that have infinite variance is a finite parameter model which exhibits both short/long-range dependence and high-variability.

Furthermore, in [253] the following relation is developed between the Hurst parameter H and the parameter d: $H = d + 1/\alpha$, and the necessary condition for the FARIMA process with stable innovations to converge is $d < 1 - 1/\alpha$.

5.4 A Fractional Second-Order Filter

The fractional-order filters have great application potential in signal modeling, filter design, controller design and nonlinear system identification [185, 312]. Due to the difficulties in analytically calculating and digitally realizing the fractional-order filter, however, the fractional-order filters have not been systematically studied. Some efforts have been made to study some simple fractional-order filters, such as the filter with the transfer function $1/s^\alpha$, $1/(s+m)^\alpha$ [55, 57], and some fractional-order filters with classic architectures. The classic fractional-order filter architecture $1/(s^{3\alpha}+2s^{2\alpha}+2s^\alpha+1)$ was presented and the performance evaluated relative to an ideal amplitude spectrum evaluated in [197]. The noise analysis of a simple single stage low-pass filter with a fractional-order capacitor was analyzed using stochastic differential equations in [157].

The key step towards the application of the fractional-order filter is its numerical discretization. The conventional discretization method for a fractional-order filter is the frequency-domain fitting technique (indirect method). In indirect discretization methods [227], two steps are required, i.e., frequency domain fitting in continuous time domain first, and then discretizing the fit s-transfer function. Other frequency-domain fitting methods can also be used but without guaranteeing the stable minimum-phase discretization [59]. In this section, the direct discretization method will be used. An effective impulse response invariant discretization method was discussed in [59, 62, 63, 182]. The method is a technique for designing discrete-time infinite impulse response (IIR) filters from continuous-time fractional-order filters in which the impulse response of the continuous-time fractional-order filter is sampled to produce the impulse response of the discrete-time filter. For more discussions of discretization methods, we cite [17, 59, 62, 63, 95, 182, 225, 241, 242, 311].

In this section, we first focus on the inverse Laplace transform of $(s^2+as+b)^{-\gamma}$ by cutting the complex plane and computing the complex integrals. The physical realization of $(s^2+as+b)^{-\gamma}$ can be illustrated as the type II fractional Langevin equation describing the fractional oscillator process with two indices [163, 173]. The centered stationary formula discussed in [173] can be partly extended by using the discussions in this section. For other previous works, we cite [225, 241, 242]. The derived results can be easily computed in MATLAB and applied to obtain the asymptotic properties of the continuous impulse responses. Moreover, a direct discretization method is proposed by using the digital impulse responses. The results are compared in both time and frequency domains.

5.4.1 *Derivation of the Analytical Impulse Response of* $(s^2+as+b)^{-\gamma}$

In this section, the inverse Laplace transform of $(s^2+as+b)^{-\gamma}=\mathcal{L}\{g(t)\}$ is derived by using the complex integral which can lead to some useful asymptotic properties of $g(t)$.

Fig. 5.27 The Hankel path

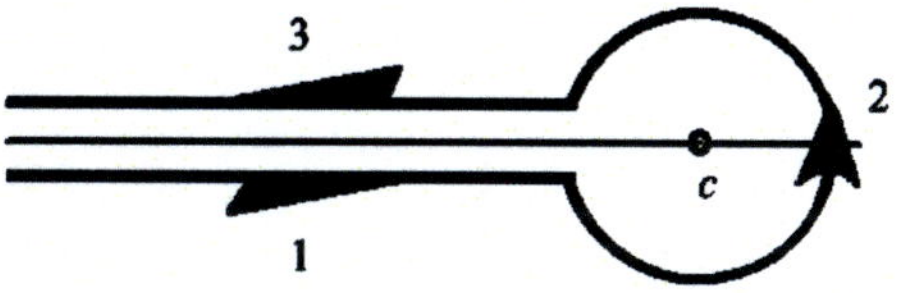

Let

$$G(s)=\frac{1}{(s^2+as+b)^\gamma}, \tag{5.63}$$

where $a, b \geq 0$, $\gamma > 0$ and $\mathcal{L}\{g(t)\} = G(s)$. It can be seen that there are two poles of $G(s)$, $s_1 = \frac{-a-\sqrt{a^2-4b}}{2}$ and $s_2 = \frac{-a+\sqrt{a^2-4b}}{2}$. It follows that

$$G(s)=\frac{1}{(s-s_1)^\gamma}\cdot\frac{1}{(s-s_2)^\gamma}.$$

Let $c \in \{s_1, s_2\}$,

$$\mathcal{L}^{-1}\left\{\frac{1}{(s-c)^\gamma}\right\}=\frac{1}{2\pi i}\int_{\sigma-i\infty}^{\sigma+i\infty}\frac{e^{st}}{(s-c)^\gamma}\mathrm{d}s. \tag{5.64}$$

When $\gamma \in \{\gamma | \gamma > 0, \gamma \neq 1, 2, 3, \ldots\}$, we have $s = c$ and $s = \infty$ which are the two branch points of $e^{st}(s-c)^{-\gamma}$. It follows that (5.64) is equivalent to the complex path integral shown in Fig. 5.27: a curve (Hankel path) which starts from $-\infty$ along the lower side of $s = \mathrm{Im}\{c\}$, encircles the circular disc $|s-c| = \epsilon \to 0$, in the positive sense, and ends at $-\infty$ along the upper side of $s = \mathrm{Im}\{c\}$.

Along path 1, by letting $s - c = xe^{-i\pi}$, where $x \in (0, \infty)$, we have

$$\begin{aligned}\int_1\frac{e^{st}\mathrm{d}s}{(s-c)^\gamma}&=\int_\infty^0\frac{e^{(c-x)t}\mathrm{d}(-x)}{x^\gamma e^{-i\gamma\pi}}=e^{ct+i\gamma\pi}\int_0^\infty\frac{e^{-xt}\mathrm{d}x}{x^\gamma}\\&=\Gamma(1-\gamma)t^{\gamma-1}e^{ct+i\gamma\pi}.\end{aligned} \tag{5.65}$$

Moreover, along path 3, let $s - c = xe^{i\pi}$, where $x \in (0, \infty)$. One obtains

$$\int_3\frac{e^{st}\mathrm{d}s}{(s-c)^\gamma}=-\Gamma(1-\gamma)t^{\gamma-1}e^{ct-i\gamma\pi}. \tag{5.66}$$

Finally, it follows from

$$\int_2\frac{e^{st}\mathrm{d}s}{(s-c)^\gamma}=\lim_{\epsilon\to0}\int_{-\pi}^{\pi}\frac{e^{\epsilon e^{i\theta}t}\cdot i\epsilon e^{i\theta}}{(\epsilon e^{i\theta}-c)^\gamma}\mathrm{d}\theta=0$$

that

$$\begin{aligned}g(t)&=\frac{1}{2\pi i}\left[\int_1+\int_3\right]\frac{e^{st}\mathrm{d}s}{(s-c)^\gamma}\\&=\frac{\sin(\gamma\pi)}{\pi}\Gamma(1-\gamma)t^{\gamma-1}e^{ct}=\frac{t^{\gamma-1}e^{ct}}{\Gamma(\gamma)},\end{aligned} \tag{5.67}$$

where $\Gamma(\gamma)\Gamma(1-\gamma) = \frac{\pi}{\sin(\gamma\pi)}$ is used in the above equation.

Based on the above discussion, we arrive at the following theorem.

Theorem 5.13 *Suppose $\gamma > 0$ and the complex number c satisfying* $\mathrm{Re}\{c\} \leq 0$. *We have*

$$\mathcal{L}^{-1}\left\{\frac{1}{(s-c)^{\gamma}}\right\} = \frac{t^{\gamma-1}e^{ct}}{\Gamma(\gamma)}, \tag{5.68}$$

and

$$\left|\mathcal{L}^{-1}\left\{\frac{1}{(s-c)^{\gamma}}\right\}\right| \leq \frac{t^{\gamma-1}}{\Gamma(\gamma)}, \tag{5.69}$$

where $t \geq 0$ and $|\cdot|$ denotes the absolute value of $\cdot$.

Proof (5.68) can be derived by using (5.67) and the frequency shifting property of Laplace transform. Moreover, it follows from $\mathrm{Re}\{c\} \leq 0$ that

$$\left|\mathcal{L}^{-1}\left\{\frac{1}{(s-c)^{\gamma}}\right\}\right| = \left|\frac{t^{\gamma-1}e^{ct}}{\Gamma(\gamma)}\right| \leq \frac{t^{\gamma-1}}{\Gamma(\gamma)},$$

where $t \geq 0$. □

Corollary 5.14 *In* (5.63), *when $a^2 - 4b = 0$ and $c = -\frac{a}{2}$, we have*

$$g(t) = \frac{t^{2\gamma-1}e^{-at/2}}{\Gamma(2\gamma)}, \tag{5.70}$$

where $\gamma > 0$.

Proof This conclusion can be proved by Theorem 5.13. □

Corollary 5.15 *In* (5.63), *when $a^2 - 4b > 0$, we have*

$$g(t) = \frac{e^{s_2 t}}{\Gamma(\gamma)}\, {}_0D_t^{-\gamma}\left[t^{\gamma-1}e^{(s_1-s_2)t}\right] \leq \frac{t^{2\gamma-1}e^{s_2 t}}{\Gamma(2\gamma)}, \tag{5.71}$$

where $\gamma > 0$, $t \geq 0$, $s_1 = \frac{-a-\sqrt{a^2-4b}}{2}$, $s_2 = \frac{-a+\sqrt{a^2-4b}}{2} \leq 0$ and D denotes the Riemann-Liouville fractional operator.

Proof It follows from Theorem 5.13 that

$$g(t) = \left[\frac{t^{\gamma-1}e^{s_1 t}}{\Gamma(\gamma)}\right] * \left[\frac{t^{\gamma-1}e^{s_2 t}}{\Gamma(\gamma)}\right],$$

where $*$ denotes the convolution on $[0, t]$. Therefore,

$$g(t) = \frac{e^{s_2 t}}{\Gamma(\gamma)^2} \int_0^t \tau^{\gamma-1}(t-\tau)^{\gamma-1} e^{(s_1-s_2)\tau} \mathrm{d}\tau \tag{5.72}$$

$$= e^{s_2 t}\, {}_0D_t^{-\gamma}\left[\frac{t^{\gamma-1}e^{(s_1-s_2)t}}{\Gamma(\gamma)}\right] \tag{5.73}$$

$$\leq e^{s_2 t}\, {}_0D_t^{-\gamma}\frac{t^{\gamma}}{\Gamma(\gamma)} = \frac{t^{2\gamma-1}e^{s_2 t}}{\Gamma(2\gamma)}, \tag{5.74}$$

where $s_1 \leq s_2 \leq 0$ and $t \geq 0$. □

Corollary 5.16 *In* (5.63), *when* $a^2 - 4b < 0$, *we have*

$$g(t) = e^{s_2 t}\, {}_0D_t^{-\gamma}\left[\frac{t^{\gamma-1}e^{(s_1-s_2)t}}{\Gamma(\gamma)}\right], \tag{5.75}$$

and

$$|g(t)| \leq \frac{t^{2\gamma-1}e^{-at/2}}{\Gamma(2\gamma)}, \tag{5.76}$$

where $\gamma > 0$, $t \geq 0$, $s_1 = \frac{-a-i\sqrt{4b-a^2}}{2}$ *and* $s_2 = \frac{-a+i\sqrt{4b-a^2}}{2}$.

Proof It follows from the same proof in Corollary 5.15 that

$$g(t) = e^{s_2 t}\, {}_0D_t^{-\gamma}\left[\frac{t^{\gamma-1}e^{(s_1-s_2)t}}{\Gamma(\gamma)}\right].$$

Applying $|\cdot|$ to the above equation yields

$$|g(t)| \leq |e^{s_2 t}|\, {}_0D_t^{-\gamma}\left[\frac{t^{\gamma-1}|e^{(s_1-s_2)t}|}{\Gamma(\gamma)}\right] \leq \frac{t^{2\gamma-1}e^{-at/2}}{\Gamma(2\gamma)},$$

where $t \geq 0$ and $|\cdot|$ denotes the magnitude of complex number $\cdot$. □

Theorem 5.17 *Let* $G(s) = (s^2 + as + b)^{-\gamma}$, *where* $a, b \geq 0$ *and* $\gamma > 0$, *we have*

$$|g(t)| \leq \frac{t^{2\gamma-1}}{\Gamma(2\gamma)}, \tag{5.77}$$

where $t \geq 0$ *and* $g(t) = \mathcal{L}^{-1}\{G(s)\}$.

Proof This conclusion can be proved by using Theorem 5.13 and Corollaries 5.14, 5.15 and 5.16. □

The impulse response $g(t)$ obtained in this section is associated with the impulse response invariant discretization method to be used in the following section.

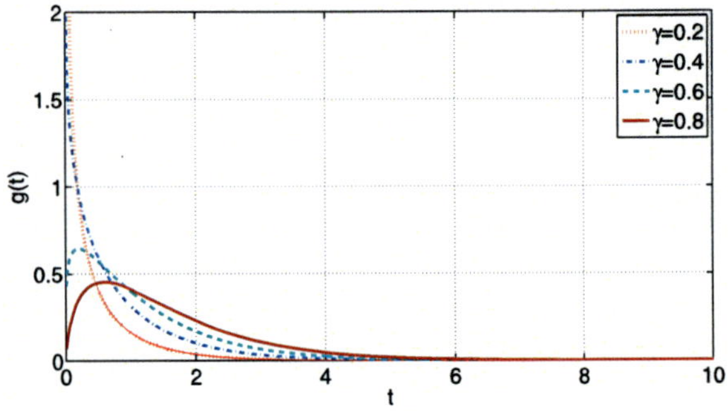

Fig. 5.28 The plots of $g(t)$, where $a = 2$, $b = 1$ and $\gamma \in \{0.2, 0.4, 0.6, 0.8\}$

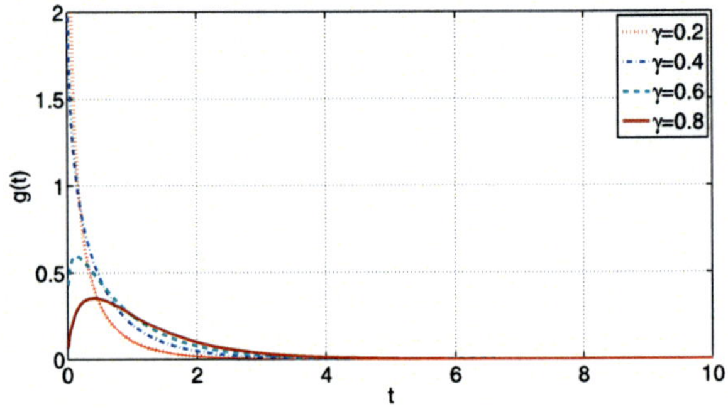

Fig. 5.29 The plots of $g(t)$, where $a = 3$, $b = 2$ and $\gamma \in \{0.2, 0.4, 0.6, 0.8\}$

5.4.2 Impulse Response Invariant Discretization of $(s^2 + as + b)^{-\gamma}$

Based on the obtained analytical impulse response function $g(t)$, given sampling period T_s, it is straightforward to perform the inverse response invariant discretization of $(s^2 + as + b)^{-\gamma}$ by using the Prony technique [95, 231, 273], which is an algorithm for finding an IIR filter with a prescribed time domain impulse response. It has applications in filter design, exponential signal modeling, and system identification (parametric modeling) [231, 273].

The plots of $g(t)$ for different a, b and γ are shown in Figs. 5.28–5.31. Specifically, when $a = 2$ and $b = 1$, it can be verified that $a^2 - 4b = 0$. The plots for different $\gamma \in \{0.2, 0.4, 0.6, 0.8\}$ are shown in Fig. 5.28. When $a = 3$ and $b = 2$, it can be verified that $a^2 - 4b > 0$. The plots for different $\gamma \in \{0.2, 0.4, 0.6, 0.8\}$ are shown in Fig. 5.29. When $a = 1$ and $b = 1$, it can be verified that $a^2 - 4b < 0$. The plots for different $\gamma \in \{0.2, 0.4, 0.6, 0.8\}$ are shown in Fig. 5.30. It can be seen that the appearances of complex poles lead to the oscillations of $g(t)$. When $a = 0$ and $b = 1$, it can be verified that $a^2 - 4b < 0$. The plots for different $\gamma \in \{0.2, 0.4, 0.6, 0.8\}$ are shown in Fig. 5.31.

Remark 5.18 Recall Corollary 5.16 and compare Fig. 5.31 with Fig. 5.30, we can see that the decreasing speed of $|g(t)|$ in Fig. 5.31 is much slower than that in

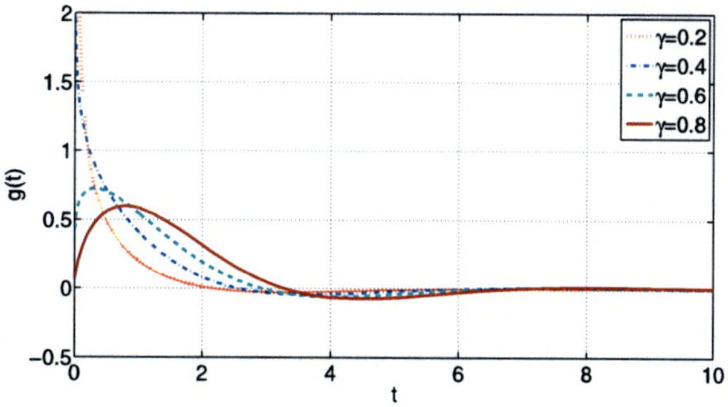

Fig. 5.30 The plots of $g(t)$, where $a = 1$, $b = 1$ and $\gamma \in \{0.2, 0.4, 0.6, 0.8\}$

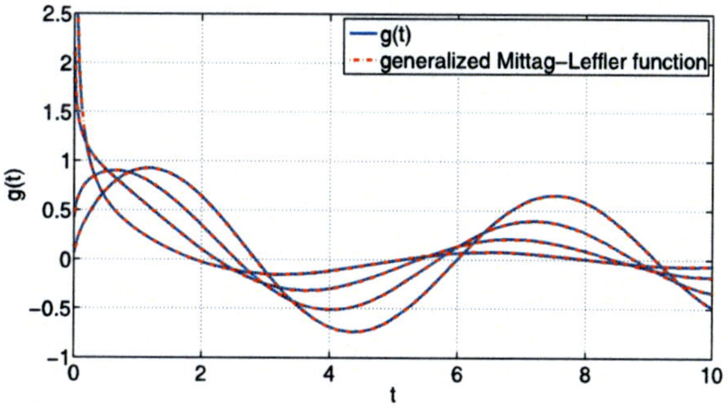

Fig. 5.31 The comparisons of $g(t)$ and generalized Mittag-Leffler function, where $a = 0$, $b = 1$ and $\gamma \in \{0.2, 0.4, 0.6, 0.8\}$

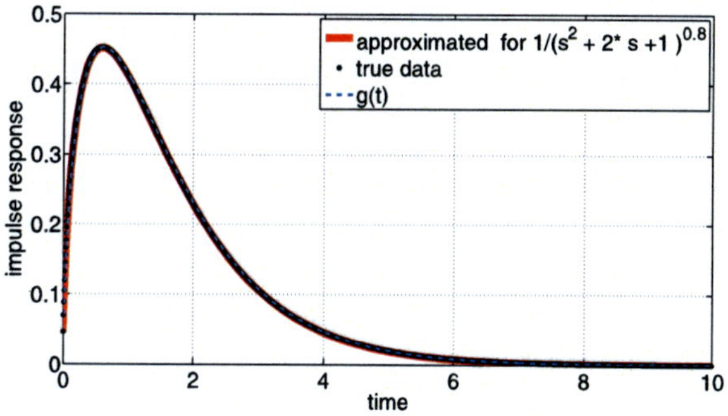

Fig. 5.32 The discrete and continuous impulse responses for $a = 2$, $b = 1$ and $\gamma = 0.8$

Fig. 5.30. Moreover, when $a = 0$, $g(t)$ is equivalent to a special case of generalized Mittag-Leffler function $t^{2\gamma-1}E^{\gamma}_{2,2\gamma}(-t^2)$. The plots of this generalized Mittag-Leffler function for different γ are also shown in Fig. 5.31 which coincide with $g(t)$.

Remark 5.19 It follows from the Laplace initial value theorem that $g(0) = 0$, $g(0) = +\infty$ and $g(0) = 1$ are corresponding to $\gamma \in (\frac{1}{2}, +\infty)$, $\gamma \in (0, \frac{1}{2})$ and $\gamma = \frac{1}{2}$, respectively.

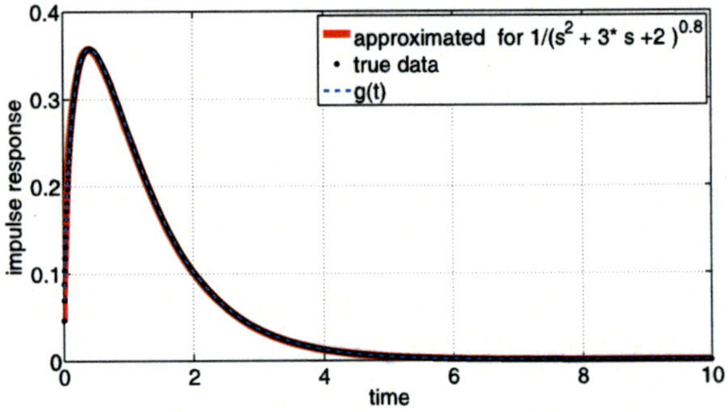

Fig. 5.33 The discrete and continuous impulse responses for $a = 3$, $b = 2$ and $\gamma = 0.8$

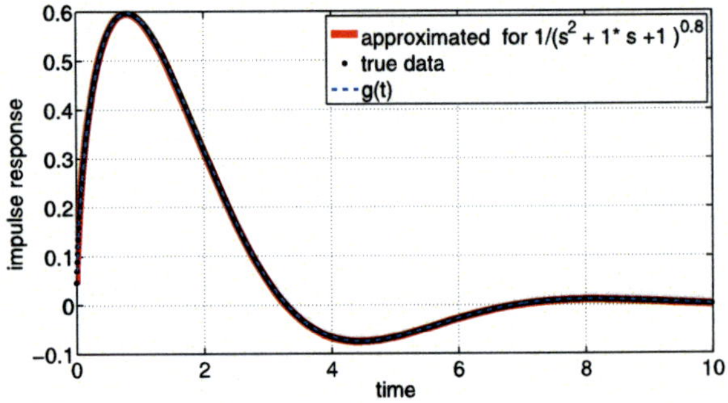

Fig. 5.34 The discrete and continuous impulse responses for $a = 1$, $b = 1$ and $\gamma = 0.8$

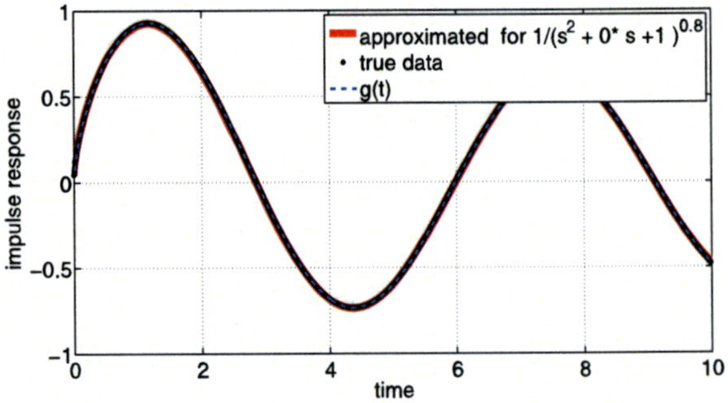

Fig. 5.35 The discrete and continuous impulse responses for $a = 0$, $b = 1$ and $\gamma = 0.8$

Remark 5.20 The centered stationary formula and the equation (2.8) discussed in [173], where $\alpha = 1/2$, are special cases of the fractional-order filters discussed in this section.

Moreover, the discrete and continuous impulse responses are shown in Figs. 5.32, 5.33, 5.34 and 5.35.

Now, let us consider how to discretize the $G(s)$ given sampling period T_s. Our goal is to get a discretized version of $G(s)$, denoted by $G_d(z^{-1})$ with a requirement that $G_d(z^{-1})$ and $G(s)$ have the same impulse response. Since the analytical impulse response of $G(s)$ had already been derived in Sect. 5.4.1, it is relatively straightforward to obtain the impulse response invariant discretized version of $G(s)$

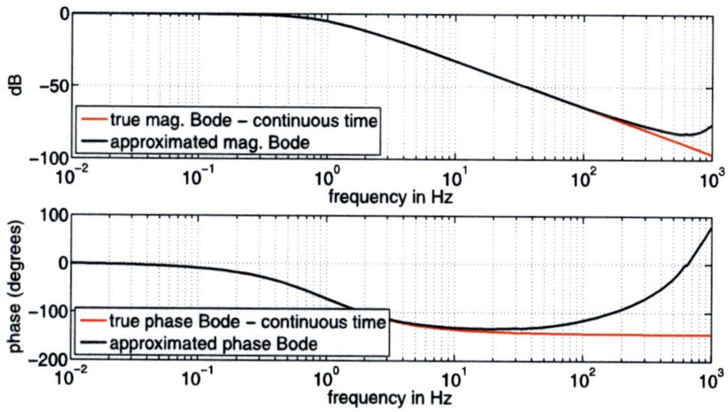

Fig. 5.36 Frequency responses for $a = 2$, $b = 1$ and $\gamma = 0.8$

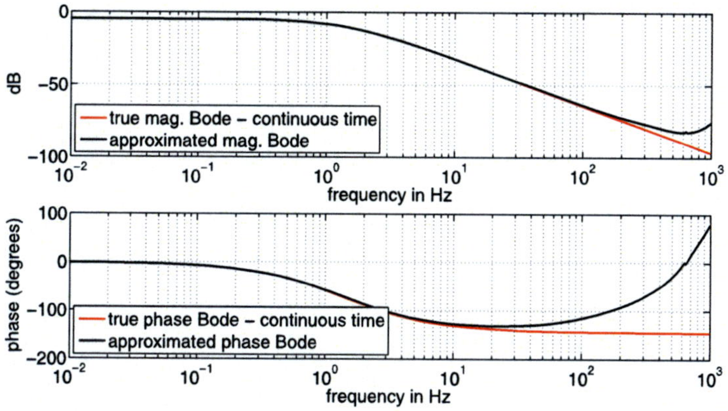

Fig. 5.37 Frequency responses for $a = 3$, $b = 2$ and $\gamma = 0.8$

via the well-known Prony technique [54, 55, 57, 58, 273]. In other words, the discretization impulse response can be obtained by using the continuous time impulse response as follows:

$$g(n) = T_s g(nT_s), \tag{5.78}$$

where $n = 0, 1, 2, \ldots$ and T_s is the sampling period.

Figures 5.36, 5.37, 5.38 and 5.39 show the magnitude and phase of the frequency response of the approximate discrete-time IIR filters and the continuous-time fractional-order filters under four different cases, where γ satisfies the convergent condition $\lim_{s\to\infty} s(s^2 + as + b)^{-\gamma} = 0$. The approximate discrete-time IIR filters can accurately reflect the time domain characteristics of continuous-time fractional-order for any a, b and γ. For frequency responses, the impulse response invariant discretization method works well under all the four cases for the band-limited continuous-time fractional-order filters. Note here, in Fig. 5.39, the two

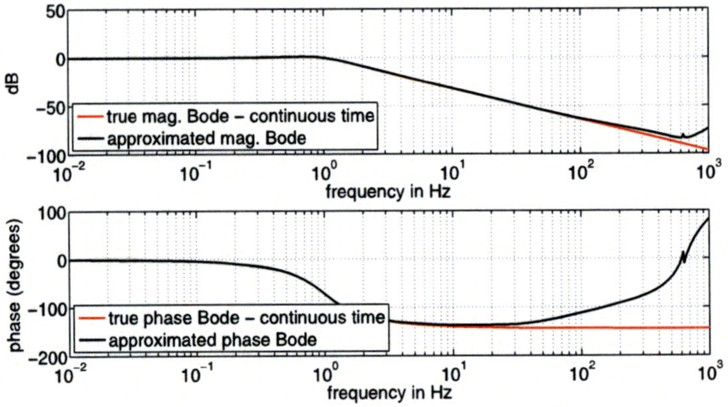

Fig. 5.38 Frequency responses for $a=1, b=1$ and $\gamma=0.8$

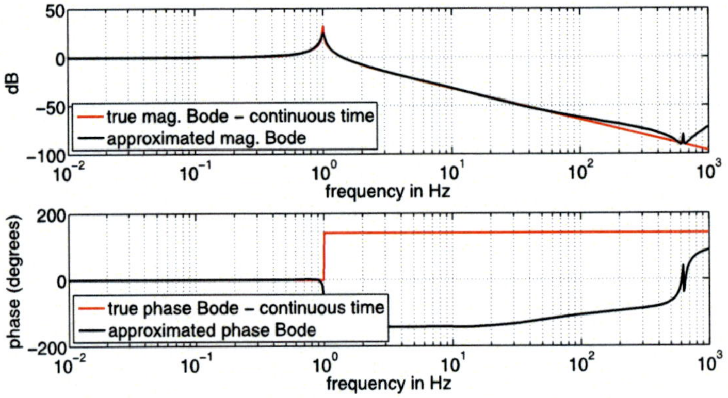

Fig. 5.39 Frequency responses for $a=0, b=1$ and $\gamma=0.8$

curves on $\omega \geq 10^0$, where $s=i\omega$ and $i=\sqrt{-1}$, are very different. Because, when $a=0$ and $b=1$, the two poles of $\frac{1}{(s^2+1)^\gamma}$ are on the imaginary axis. In other words, the red line in Fig. 5.39 is not accurate for large ω due to the direct computations of real and imaginary parts of $\frac{1}{(s^2+1)^\gamma}$, where $s=i\omega$. Overall, the impulse response invariant discretization method can accurately describe the fractional-order filter $(s^2+as+b)^{-\gamma}$. Using the approximate discrete-time IIR filters we can make full use of the discussed fractional-order filter. Moreover, in Figs. 5.36–5.39, $\gamma=0.8$, $T_s=0.01$, the order of $G_d(z^{-1})$ is 5, $G_d(z^{-1})$ for different cases are shown below:

$$G_d(z^{-1}) = \frac{a_1 z^5 + a_2 z^4 + a_3 z^3 + a_4 z^2 + a_5 z + a_6}{z^5 + b_1 z^4 + b_2 z^3 + b_3 z^2 + b_4 z + b_5}, \tag{5.79}$$

and a_i $(i=1,2,\ldots,6)$ and b_i $(i=1,2,\ldots,5)$ are shown in Tables 5.4 and 5.5.

Lastly, the above discussions are also valid for $\gamma \geq 1$.

Table 5.4 The values of a_i ($i = 1, 2, \ldots, 6$)

	$a = 2, b = 1$	$a = 3, b = 2$	$a = 1, b = 1$	$a = 0, b = 1$
a_1	0.0006991	0.0006956	0.0007028	0.0007044
a_2	0.002133	0.002155	0.002293	0.002281
a_3	0.002315	0.002374	0.002616	0.002397
a_4	0.0009995	0.001036	0.00108	0.0005399
a_5	9.904e−005	9.815e−005	2.852e−005	0.0005101
a_6	1.985e−005	2.38e−005	8.303e−005	0.0002292

Table 5.5 The values of b_i ($i = 1, 2, \ldots, 5$)

	$a = 2, b = 1$	$a = 3, b = 2$	$a = 1, b = 1$	$a = 0, b = 1$
b_1	4.554	4.595	4.807	4.922
b_2	8.267	8.419	9.235	9.69
b_3	7.471	7.686	8.864	9.536
b_4	3.361	3.494	4.251	4.692
b_5	0.6015	0.6327	0.8145	0.9232

5.5 Analogue Realization of Constant-Order Fractional Systems

5.5.1 *Introduction of Fractional-Order Component*

The transfer function of the fractional-order integrator is represented in the frequency domain by

$$G(s) = s^{-\alpha}, \tag{5.80}$$

where $s = j\omega$ is the complex frequency and α is a positive real number and $0 < \alpha < 1$. The transfer function of the fractional-order differentiator under zero initial condition is represented in the frequency domain by

$$G(s) = s^{\alpha}. \tag{5.81}$$

The frequency responses for the fractional-order integrator or differentiator are different from the integer-order ones.

Motivated by the application of a fractional-order system, and by the need for analogue realization of the fractional-order $PI^{\lambda}D^{\mu}$ controller [238], many efforts have been made to construct the analogue fractional-order integrator and differentiator [89, 131, 137, 155]. The fractional-order differentiator s^{μ}, integrator $s^{-\lambda}$ and the fractional $PI^{\lambda}D^{\mu}$ controller were studied in [50]. Podlubny proposed an approach to designing analogue circuits by implementing fractional-order controllers in [236]. These analogue realization methods of the fractional-order operator are based on the resistor, capacitor or inductor networks. Different from the above "fractance" device realization methods, physical experiments in this section were based on an electrical element named 'Fractor', which was manufactured by Bohannan [27, 28]. In

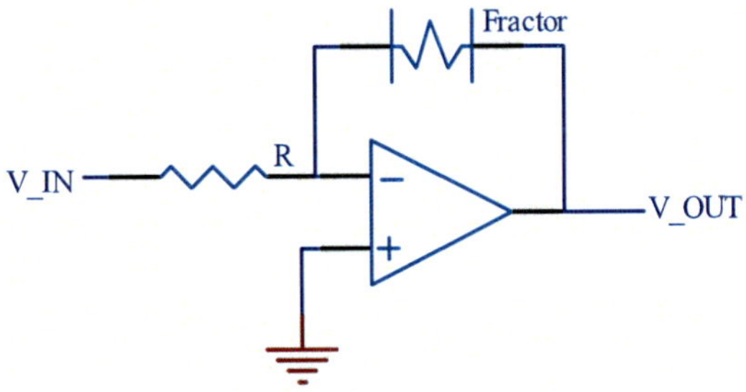

Fig. 5.40 Schematic circuit diagram for a fractional-order integrator

experiments, an analogue fractional-order integrator circuit was constructed using operational amplifier and a Fractor.

5.5.2 Analogue Realization of Fractional-Order Integrator and Differentiator

The impedance behavior of a Fractor is accurately described as the following inverse power-law form [28]

$$Z_F(\omega) = \frac{K}{(j\omega\tau)^\lambda}, \quad 0 < \lambda < 1, \tag{5.82}$$

where K is the impedance magnitude at a calibration frequency $\omega_0 = 1/\tau$, and λ is a non-integer exponent. The phase shift is related to the exponent by $\phi = -90^\circ \times \lambda$.

An analogue fractional-order integrator can be designed using the standard integrator amplifier circuit. Figure 5.40 presents the schematic circuit diagram of a fractional-order integrator using an operational amplifier, in which the Fractor with impedance Z_F is connected as a feedback in the operational amplifier. The gain of the fractional-order integrator in Fig. 5.40 is represented in the frequency domain as the ratio of the feedback impedance to the input impedance:

$$G(\omega) = \frac{V_{\mathrm{OUT}}}{V_{\mathrm{IN}}} = -\frac{Z_F(\omega)}{Z_R(\omega)} = -\frac{K}{R(j\omega\tau)^\lambda}, \tag{5.83}$$

where R is the resistor. Rewriting (5.83) in the Laplace domain, $s = j\omega$, we get

$$G(s) = -\frac{K}{R(s\tau)^\lambda}. \tag{5.84}$$

Figure 5.41 illustrates the experiment setup for the analogue realization of fractional integrator. The small cube at the bottom of the photo is the Fractor.

The frequency response for the fractional-order integrator with $\lambda \approx 0.9$ is presented in Fig. 5.42. The frequency response was measured using HP 35665A Dynamic Signal Analyzer (DSA) which measures both magnitude and phase.

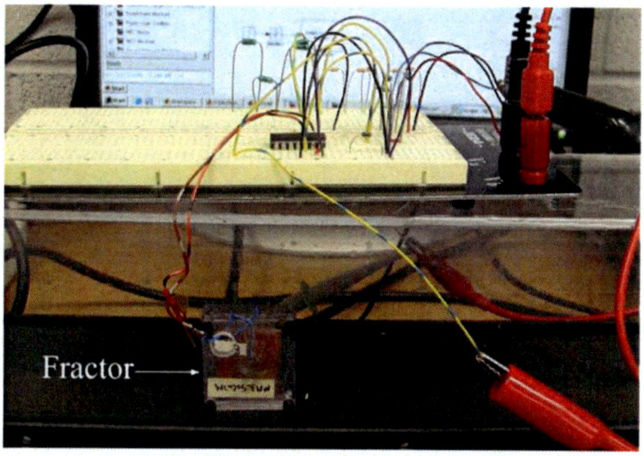

Fig. 5.41 Experiment setup for fractional-order integrator

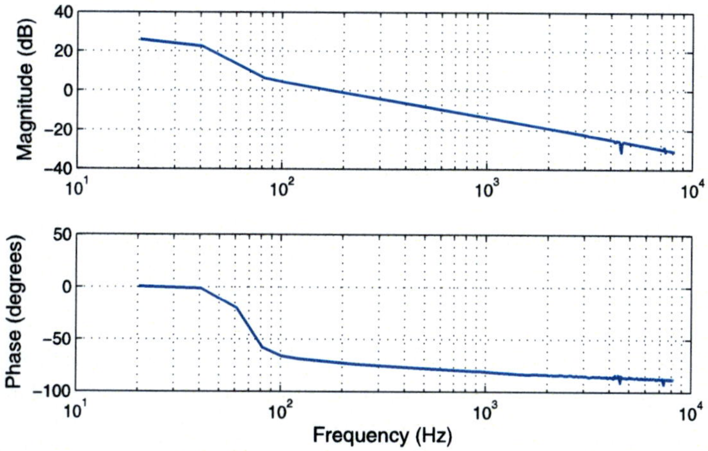

Fig. 5.42 Frequency response for the fractional-order integrator with $\lambda \approx 0.9$

Similar to the realization of fractional-order integrator, the fractional-order differentiator can be achieved by exchanging the positions of Fractor and resistor. The Fractor can be put to the input terminal of the operational amplifier as in Fig. 5.43. The gain of the fractional-order differentiator can be represented in the frequency domain as

$$G(\omega) = \frac{V_{\text{OUT}}}{V_{\text{IN}}} = -\frac{Z_R(\omega)}{Z_F(\omega)} = -\frac{R(j\omega\tau)^{\lambda}}{K}. \tag{5.85}$$

Rewriting (5.85) in the Laplace domain, one has

$$G(s) = -\frac{R(s\tau)^{\lambda}}{K}. \tag{5.86}$$

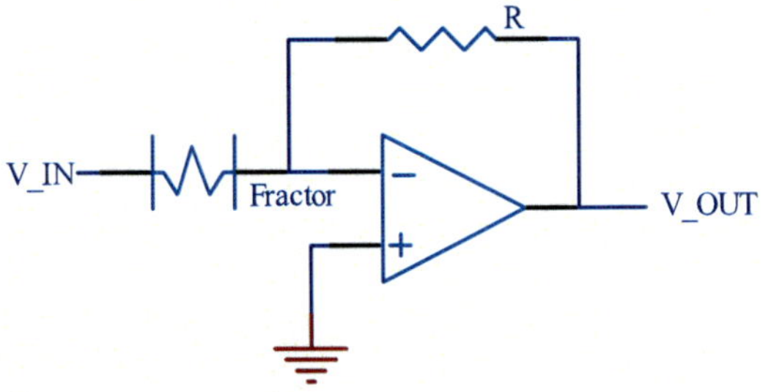

Fig. 5.43 Schematic circuit diagram for a fractional-order differentiator

As a side remark, we can actually cascade another differentiator of first-order to the output terminal of a fractional-order integrator to realize a fractional-order differentiator of order $1-\lambda$.

5.6 Chapter Summary

In this chapter, we introduced the constant-order FOSP techniques. Section 5.1 provided some realization methods for the fractional-order differentiator and integrator. Similar to integer-order signal processing techniques, the constant-order FOSP techniques include the simulation of constant-order fractional processes, constant-order fractional system modeling, fractional-order filter, and analogue realization of constant-order fractional systems. The relationship between constant-order fractional processes and constant-order fractional systems was investigated. Based on this relationship, the fGn and fractional stable noise can both be simulated using the constant-order fractional integrator. In order to capture the LRD property of the constant-order fractional processes, some constant-order fractional models, including FARIMA, FIGARCH and FARIMA with stable innovations were introduced. In addition, a fractional second-order filter $G(s)=(s^2+as+b)^{-\gamma}$ and its asymptotic properties were studied. The impulse response invariant discretization method was used to obtain a discrete time approximate IIR filter and frequency responses which were compared with the corresponding continuous cases. At the end of the chapter, the analogue realization of the constant-order fractional integrator and differentiator was provided to meet the needs of practical applications.

Chapter 6
Variable-Order Fractional Signal Processing

6.1 Synthesis of Multifractional Processes

6.1.1 Synthesis of mGn

Similar to the relationship of wGn, fGn and fBm, the relationship of wGn, mGn and mBm can be established by replacing the constant Hurst exponent by local Hölder exponent $H(t)$, and replacing the fractional-order α of integral in (5.42) by $\alpha(t)$. Lim generalized the Riemann-Liouville type fBm to the Riemann-Liouville type mBm, which is defined as [172]

$$X_+(t) = \frac{1}{\Gamma(H(t)+1/2)} \int_0^t (t-\tau)^{H(t)-1/2} \omega(\tau) \mathrm{d}\tau, \tag{6.1}$$

where $\omega(t)$ is the wGn. According to the relationship between wGn and fGn, we can extend the fGn to mGn with the help of local Hölder exponent $H(t)$ and $\alpha(t)$. Therefore, we can consider mGn as the output of a variable-order fractional integrator with wGn as the input. The mGn $Y_{H(t)}(t)$ can be described as

$$\begin{aligned} Y_{H(t)}(t) &= {}_0D_t^{-\alpha(t)} \omega(t) \\ &= \frac{1}{\Gamma(\alpha(t))} \int_0^t (t-\tau)^{\alpha(t)-1} \omega(\tau) \mathrm{d}\tau, \quad 0 < \alpha(t) < 1/2, \end{aligned} \tag{6.2}$$

where $H(t) = 1/2 + \alpha(t)$, and $\omega(t)$ is the wGn. According to the definition of the mGn, mBm is the integration of mGn, so the mBm is the $(\alpha(t)+1)$th integration of wGn. Assume that $\omega(t) = 0$ when $t < 0$. Then, the mBm can be described as

$$\begin{aligned} B_{H(t)}(t) &= {}_0D_t^{-1-\alpha(t)} \omega(t) \\ &= \frac{1}{\Gamma(H(t)+1/2)} \int_0^t (t-\tau)^{H(t)-1/2} \omega(\tau) \mathrm{d}\tau, \quad 1/2 < H(t) < 1, \end{aligned} \tag{6.3}$$

H. Sheng et al., *Fractional Processes and Fractional-Order Signal Processing*, Signals and Communication Technology,
DOI 10.1007/978-1-4471-2233-3_6, © Springer-Verlag London Limited 2012

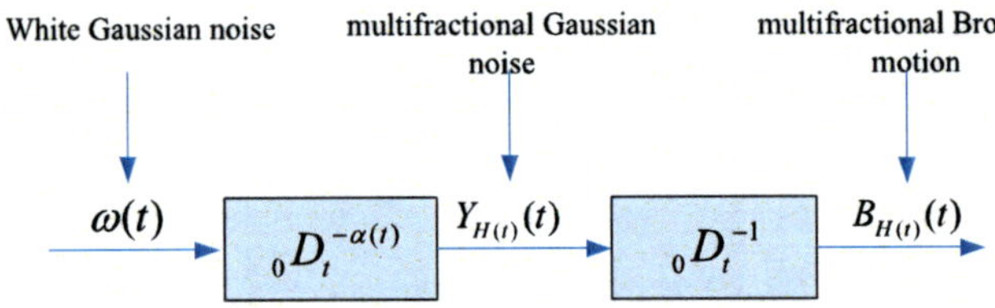

Fig. 6.1 Relationship of wGn, mGn and mBm

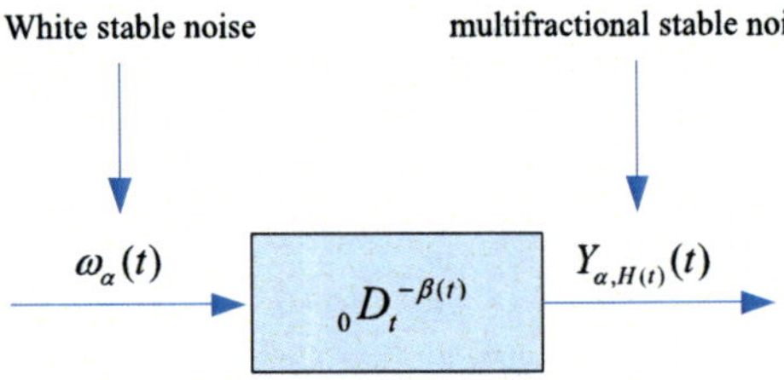

Fig. 6.2 Relationship between white stable noise and multifractional α-stable noise

where $\omega(t)$ is the wGn. (6.3) is the same as the definition of mBm introduced by Lim [172] based on the Riemann-Liouville fractional integral. The relationship of wGn, mGn and mBm is presented in Fig. 6.1, where $\omega(t)$ is a white Gaussian noise, $Y_{H(t)}(t)$, is the mGn, and $B_{H(t)}(t)$ is the mBm. Therefore, we can use the variable-order integration of wGn to synthesize the mGn. The variable-order integration can be numerically calculated based on the definition. In this synthesis method, we used the algorithm and the related MATLAB® code in [289] for the numerical solution of variable-order integration. The discrete function to be integrated is a wGn, and the variable-order is a discrete time-dependent Hölder exponent, so the numerical result is an mGn.

It was introduced in [235] that the fractional α-stable processes can be viewed as the output of a fractional integrator driven by a white α-stable noise. In the same way, the multifractional α-stable noise, the generalization of the fractional α-stable processes, can also be generated using variable-order fractional operators. If the input, a white Gaussian noise $\omega(t)$, is replaced by a white α-stable noise $\omega_\alpha(t)$, then the output of the variable-order integrator is a multifractional α-stable noise. Figure 6.2 illustrates the relationship between white stable noise and multifractional α-stable noise. For the synthesis method of multifractional α-stable noise, we can also use the algorithm for the numerical solution of variable-order integration in [289]. The discrete function to be integrated is a white α-stable noise, and the variable-order is a discrete time-dependent Hölder exponent. So, the numerical result is a multifractional α-stable noise.

Multifractional α-stable noise is the generalization of mGn, since the mGn is the special case when $\alpha = 2$. The multifractional α-stable noise has broader application areas. Based on multifractional α-stable noise, more accurate models of processes with local scaling characteristic and heavy-tailed distribution can be built.

Fig. 6.3 Examples of synthetic mGn with $H_1(t) = 0.7$: (**a**) mGn with $H_1(t) = 0.7$; (**b**) $H_1(t) = 0.7$ (*blue line*) and estimated $\widehat{H_1}(t)$ of mGn (*red line*)

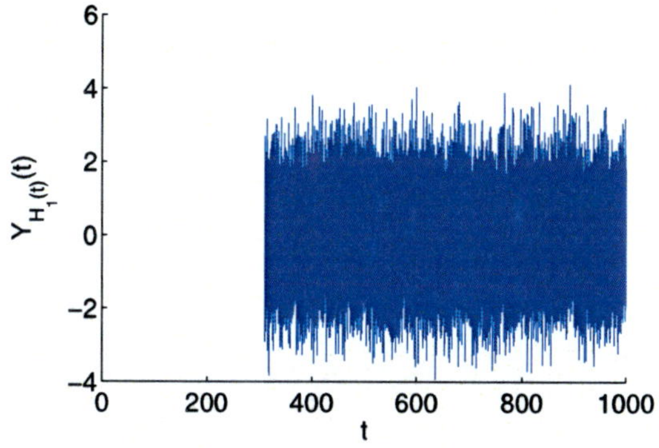

(a) mGn with $H_1(t) = 0.7$

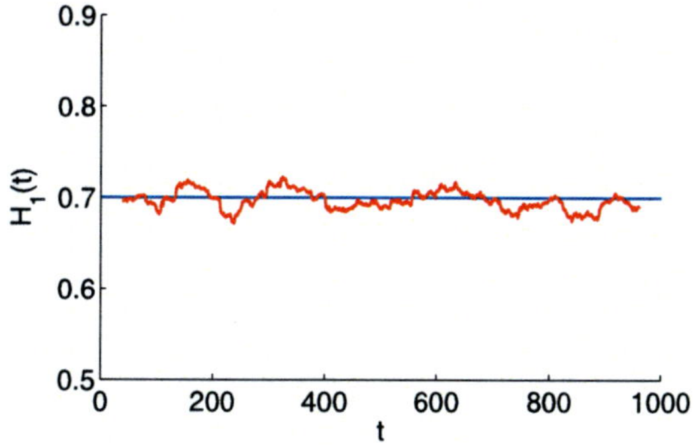

(b) Estimated $\widehat{H_1}(t)$ of mGn

6.1.2 Examples of the Synthesized mGns

The examples of mGns and multifractional α-stable noise are shown in Figs. 6.3, 6.4 and 6.5. The mGn with $H_1(t) = 0.7$ is shown in Fig. 6.3(a). The $H(t)$ (blue line) and the estimated Hölder exponents $\widehat{H}(t)$ (red line) are illustrated in Fig. 6.3(b). The mGn with $H_2(t) = at + b$, and multifractional α-stable noise ($\alpha = 1.8$) with $H_2(t) = at + b$ are shown in Fig. 6.4(a) and Fig. 6.5(a), respectively, where $a = 4 \times 10^{-4}$ and $b = 0.5$. The sample paths (blue line) and the estimated local Hölder exponents $\widehat{H}(t)$ (red line) of these two stochastic processes are shown in Fig. 6.4(b) and Fig. 6.5(b), respectively. The local Hölder exponents $\widehat{H}(t)$ are estimated using sliding windowed Koutsoyiannis' method [153], which is an iterative method to determine the Hurst exponent. For sliding windowed Koutsoyiannis' method, the time series is truncated by a sliding window with constant width, and the Hurst parameter of each truncated time series is estimated using Koutsoyiannis' method. From Fig. 6.5(b) we can see that the estimation results were affected by the heavy-tailed distribution of the multifractional α-stable process. The estimated Hölder exponents of these three time series are close to the actual value $H(t)$, which testifies the validity of the synthesis method.

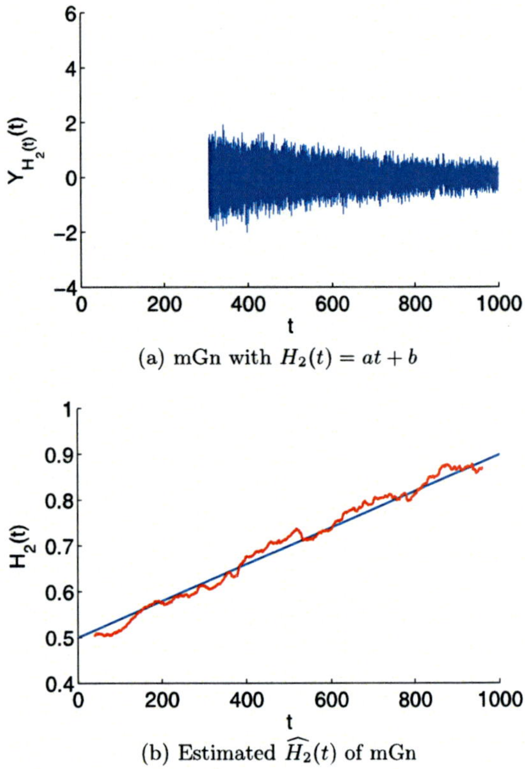

Fig. 6.4 Examples of synthetic mGn with $H_2(t) = at + b$: (**a**) mGn with $H_2(t) = at + b$; (**b**) $H_2(t) = at + b$ (*blue line*) and estimated $\widehat{H_2}(t)$ of mGn (*red line*)

6.2 Variable-Order Fractional System Modeling

6.2.1 Locally Stationary Long Memory FARIMA(p, d_t, q) Model

Similar to the modeling of constant-order fractional systems, the variable-order fractional systems can be modeled by generalizing the standard long memory modeling, assuming that the long memory parameter is time-varying.

A locally stationary long memory FARIMA(p, d_t, q) process is defined as [30]

$$\Phi(B)(1 - B)^{d_t} X_t = \Theta(B)\epsilon_t, \tag{6.4}$$

where ϵ_t is a wGn and d_t is a time-varying parameter, $d_t \in (-0.5, 0.5)$. B is the backshift operator, defined by $BX_t = X_{t-1}$,

$$\Phi(B) = 1 - \Phi_1 B - \Phi_2 B^2 - \cdots - \Phi_p B^p, \tag{6.5}$$

$$\Theta(B) = 1 + \Theta_1 B + \Theta_2 B^2 + \cdots + \Theta_q B^q. \tag{6.6}$$

Fig. 6.5 Examples of synthetic multifractional α-stable noise ($\alpha = 1.8$) with $H_2(t) = at + b$:
(**a**) multifractional α-stable noises ($\alpha = 1.8$) with $H_2(t) = at + b$;
(**b**) $H_2(t) = at + b$ (*blue line*) and estimated $\widehat{H_2}(t)$ of multifractional α-stable noise (*red line*)

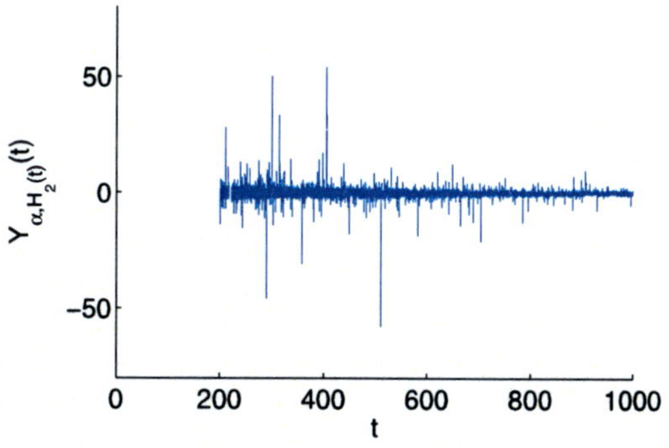

(a) Multifractional α-stable noise with $H_2(t) = at + b$

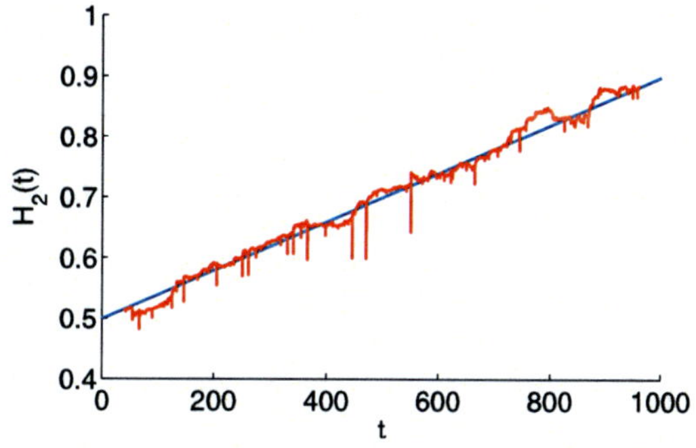

(b) Estimated $\widehat{H_2}(t)$ of multifractional α-stable noise

$(1 - B)^{d_t}$ can be defined by

$$(1 - B)^{d_t} = \sum_{k=0}^{\infty} \frac{\Gamma[k - d_t]}{\Gamma(k+1)\Gamma[-d_t]} B^k, \tag{6.7}$$

where Γ denotes the Gamma function. Many efforts have been made to investigate the time-varying long memory parameter FARIMA model based variable-order fractional system. Ray and Tsay (2002) propose a random persistence-shift FARIMA model where the parameter d_t is allowed to change randomly over time as [243]

$$d_t = d_0 + \sum_{j=1}^{t} \delta_j \beta_j = d_{t-1} + \delta_t \beta_t, \tag{6.8}$$

where the δ_ts are independent and identically distributed Bernoulli random variables, and β_t is a sequence of random observations from a known distribution.

6.2.2 Locally Stationary Long Memory FARIMA(p, d_t, q) with Stable Innovations Model

By assuming that the innovations of the locally stationary long memory FARIMA process follow the α-stable distribution, we can deal with the models which exhibit both local memory and heavy-tailed distribution. Infinite variance α-stable distributions are a rich class of distributions with numerous applications in telecommunications, engineering, finance, insurance, physics etc. Locally stationary long memory FARIMA process with stable innovations can be described as the output of a variable-order fractional system driven by a white stable noise.

Let X_t be a time series, and consider the model

$$\Phi(B)(1-B)^{d_t}X_t = \Theta(B)\xi_t, \tag{6.9}$$

where $d_t \in (-0.5, 0.5)$ is a time-varying parameter, and $\Phi(B) = 1 - \Phi_1 B - \Phi_2 B^2 - \cdots - \Phi_p B^p$; $\Theta(B) = 1 + \Theta_1 B + \Theta_2 B^2 + \cdots + \Theta_p B^q$ are stationary autoregressive and invertible moving average operators respectively, and ξ_t is a sequence of white stable random variables.

6.2.3 Variable Parameter FIGARCH Model

A new variable parameter FIGARCH model, a natural extension of the conventional FIGARCH model, was studied in [21]. In the variable parameter FIGARCH model, the mean and variance parameters are made dependent upon a latent state variable s_t, given by [21]

$$r_t = \mu_{s_t} + \epsilon_t, \; \epsilon_t \sim N(0, \sigma_{s_t}^2), \tag{6.10}$$

$$\sigma_{s_t}^2 = \omega_{s_t} + \beta_{1,s_t}\tilde{\sigma}_{t-1}^2 + \left[1 - \beta_{1,s_t}L - (1-\phi_{1,s_t}L)(1-L)^{d,s_t}\right]\tilde{\epsilon}_t^2, \tag{6.11}$$

where μ_t represents the regression function for the conditional mean, and L is the lag operator.

6.3 Analogue Realization of Variable-Order Fractional Systems

6.3.1 Physical Experimental Study of Temperature-Dependent Variable-Order Fractional Integrator and Differentiator

It has been demonstrated that some complex physical phenomena show variable-order fractional integrator or differentiator properties. It has also been indicated that the stress-stain behavior of viscoelastic materials with changing strain level can

Table 6.1 Fractor characteristics comparison

Fractor No.	λ-2008	λ-2009	λ-2010
0	0.5175	0.996	0.9709
1	0.5645	0.715	0.9637
2	0.7425	N/A	0.9366
3	0.7900	N/A	0.9809
4	0.9440	0.990	0.9804
5	0.8630	0.905	0.9617
6	0.9190	0.922	N/A
7	0.8680	N/A	0.9552
8	0.7460	0.771	0.9324
9	0.3850	0.832	0.9507
10	0.6495	0.772	0.9597

be characterized by variable-order fractional differential equations [274]. Glockle showed that the relaxation processes and reaction kinetics of proteins under different temperatures show variable-order fractional dynamic properties [98]. The time-dependent and space-dependent variable-order differential operators in anomalous diffusion modeling were studied in [290]. However, how to physically implement or realize an element of the variable-order fractional integrator and differentiator using an analogue circuit is not reported. In this section, we will present some Fractor based physical experimental results of temperature-dependent variable-order fractional operators.

The orders of eleven Fractors in the years 2008 and 2009 were tested in [210]. The orders of these eleven Fractors were changing over time due to aging. In order to investigate the time-dependent variable-order physical properties of the Fractor, we tested the order of these eleven Fractors in the year 2010. The time varying orders of the eleven Fractors in the years 2008–2010 were provided in Table 6.1. It can be seen that the orders of these Fractors become bigger and closer to one with the lapse of time.

What interests us most here is the influence of temperature on the order of the Fractor. So, in this section, we focus on physical realization of the variable-order fractional operator, and the relationship between the order and the temperature of the Fractor. Our physical experiments are based on the research in [96], which studied the electrical conductivity of $LiN_2H_5SO_4$ in the temperature range from 295 K to about 470 K. It was found that the electric properties and the crystal structures are influenced by temperature variations. If the analogue fractional-order operator in the circuit of Fig. 5.40 or Fig. 5.43 is put into an environment with a changing temperature, the order of integrator or differentiator should change with the variation of temperature. In order to accurately control the temperature of the Fractor, the Heat-Flow Experiment (HFE) platform was adopted [240]. The HFE system consists of a duct equipped with a heater and a blower at one end, and three temperature sensors located along the interior of the duct. The fan speed and the power delivered to the

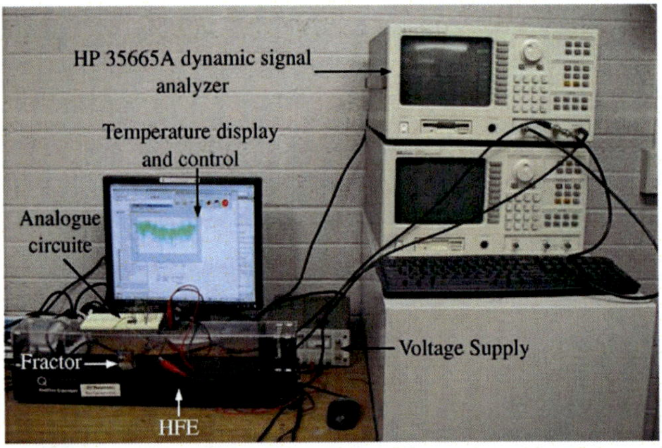

Fig. 6.6 Experiment for the analogue realization of temperature-dependent variable-order fractional integrator

heater can be controlled using an analog signal. Fast settling platinum temperature transducers are used to measure the temperature [240]. Figure 6.6 shows the experiment setup for the analogue realization of the temperature-dependent variable-order fractional integrator.

Different from fractional-order integrator and differentiator, the temperature-dependent variable-order fractional integrator and differentiator cannot be simply described in the frequency domain. In the time domain, the current-voltage relationship for a time dependent variable-order fractional derivative model of a Fractor with order $\lambda(t)$ can be expressed as

$$I(t) = \frac{\tau^{\lambda(t)}}{K} D_t^{\lambda(t)} V(t), \tag{6.12}$$

where $0 < \lambda(t) < 1$. In a temperature-dependent variable-order fractional integrator circuit, the order of the Fractor is a function of the temperature variable $T(t)$, where T is the function of the time variable t. Therefore, the current-voltage relationship for the temperature-dependent variable-order fractional integrator in the circuit is

$$I(t) = \frac{\tau^{\lambda(T(t))}}{K} D_t^{\lambda(T(t))} V(t). \tag{6.13}$$

Since the node voltage of the integrating operational amplifier at its inverting input terminal is zero (virtual earth), the current I flowing through the input resistor is given as

$$I(t) = \frac{V_{\mathrm{IN}}(t)}{R}. \tag{6.14}$$

The current flowing through the feedback Fractor is given as (6.13). Assuming that the input impedance of the operational amplifier is infinite, no current flows

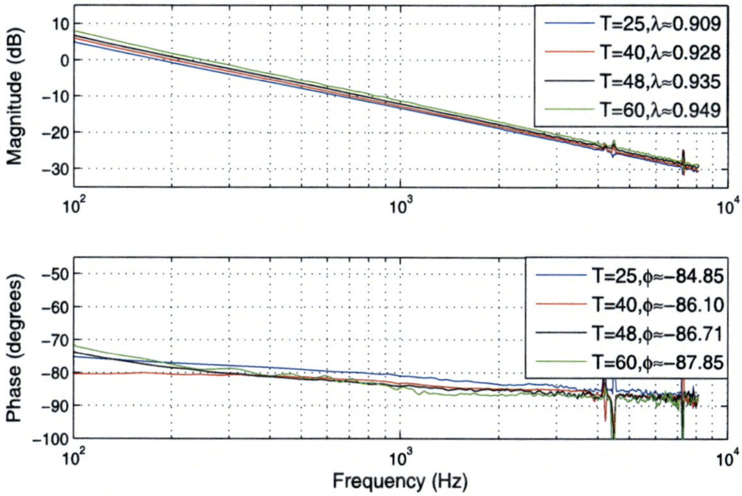

Fig. 6.7 Frequency responses for the temperature-dependent variable-order fractional integrator with $T = 25°\text{C}$, $T = 40°\text{C}$, $T = 48°\text{C}$ and $T = 60°\text{C}$

into the operational amplifier terminal. Therefore, the nodal equation at the inverting input terminal is given as:

$$\frac{V_{\text{IN}}(t)}{R} = -\frac{\tau^{\lambda(T(t))}}{K} D_t^{\lambda(T(t))} V_{\text{OUT}}(t). \tag{6.15}$$

Then, we have an ideal voltage output for the integrator amplifier as

$$V_{\text{OUT}}(t) = -D_t^{-\lambda(T(t))} \frac{K\, V_{\text{IN}}(t)}{R\tau^{\lambda(T(t))}}. \tag{6.16}$$

Figure 6.7 illustrates the frequency responses for the temperature-dependent variable-order fractional integrator with $T = 25°\text{C}$, $T = 40°\text{C}$, $T = 48°\text{C}$ and $T = 60°\text{C}$, respectively. The frequency responses were measured using HP 35665A Dynamic Signal Analyzer. It can be seen that the order $\lambda(T(t))$ changes with the temperature $T(t)$. The reason why the phases are not exactly equal to $-90° \times \lambda(T)$ might be due to the experiment errors or measurement errors.

Figure 6.8 illustrates the relationship between the order λ and the temperature $T(t)$ in the range of 25°C to 60°C. In Fig. 6.8, the y-axis is the order λ, and x-axis is the temperature $T(t)$. The values in the brackets on x-axis provide the time t. The unit of time t is a minute, and the time interval of the measurement is 20 minutes. The relationship between λ and $T(t)$ is approximately linear. Therefore, the variable-order fractional integrator can be realized by precise control of the temperature.

Similar to the realization of a temperature-dependent variable-order fractional integrator, the temperature-dependent variable-order fractional differentiator can be achieved by putting the circuit of Fig. 5.43 into a controlled environment with

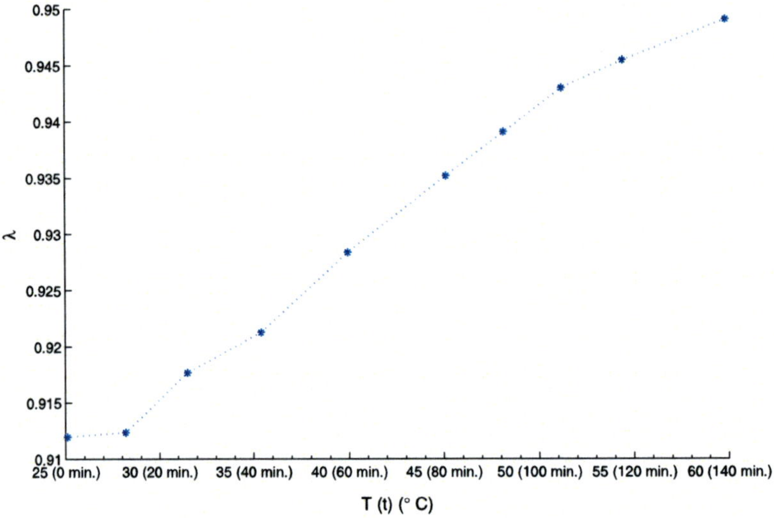

Fig. 6.8 The relation of order λ and $T(t)$

changing temperature. The voltage output for the variable-order fractional differentiator amplifier is then

$$V_{\rm OUT}(t) = -\frac{R\tau^{\lambda(T(t))}}{K} D_t^{\lambda(T(t))} V_{\rm IN}(t). \tag{6.17}$$

6.3.2 Application Examples of Analogue Variable-Order Fractional Systems

Based on the variable-order fractional integrator and differentiator, variable-order fractional $PI^{\lambda(t)}D^{\mu(t)}$ can be proposed as a generalization of the constant-order $PI^{\lambda}D^{\mu}$ controller. The internal structure of the variable-order fractional controller consists of the parallel connection, the proportional, integration, and the derivative parts. In time domain, the variable-order fractional controller has the form

$$u(t) = K_p e(t) + K_i D_t^{-\lambda(t)} e(t) + K_d D_t^{\mu(t)} e(t). \tag{6.18}$$

Figure 6.9 shows the structure of the variable-order fractional $PI^{\lambda(t)}D^{\mu(t)}$ controller.

When $\lambda(t)$ and $\mu(t)$ are constants, we obtain the commonly used constant-order fractional $PI^{\lambda}D^{\mu}$ controller. When $\lambda(t) = 1$ and $\mu(t) = 1$, we obtain the classical integer order PID controller. Based on the $PI^{\lambda(t)}D^{\mu(t)}$ controller, $PI^{\lambda(t)}$ and $PD^{\mu(t)}$ controllers can be adopted to better adjust the dynamic properties of the control system.

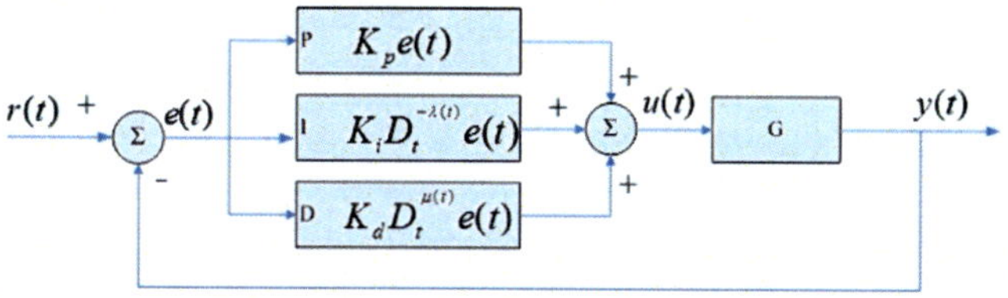

Fig. 6.9 Structure of the variable-order fractional $PI^{-\lambda(t)}D^{\mu(t)}$ controller

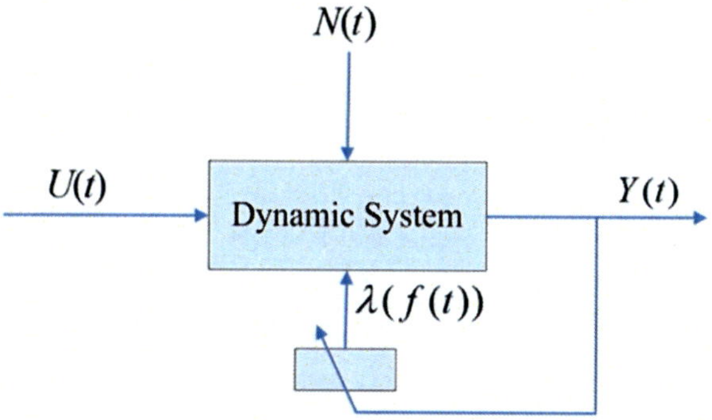

Fig. 6.10 Dynamic-order system

The variable-order fractional integrator and differentiator can also be used to analyze or simulate complex dynamic system. Figure 6.10 presents an interesting dynamic-order system. The order of the dynamic system is influenced by the feedback of the output of the system.

Besides the above mentioned potential applications, the variable-order fractional integrator and differentiator should find wide use in the field of physics, control, and signal processing.

6.4 Chapter Summary

This chapter introduced variable-order fractional signal processing techniques. The simulation of multifractional processes was realized by replacing the constant-order fractional integrator with a variable-order integrator. So, the generated multifractional processes exhibit the local memory property. Similarly, variable-order fractional system models were built by replacing the constant-order long memory parameter d with a variable-order local memory parameter d_t. The variable-order fractional system models can characterize the local memory of the fractional processes. A physical experimental study of the temperature-dependent variable-order fractional integrator and differentiator was introduced at the end of this chapter. The experiments were based on an analogue component named Fractor and the HFE temperature control platform. The experimental results show that the order λ of the fractional operator is the function of the temperature variable $T(t)$. The frequency responses of the variable-order fractional integrator and differentiator were measured using HP 35665A Dynamic Signal Analyzer. Some potential applications of the variable-order fractional integrator and differentiator are briefly discussed. It is expected that variable-order fractional signal processing techniques will be widely used in various fields of research and real world applications.

Chapter 7
Distributed-Order Fractional Signal Processing

The idea of using the distributed-order differential equation was first proposed by M. Caputo in 1969 [45], and a class of distributed-order differential equations were solved by him in 1995 [46]. These distributed-order equations were introduced in constitutive equations of dielectric media [46] and in diffusion equations [15]. Later, in 2002 [180], the authors studied the rheological properties of composite materials. Distributed-order fractional kinetics was discussed in 2004 [276]. In 2006 [307], the multi-dimensional random walk models were shown to be governed by distributed-order differential equations. These ultraslow and lateral diffusion processes were discussed in 2008 [141].

The theories of the distributed-order equations are classified as follows: distributed-order equations [12, 13, 15, 46], distributed-order system identification [111, 276, 280], special functions in distributed-order calculus [10, 47, 189, 191], numerical methods [53, 81, 290, 291] and so on [9, 141]. Moreover, there are also three surveys [179, 180, 307] and three theses [26, 69, 303] discussing the theories and applications of the distributed-order operators. It can be seen that both integer and fractional-order systems are the special cases of distributed-order systems [180]. Particularly, the distributed-order operator becomes a more precise tool to explain and describe some real physical phenomena such as the complexity of nonlinear systems [2, 11–13, 81, 111, 179, 180, 189, 276], networked structures [48, 180, 325], nonhomogeneous phenomena [47, 53, 141, 280, 290, 291, 307], multi-scale and multi-spectral phenomena [9, 26, 69, 190, 191, 303], etc. However, the time domain analysis of the distributed-order operator is still immature, and is in urgent need of further development. In this chapter, the distributed-order integrator/differentiator, distributed-order low-pass filter, and distributed parameter low-pass filter are studied in time domain. Moreover, the discretization method is used to get the digital impulse responses of these distributed-order fractional filters. The results are verified in both time and frequency domains.

H. Sheng et al., *Fractional Processes and Fractional-Order Signal Processing*,
Signals and Communication Technology,
DOI 10.1007/978-1-4471-2233-3_7,

7.1 Distributed-Order Integrator/Differentiator

Motivated by the applications of the distributed-order operators in control, filtering and signal processing, a distributed-order integrator/differentiator is derived step by step in this section. Firstly, the classical integrator can be rewritten as

$$\frac{1}{s} = \int_{-\infty}^{\infty} \delta(\alpha - 1)\frac{1}{s^\alpha}\mathrm{d}\alpha, \tag{7.1}$$

where $\delta(\cdot)$ denotes the Dirac-Delta function and $\frac{1}{s^\alpha}$ is the fractional-order integrator/differentiator with order $\alpha \in \mathbb{R}$. Moreover, the summation of a series of fractional-order integrators/differentiators can be expressed as

$$\sum_k \frac{1}{s^{\alpha_k}} = \int_{-\infty}^{\infty} \left[\sum_k \delta(\alpha - \alpha_k)\right] \frac{1}{s^\alpha}\mathrm{d}\alpha, \tag{7.2}$$

where k can belong to any countable or noncountable set. Now, it is straightforward to replace $[\sum_k \delta(\alpha - \alpha_k)]$ by a weighted kernel $w(\alpha)$. It follows that the right side of the above equation becomes

$$\int_{-\infty}^{\infty} w(\alpha)\frac{1}{s^\alpha}\mathrm{d}\alpha, \tag{7.3}$$

where $w(\alpha)$ is independent of time, and the above equation defines a distributed-order integrator/differentiator. Particularly, when $w(\alpha)$ is a piecewise constant function,

$$\int_{-\infty}^{\infty} w(\alpha)\frac{1}{s^\alpha}\mathrm{d}\alpha = w(\alpha_l)\int_{a_l}^{b_l} \frac{1}{s^\alpha}\mathrm{d}\alpha, \tag{7.4}$$

where a_l, b_l are real numbers, $\alpha_l \in (a_l, b_l)$ and $w(\alpha)$ is a constant on $\alpha \in (a_l, b_l)$. Based on the above discussions, without loss of generality, we focus on the uniform distributed-order integrator/differentiator $\int_a^b \frac{1}{s^\alpha}\mathrm{d}\alpha$, where $a < b$ are arbitrary real numbers.

In order to apply the distributed-order integrator/differentiator, the numerical discretization method is needed. This finds applications in signal modeling, filter design, controller design [185] and nonlinear system identification [2, 111]. The numerical discretization of the distributed-order integrator/differentiator, the key step towards application, can be realized in two ways: direct methods and indirect methods. In indirect discretization methods [59, 227], two steps are required, i.e., frequency domain fitting in continuous time domain first and then discretizing the fit s-transfer function [59]. Other frequency-domain fitting methods can also be used but without guaranteeing the stable minimum-phase discretization [59]. In this section, the direct discretization method will be used by an effective impulse response invariant discretization method discussed in [59, 62, 63, 171, 182]. In the above-mentioned references, the authors developed a technique for designing the discrete-time IIR filters from continuous-time fractional-order filters, in which the impulse response of the continuous-time fractional-order filter is sampled to produce the impulse response of the discrete-time filter. The detailed techniques of the impulse response invariant discretization method will be introduced in Sect. 7.1.2. For more discussions of the discretization methods, we cite [17, 95, 225, 241, 242, 311].

7.1.1 Impulse Response of the Distributed-Order Integrator/Differentiator

For the distributed-order integrator/differentiator

$$\int_a^b \frac{1}{s^\alpha}\mathrm{d}\alpha = \frac{s^{-a}-s^{-b}}{\ln(s)}, \tag{7.5}$$

where $a < b$ are arbitrary real numbers, its inverse Laplace transform is written as

$$\mathcal{L}^{-1}\left\{\int_a^b \frac{1}{s^\alpha}\mathrm{d}\alpha\right\} = \frac{1}{2\pi i}\int_{\sigma-i\infty}^{\sigma+i\infty} e^{st}\frac{s^{-a}-s^{-b}}{\ln(s)}\mathrm{d}s, \tag{7.6}$$

where $\sigma > 0$. It can be seen that there are two branch points of (7.5), $s = 0$ and $s = \infty$. Therefore, we can cut the complex plane by connecting the branch points along the negative real domain, so that the path integral in (7.6) is equivalent to the path integral along the Hankel path[1] (Fig. 5.27). The Hankel path starts from $-\infty$ along the lower side of real (horizontal) axis, encircles the circular disc $|s| = \epsilon \to 0$, in the positive sense, and ends at $-\infty$ along the upper side of real axis. Moreover, it can also be proved that the path integral of $\frac{e^{st}(s^{-a}-s^{-b})}{\ln(s)}$ along $s \to 0$ equals zero for $b \le 1$, and that there are no poles in the single value analytical plane. Therefore, by substituting $s = -xe^{-i\pi}$ and $s = xe^{i\pi}$, where $x \in (0, +\infty)$, we have, for an arbitrary $\sigma > 0$ and $b \le 1$,

$$\begin{aligned}
\mathcal{L}^{-1}\left\{\int_a^b \frac{1}{s^\alpha}\mathrm{d}\alpha\right\} =& \frac{1}{2\pi i}\int_{\sigma-i\infty}^{\sigma+i\infty} \frac{e^{st}(s^{-a}-s^{-b})}{\ln(s)}\mathrm{d}s \\
=& \frac{1}{2\pi i}\int_0^\infty \frac{e^{-xt}(x^{-a}e^{a\pi i}-x^{-b}e^{b\pi i})}{\ln(x)-i\pi}\mathrm{d}x \\
&- \frac{1}{2\pi i}\int_0^\infty \frac{e^{-xt}(x^{-a}e^{-a\pi i}-x^{-b}e^{-b\pi i})}{\ln(x)+i\pi}\mathrm{d}x \\
=& \frac{1}{\pi}\int_0^\infty \frac{e^{-xt}}{(\ln(x))^2+\pi^2}\big[x^{-a}(\sin(a\pi)\ln(x)+\pi\cos(a\pi)) \\
&- x^{-b}(\sin(b\pi)\ln(x)+\pi\cos(b\pi))\big]\mathrm{d}x.
\end{aligned} \tag{7.7}$$

[1] It follows from the residue of $\frac{e^{st}(s^{-a}-s^{-b})}{\ln(s)}$ which equals zero at $s = \infty$, that the path integral of it along $s \to \infty$ is vanished for $b \le 1$.

Based on the above discussions, we arrive at the following theorem.

Theorem 7.1 *For any $a < b \le 1$, we have*

$$\mathcal{L}^{-1}\left\{\int_a^b \frac{1}{s^\alpha}\mathrm{d}\alpha\right\} = \frac{1}{\pi}\int_0^\infty \frac{e^{-xt}}{(\ln(x))^2+\pi^2}\big[x^{-a}(\sin(a\pi)\ln(x)+\pi\cos(a\pi)) - x^{-b}(\sin(b\pi)\ln(x)+\pi\cos(b\pi))\big]\mathrm{d}x. \tag{7.8}$$

Especially, when $0 \le a < b \le 1$, it can be derived that

$$\left|\mathcal{L}^{-1}\left\{\int_a^b \frac{1}{s^\alpha}\mathrm{d}\alpha\right\}\right| \le \frac{1}{\pi^2}\left(\frac{M_1 t^{a-1}}{|a-1|} + \frac{M_2 t^{b-1}}{|b-1|}\right), \tag{7.9}$$

where M_1 and M_2 are finite positive constants.

Proof The first equation in this theorem is the same as (7.7). Moreover, by using (7.7), it can be easily proved that

$$\left|\mathcal{L}^{-1}\left\{\int_a^b \frac{1}{s^\alpha}\mathrm{d}\alpha\right\}\right| \le \frac{1}{\pi^2}\int_0^\infty e^{-xt}(x^{-a}+x^{-b})\mathrm{d}x = \frac{1}{\pi^2}\left(\frac{M_1 t^{a-1}}{|a-1|} + \frac{M_2 t^{b-1}}{|b-1|}\right),$$

where $M_1 = \int_0^\infty e^{-\tau^{1/(1-a)}}\mathrm{d}\tau$ and $M_2 = \int_0^\infty e^{-\tau^{1/(1-b)}}\mathrm{d}\tau$ are finite positive constants for any $0 \le a < b \le 1$. □

Based on the above discussions we can get the time domain expression of the impulse response of the distributed-order integrator/differentiator for any $a < b \le 1$. Note here, for $a < b \le 1$, (7.7) can be easily computed by using "*quadgk*" in MATLAB®, which will be used in the discretization method. Moreover, in order to extend a and b to the whole real axis, we can use the following properties.

Property 7.2 *It can be proved that*

(A) $s^c \int_a^b \frac{1}{s^\alpha}\mathrm{d}\alpha = \int_{a-c}^{b-c} \frac{1}{s^\alpha}\mathrm{d}\alpha$, *where $c \in \mathbb{R}$.*

(B) $\int_a^b \frac{1}{s^\alpha}\mathrm{d}\alpha = s\int_{a+1}^{1} \frac{1}{s^\alpha}\mathrm{d}\alpha + \int_0^b \frac{1}{s^\alpha}\mathrm{d}\alpha$, *where $a \in [-1, 0)$ and $b \in [0, 1]$.*

(C) $\int_a^b \frac{1}{s^\alpha}\mathrm{d}\alpha = (s^{-1} + \cdots + s^{-N})s^{-[a]}\int_{a-[a]-1}^{a-[a]} \frac{1}{s^\alpha}\mathrm{d}\alpha + s^{-(N+[a]+1)} \times \int_{a-[a]-1}^{b-(N+[a]+1)} \frac{1}{s^\alpha}\mathrm{d}\alpha$, *where $b - a > 1$, $N = [b - a]$ and $[*]$ denotes the integer part of $*$.*

(D) $\int_{\tilde{a}}^{\tilde{b}} s^\alpha\mathrm{d}\alpha = s\int_a^b \frac{1}{s^\alpha}\mathrm{d}\alpha$, *where $\tilde{a} < \tilde{b}$, $a = 1 - \tilde{b}$ and $b = 1 - \tilde{a}$.*

(E) *For the distributed integrator/differentiator $\int_a^b \frac{w(\alpha)}{s^\alpha}\mathrm{d}\alpha$, where $w(\alpha)$ is a piecewise function. Then it can be converted to the summation of uniform distributed integrator/differentiator.*

Theorem 7.3 *Any distributed-order integrator/differentiator can be composed by the distributed-order integrator for $0 \le a < b \le 1$, integrator $\frac{1}{s}$ and differentiator s.*

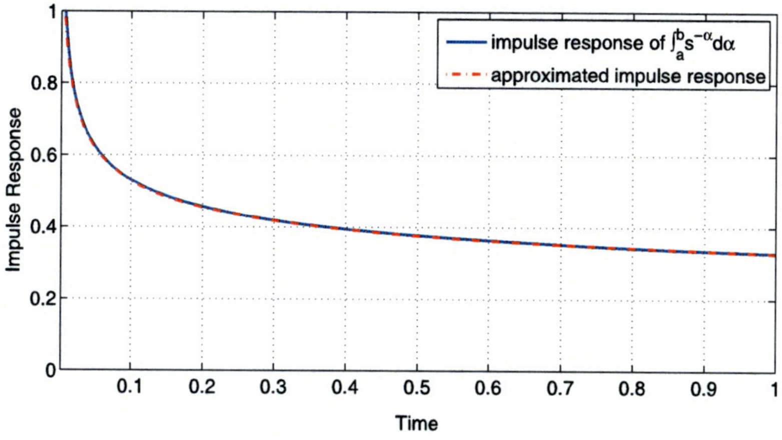

Fig. 7.1 The impulse responses of the approximate discrete-time IIR filter and the continuous-time distributed-order filter when $a = 0.6$, $b = 1$ and $T_s = 0.001$ second

Proof This theorem can be proved by Property 7.2. □

7.1.2 Impulse Response Invariant Discretization of DOI/DOD

The impulse response invariant discretization method converts analog filter transfer functions to digital filter transfer functions in such a way that the impulse responses are the same (invariant) at the sampling instants. Thus, if $g(t)$ denotes the impulse-response of an analog (continuous-time) filter, then the digital (discrete-time) filter given by the impulse-invariant method will have impulse response $g(nT_s)$, where T_s denotes the sampling period in seconds. Moreover, the frequency response of digital filter is an aliased version of the analog filter's frequency response [275].

Impulse invariance-based IIR-type discretization method is a simple and efficient numerical discretization method for the approximation of fractional-order filter [54, 55, 57, 58]. The method not only can accurately approximate the fractional-order filter in time domain but also fit the frequency response very well in the low frequency band in the frequency domain [167]. Figures 7.1 and 7.2 show the impulse responses and the frequency response of the frequency response of the approximated discrete-time IIR filter and the continuous-time fractional-order filter when $a = 0.6$, $b = 1$ and $T_s = 0.001$ second, respectively. The transfer function of the approximated IIR filter is

$$\frac{0.00167 - 0.006112z^{-1} + 0.008409z^{-2} - 0.005208z^{-3} + 0.00129z^{-4} - 4.785 \cdot 10^{-5}z^{-5}}{1 - 4.488z^{-1} + 8.004z^{-2} - 7.082z^{-3} + 3.104z^{-4} - 0.5383z^{-5}}. \tag{7.10}$$

For frequency response, the impulse response invariant discretization method works well for the band-limited (1–100 Hz) continuous-time fractional-order filters. This

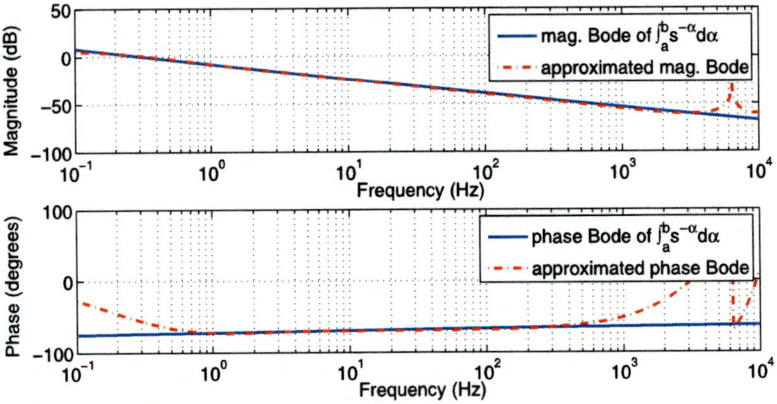

Fig. 7.2 The frequency response of the approximate discrete-time IIR filter and the continuous-time distributed-order filter when $a = 0.6$, $b = 1$ and $T_s = 0.001$ second

figure is plotted by the MATLAB code [265], where we used the MATLAB command [sr] = irid_doi(0.001, 0.6, 1, 5, 5).

Remark 7.4 The algorithm proposed in [283] permits more accurate identification when the impulse response is slowly varying. Therefore, it follows from Theorem 7.1 that the performance of "*stmcb*", an algorithm for finding an IIR filter with a prescribed time domain response given an input signal, in MATLAB is related to a and b. Particularly, when $0 \le a < b \le 1$, the approximated results are more accurate for the case when a, b are closer to 1.

It follows from Remark 7.4 that the approximated results obtained by the application of (7.7) and the discretization method have relatively good performances for $0.5 \le a < b \le 1$ in both time and frequency domains. Allowing for Theorem 7.3, and in order to extend a and b to the whole real domain, we arrive at the following property.

Property 7.5 *When* $0 \le a < b \le 0.5$, *it follows from* (A) *in Property* 7.2 *that* $\int_a^b \frac{1}{s^\alpha} d\alpha = s^{0.5-a} \int_{0.5}^{0.5+b-a} \frac{1}{s^\alpha} d\alpha$, *where* $0.5 \le 0.5 + b - a \le 1$.

Remark 7.6 It follows from Properties 7.2 and 7.5 that, for arbitrary $\tilde{a}, \tilde{b} \in \mathbb{R}$, $\int_{\tilde{a}}^{\tilde{b}} \frac{1}{s^\alpha} d\alpha$ can be divided into the combination of $s^\lambda (\lambda \in \mathbb{R})$ and $\int_a^b \frac{1}{s^\alpha} d\alpha$, where $a, b \in [0.5, 1]$.

Lastly, it can be shown in both time[2] and frequency domains that the distributed-order integrator/differentiator exhibits some intermediate properties among the

[2] This conclusion can be seen from (7.7) and Theorem 7.1.

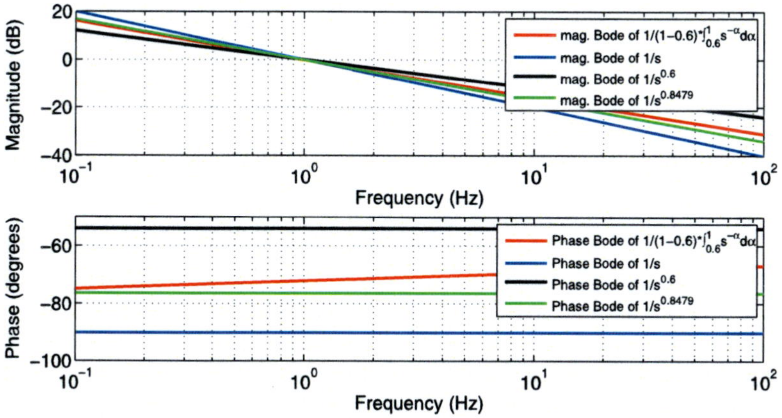

Fig. 7.3 Frequency response comparisons

integer-order and fractional-order integrators/differentiators. In the frequency domain, for example, Figure 7.3 presents the frequency responses of distributed-order integrator $\frac{1}{1-0.6}\int_{0.6}^{1} s^{-\alpha}\mathrm{d}\alpha$, integer-order integrator $\frac{1}{s}$, and fractional-order integrators $\frac{1}{s^{0.6}}$ and $\frac{1}{s^{0.8479}}$. The fractional integrator $\frac{1}{s^{0.8479}}$ was constructed by searching the best fit to the magnitude of the distributed-order integrator $\frac{1}{1-0.6}\int_{0.6}^{1} s^{-\alpha}\mathrm{d}\alpha$. It can be seen that the magnitude and phase of the frequency response of the distributed-order integrator are totally different from that of the fractional-order integrator and integer-order integrator. The phase of the distributed-order integrator is no longer a constant. The comparison study of the three types of integrators indicates that the distributed-order integrator exhibits distinct frequency response characteristics. There does not exist a so-called "mean order" equivalent constant-order integrator/differentiator for distributed-order one.

7.2 Distributed-Order Low-Pass Filter

In this section, we focus on the discussions of the uniform distributed-order low-pass filter

$$\frac{\lambda^{a+b}\ln\lambda}{\lambda^b-\lambda^a}\int_a^b \frac{1}{(s+\lambda)^\alpha}\mathrm{d}\alpha, \tag{7.11}$$

where $\lambda \geq 0$, $a < b$ are arbitrary real numbers and $\frac{\lambda^{a+b}\ln\lambda}{\lambda^b-\lambda^a}$ is the normalizing constant, such that the filter (7.11) has a unity DC gain[3].

[3]When $s=0$, DC gain of $\int_a^b \frac{1}{(s+\lambda)^\alpha}\mathrm{d}\alpha = \int_a^b \frac{1}{\lambda^\alpha}\mathrm{d}\alpha = \frac{1}{\ln\lambda}(\frac{1}{\lambda^a}-\frac{1}{\lambda^b})$. So, unity gain requires gain scaling factor $\frac{\lambda^{a+b}\ln\lambda}{\lambda^b-\lambda^a}$.

Firstly, the classical first order low-pass filter can be rewritten as

$$\frac{1}{Ts+1} = \int_{-\infty}^{\infty} \delta(\alpha - 1)\frac{1}{(Ts+1)^{\alpha}}\mathrm{d}\alpha, \tag{7.12}$$

where $\delta(\cdot)$ denotes the Dirac-Delta function and $\frac{1}{(Ts+1)^{\alpha}}$ is a fractional-order low-pass filter with order $\alpha \in \mathbb{R}$. Moreover, the summation of a series of fractional-order low-pass filters can be expressed as

$$\sum_k \frac{1}{(T_k s+1)^{\alpha_k}} = \int_{-\infty}^{\infty} \left[\sum_k \delta(\alpha - \alpha_k)\right]\frac{1}{(T_k s+1)^{\alpha}}\mathrm{d}\alpha, \tag{7.13}$$

where k can belong to any countable or non countable set. Now, it is straightforward to replace $[\sum_k \delta(\alpha - \alpha_k)]$ by a weighted kernel $w(\alpha)$. It follows that the right side of the above equation becomes

$$\int_{-\infty}^{\infty} w(\alpha)\frac{1}{(T_w s+1)^{\alpha}}\mathrm{d}\alpha, \tag{7.14}$$

where $w(\alpha)$ and T_w are independent of time and the above equation defines a distributed-order low-pass filter. Particularly, when $w(\alpha)$ and T_w are piecewise functions,

$$\int_{-\infty}^{\infty} w(\alpha)\frac{1}{(T_w s+1)^{\alpha}}\mathrm{d}\alpha = \sum_l \frac{w(\alpha_l)}{T_l^{\alpha}}\int_{a_l}^{b_l} \frac{1}{(s+1/T_l)^{\alpha}}\mathrm{d}\alpha, \tag{7.15}$$

where a_l, b_l are real numbers, $\alpha_l \in (a_l, b_l)$ and $w(\alpha)$ and T_w are constants on $\alpha \in (a_l, b_l)$. Based on the above discussions, without loss of generality, we focus on the following uniform distributed-order low-pass filter

$$\frac{\lambda^{a+b}\ln\lambda}{\lambda^b - \lambda^a}\int_a^b \frac{1}{(s+\lambda)^{\alpha}}\mathrm{d}\alpha, \tag{7.16}$$

where $\lambda \geq 0$, $a < b$ are arbitrary real numbers and $\frac{\lambda^{a+b}\ln\lambda}{\lambda^b-\lambda^a}$ is the normalizing constant for unity DC gain.

Moreover, to enable the applications of the distributed-order low-pass filter in engineering, the numerical discretization method should be applied so that the filter can be used in signal modeling, filter design and nonlinear system identification [2, 111, 168]. Let us first derive the analytical form of the filter's impulse response.

7.2.1 Impulse Response of the Distributed-Order Low-Pass Filter

In this section, the analytical form of

$$\mathcal{L}^{-1}\left\{\int_a^b \frac{1}{(s+\lambda)^{\alpha}}\mathrm{d}\alpha\right\} \tag{7.17}$$

is derived and is in a computable form in MATLAB. This will be used in the impulse response invariant discretization in the next section.

It follows from the properties of inverse Laplace transform that

$$\mathcal{L}^{-1}\left\{\int_a^b \frac{1}{(s+\lambda)^\alpha}\mathrm{d}\alpha\right\} = e^{-\lambda t}\mathcal{L}^{-1}\left\{\int_a^b \frac{1}{s^\alpha}\mathrm{d}\alpha\right\}. \tag{7.18}$$

It has been provided that by substituting $s = -xe^{-i\pi}$ and $s = xe^{i\pi}$, where $x \in (0, +\infty)$, we have, for an arbitrary $\sigma > 0$ and $b \le 1$,

$$\begin{aligned}\mathcal{L}^{-1}\left\{\int_a^b \frac{1}{s^\alpha}\mathrm{d}\alpha\right\} =& \frac{1}{\pi}\int_0^\infty \frac{e^{-xt}}{(\ln(x))^2+\pi^2}\big[x^{-a}(\sin(a\pi)\ln(x)+\pi\cos(a\pi))\\ &- x^{-b}(\sin(b\pi)\ln(x)+\pi\cos(b\pi))\big]\mathrm{d}x.\end{aligned} \tag{7.19}$$

Theorem 7.7 *For any $a, b \in \mathbb{R}$, we have*

$$\left|\mathcal{L}^{-1}\left\{\int_a^b \frac{1}{(s+\lambda)^\alpha}\mathrm{d}\alpha\right\}\right| \le \frac{e^{-\lambda t}}{\pi^2}\left(\frac{M_1 t^{a-1}}{|a-1|} + \frac{M_2 t^{b-1}}{|b-1|}\right),$$

where M_1 and M_2 are finite positive constants.

Proof By using (7.19), it can be easily proved that

$$\begin{aligned}\left|\mathcal{L}^{-1}\left\{\int_a^b \frac{1}{(s+\lambda)^\alpha}\mathrm{d}\alpha\right\}\right| &\le \frac{e^{-\lambda t}}{\pi^2}\int_0^\infty e^{-xt}(x^{-a}+x^{-b})\mathrm{d}x\\ &= \frac{e^{-\lambda t}}{\pi^2}\left(\frac{M_1 t^{a-1}}{|a-1|} + \frac{M_2 t^{b-1}}{|b-1|}\right),\end{aligned}$$

where $M_1 = \int_0^\infty e^{-\tau^{1/(1-a)}}\mathrm{d}\tau$ and $M_2 = \int_0^\infty e^{-\tau^{1/(1-b)}}\mathrm{d}\tau$ are finite positive constants for any $a, b \in \mathbb{R}\setminus\{1\}$. When $a = 1$ or $b = 1$, it is obvious that $|\mathcal{L}^{-1}\{\int_a^b \frac{1}{(s+\lambda)^\alpha}\mathrm{d}\alpha\}| \le +\infty$. □

7.2.2 Impulse Response Invariant Discretization of DO-LPF

Now, let us consider how to discretize the $G(s)$ given sampling period T_s. Our goal is to get a discretized version of $G(s)$, denoted by $G_d(z^{-1})$ with a constraint that $G_d(z^{-1})$ and $G(s)$ have the same impulse responses. Since the analytical impulse response of $G(s)$ had already been derived in Sect. 7.2.1, it is relatively straightforward to obtain the impulse response invariant discretized version of $G(s)$ via the well-known Prony technique [54, 55, 57, 58, 273]. In other words, the discretization impulse response can be obtained by using the continuous time impulse response as follows:

$$g(n) = T_s g(nT_s), \tag{7.20}$$

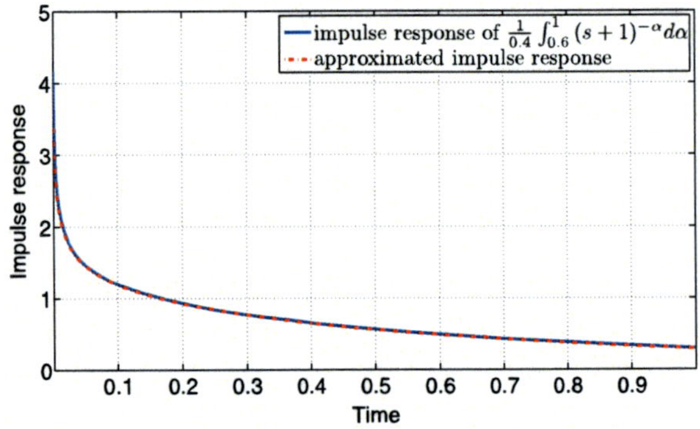

Fig. 7.4 The impulse response of $\frac{1}{0.4}\int_{0.6}^{1}\frac{1}{(s+1)^{\alpha}}\mathrm{d}\alpha$

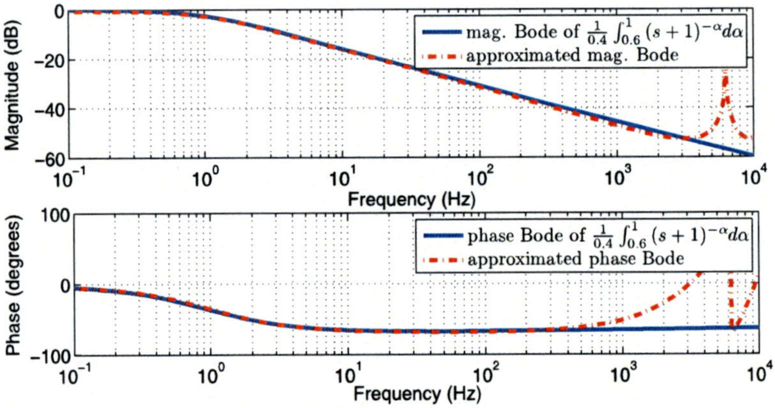

Fig. 7.5 The frequency response of $\frac{1}{0.4}\int_{0.6}^{1}\frac{1}{(s+1)^{\alpha}}\mathrm{d}\alpha$

where $n = 0, 1, 2, \ldots$ and T_s is the sampling period.

Figure 7.4 shows the magnitude and phase of the frequency response of the approximate discrete-time IIR filter and the continuous-time fractional distributed order filter $\frac{1}{0.4}\int_{0.6}^{1}\frac{1}{(s+1)^{\alpha}}\mathrm{d}\alpha$. The transfer function of the approximate IIR filter $H(z)$ is

$$\frac{.00417 - .01509z^{-1} + .02048z^{-2} - .01248z^{-3} + .003019z^{-4} - .0001022z^{-5}}{1 - 4.445z^{-1} + 7.844z^{-2} - 6.859z^{-3} + 2.967z^{-4} - .5066z^{-5}}.$$

For frequency responses, the impulse response invariant discretization method works well for the continuous-time fractional-order filters. Moreover, the continuous and discretized impulse response and frequency response are also shown in Figs. 7.4 and 7.5, where $T_s = 0.001$ second. Then, several low-pass filters are compared and shown in Fig. 7.6.

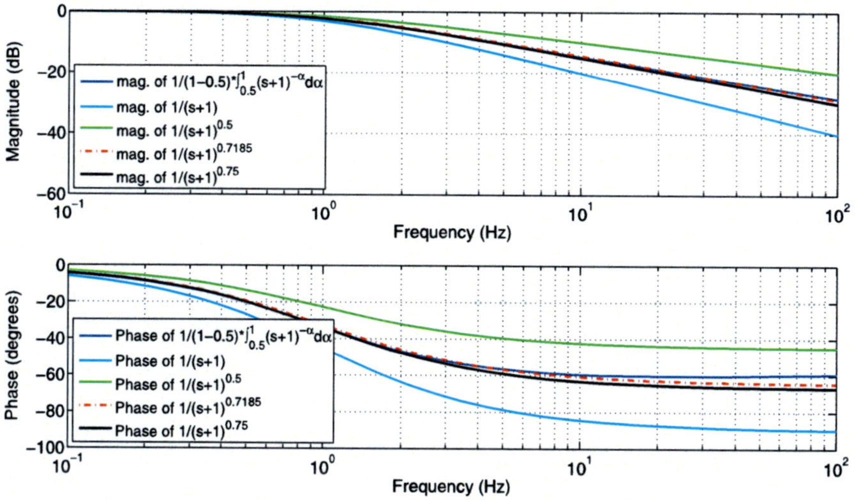

Fig. 7.6 The comparisons of distributed-order low-pass filter with several integer-order and constant-order low-pass filters

7.3 Distributed Parameter Low-Pass Filter

It is interesting to note that the classical fractional-order low-pass filter can also be rewritten as

$$\frac{1}{T_1 s^\alpha + 1} = \int_{-\infty}^{\infty} \delta(T - T_1) \frac{1}{T s^\alpha + 1} \mathrm{d}T, \tag{7.21}$$

where $T_1 > 0$ and $\delta(\cdot)$ denotes the Dirac-Delta function and $\frac{1}{T_1 s^\alpha + 1}$ is a fractional-order low-pass filter with order $\alpha > 0$. Moreover, the summation of a series of fractional-order low-pass filters can be expressed as

$$\sum_k \frac{1}{T_k s^\alpha + 1} = \int_{-\infty}^{\infty} \left[\sum_k \delta(T - T_k) \right] \frac{1}{T s^\alpha + 1} \mathrm{d}T, \tag{7.22}$$

where k can belong to any countable or non countable set and $T_k > 0$. Now, it is straightforward to replace $\sum_k \delta(T - T_k)$ by a weighted kernel $w(T)$. It follows that the right side of the above equation becomes

$$\int_{-\infty}^{\infty} w(T) \frac{1}{T s^\alpha + 1} \mathrm{d}T, \tag{7.23}$$

where $w(T)$ and α are independent of time and the above equation is leading to the fractional-order distributed parameter low-pass filter. Particularly, when $w(T)$ is a piecewise function,

$$\int_{-\infty}^{\infty} w(T) \frac{1}{T s^\alpha + 1} \mathrm{d}T = \sum_l w(T_l) \int_{a_l}^{b_l} \frac{1}{T s^\alpha + 1} \mathrm{d}T,$$

where a_l, b_l are real numbers and $w(T)$ is equal to a constant on $\alpha \in (a_l, b_l)$ for all l. Based on the above discussions, without loss of generality, we focus on the discussions of the uniform fractional-order distributed parameter low-pass filter

$$\frac{1}{b-a}\int_a^b \frac{1}{Ts^\alpha + 1}\mathrm{d}T, \tag{7.24}$$

where $0 \le a \le b$, $\alpha \in (0,1)$, and $\frac{1}{b-a}$ is the normalizing constant to ensure unity DC gain of the filter.

For applications of the fractional-order distributed parameter low-pass filter in engineering, the numerical discretization method is applied allowing potential applications in signal modeling, filter design and nonlinear system identification [2, 111]. The results are compared in both time and frequency domains.

7.3.1 Derivation of the Analytical Impulse Response of the Fractional-Order Distributed Parameter Low-Pass Filter

In this section, the inverse Laplace transform of the fractional-order distributed parameter low-pass filter is derived by using the complex integral, leading to some useful asymptotic properties of its impulse response. The filter we discuss here is the form (7.24). Let $\tau = Ts^\alpha + 1$, the above integral becomes

$$\int_a^b \frac{1}{Ts^\alpha + 1}\mathrm{d}T = \frac{1}{s^\alpha}\int_{as^\alpha+1}^{bs^\alpha+1}\frac{1}{\tau}\mathrm{d}\tau = \frac{\ln(bs^\alpha+1) - \ln(as^\alpha+1)}{s^\alpha}. \tag{7.25}$$

It follows from the definition of the inverse Laplace transform that

$$\mathcal{L}^{-1}\left\{\frac{\ln(bs^\alpha+1) - \ln(as^\alpha+1)}{s^\alpha}\right\} = \frac{1}{2\pi i}\int_{\sigma-i\infty}^{\sigma+i\infty} e^{st}\frac{\ln(bs^\alpha+1) - \ln(as^\alpha+1)}{s^\alpha}\mathrm{d}s. \tag{7.26}$$

When $a, b > 0$ and $\alpha \in (0,1)$, $s = 0$ and $s = \infty$ are the two branch points of the above integrand. (7.26) is equivalent to the complex path integral, a curve (Hankel path) which starts from $-\infty$ along the lower side of the real axis, encircles the circular disc $|s| = \epsilon \to 0$ in the positive side, and ends at $-\infty$ along the upper side of the real axis. Moreover, from the fact the residue of the above integrand equals zero at the origin, one obtains

$$\frac{1}{2\pi i}\int_{\sigma-i\infty}^{\sigma+i\infty} e^{st}\frac{\ln(bs^\alpha+1) - \ln(as^\alpha+1)}{s^\alpha}\mathrm{d}s = \frac{1}{2\pi i}\int_{\text{Hankel}} e^{st}\frac{\ln(bs^\alpha+1) - \ln(as^\alpha+1)}{s^\alpha}\mathrm{d}s \tag{7.27}$$

$$= \frac{1}{2\pi i}\left[\int_{\text{lower}} + \int_{\text{upper}}\right] e^{st} \frac{\ln(bs^\alpha + 1) - \ln(as^\alpha + 1)}{s^\alpha} \mathrm{d}s. \tag{7.28}$$

Therefore, on the lower side, substituting $s = xe^{-i\pi}$ into the above integral yields

$$\begin{aligned}
&\int_{\text{lower}} e^{st} \cdot \frac{\ln(bs^\alpha + 1) - \ln(as^\alpha + 1)}{s^\alpha} \mathrm{d}s \\
&= \int_\infty^0 e^{xe^{-i\pi}t} \cdot \frac{\ln(bx^\alpha e^{-i\alpha\pi} + 1) - \ln(ax^\alpha e^{-i\alpha\pi} + 1)}{x^\alpha e^{-i\alpha\pi}} \mathrm{d}(xe^{-i\pi}) \\
&= \int_0^\infty e^{-xt} \cdot \frac{\ln(bx^\alpha e^{-i\alpha\pi} + 1) - \ln(ax^\alpha e^{-i\alpha\pi} + 1)}{x^\alpha e^{-i\alpha\pi}} \mathrm{d}x.
\end{aligned} \tag{7.29}$$

On the upper side, set $s = xe^{i\pi}$ and we have

$$\begin{aligned}
&\int_{\text{upper}} e^{st} \cdot \frac{\ln(bs^\alpha + 1) - \ln(as^\alpha + 1)}{s^\alpha} \mathrm{d}s \\
&= \int_0^\infty e^{xe^{i\pi}t} \cdot \frac{\ln(bx^\alpha e^{i\alpha\pi} + 1) - \ln(ax^\alpha e^{i\alpha\pi} + 1)}{x^\alpha e^{i\alpha\pi}} \mathrm{d}(xe^{i\pi}) \\
&= -\int_0^\infty e^{-xt} \cdot \frac{\ln(bx^\alpha e^{i\alpha\pi} + 1) - \ln(ax^\alpha e^{i\alpha\pi} + 1)}{x^\alpha e^{i\alpha\pi}} \mathrm{d}x.
\end{aligned}$$

In the above equations,

$$\ln(ax^\alpha e^{\pm i\alpha\pi} + 1) = \ln(r_a) \pm i\theta_a,$$

where $r_a = \sqrt{a^2x^{2\alpha} + 2ax^\alpha \cos(\alpha\pi) + 1}$ and $\theta_a = \mathrm{Arg}\{ax^\alpha \cos(\alpha\pi) + 1 + iax^\alpha \times \sin(\alpha\pi)\}$. Similarly,

$$\ln(bx^\alpha e^{\pm i\alpha\pi} + 1) = \ln(r_b) \pm i\theta_b,$$

where $r_b = \sqrt{b^2x^{2\alpha} + 2bx^\alpha \cos(\alpha\pi) + 1}$ and $\theta_b = \mathrm{Arg}\{bx^\alpha \cos(\alpha\pi) + 1 + ibx^\alpha \times \sin(\alpha\pi)\}$. Therefore,

$$\begin{aligned}
&\frac{1}{2\pi i}\int_{\sigma - i\infty}^{\sigma + i\infty} e^{st} \frac{\ln(bs^\alpha + 1) - \ln(as^\alpha + 1)}{s^\alpha} \mathrm{d}s \\
&= \frac{1}{\pi}\int_0^\infty \frac{(\ln(r_b) - \ln(r_a))\sin(\alpha\pi) + (\theta_a - \theta_b)\cos(\alpha\pi)}{x^{2\alpha}e^{xt}} \mathrm{d}x.
\end{aligned}$$

Theorem 7.8 *Let* $\mathcal{L}\{g(t)\} = \int_a^b \frac{1}{Ts^\alpha + 1}\mathrm{d}T$, *where* $0 \le a \le b$ *and* $\alpha \in (0, 1)$. *We have*

$$g(t) = \frac{1}{\pi}\int_0^\infty \frac{(\ln(r_b) - \ln(r_a))\sin(\alpha\pi) + (\theta_a - \theta_b)\cos(\alpha\pi)}{x^{2\alpha}e^{xt}} \mathrm{d}x \tag{7.30}$$

and

$$g(t) \leq \frac{(b^2-a^2)\Gamma(1+\alpha)t^{-\alpha-1}}{2\pi \sin(\alpha\pi)}. \tag{7.31}$$

Proof (7.30) has been proved in the previous part of this session.

For the asymptotic property of $g(t)$, we need to derive the integral expressions of $t^{\alpha-1}E_{\alpha,\alpha}(-\frac{t^\alpha}{T})$. It can be seen that the Laplace transform of $t^{\alpha-1}E_{\alpha,\alpha}(-\frac{t^\alpha}{T})$ is $\frac{1}{s^\alpha+\frac{1}{T}}$, which has two branch points $s=0$ and $s=\infty$. By cutting the complex plane with the same Hankel path (Fig. 5.27) described in the previous discussion, it follows from the fact that the path integral around the origin equals zero, that

$$t^{\alpha-1}E_{\alpha,\alpha}\left(-\frac{t^\alpha}{T}\right) = \frac{1}{\pi}\int_0^\infty \frac{x^\alpha \sin(\alpha\pi)e^{-xt}\mathrm{d}x}{x^{2\alpha}+\frac{2\sin(\alpha\pi)}{T}x^\alpha+\frac{1}{T^2}}.$$

It then follows from $x^{2\alpha}+\frac{2\sin(\alpha\pi)}{T}x^\alpha+\frac{1}{T^2} \geq \frac{\sin^2(\alpha\pi)}{T^2}$ that

$$0 \leq t^{\alpha-1}E_{\alpha,\alpha}\left(-\frac{t^\alpha}{T}\right) \leq \frac{T^2}{\pi\sin(\alpha\pi)}\int_0^\infty x^\alpha e^{-xt}\mathrm{d}x = \frac{T^2\Gamma(1+\alpha)t^{-\alpha-1}}{\pi\sin(\alpha\pi)}, \tag{7.32}$$

where the definition of the Gamma function is used in (7.32). Then, clearly,

$$g(t) = \int_a^b \frac{1}{T}t^{\alpha-1}E_{\alpha,\alpha}\left(-\frac{t^\alpha}{T}\right)\mathrm{d}T \leq \frac{(b^2-a^2)\Gamma(1+\alpha)t^{-\alpha-1}}{2\pi\sin(\alpha\pi)}.$$

□

7.3.2 Impulse Response Invariant Discretization of FO-DP-LPF

Based on the obtained analytical impulse response function $g(t)$, given sampling period T_s, it is straightforward to perform the inverse response invariant discretization of

$$\frac{1}{b-a}\int_a^b \frac{1}{Ts^\alpha+1}\mathrm{d}T,$$

by using the well-known Prony technique [273] similar to the procedures presented in [95].

Figures 7.7 and 7.8 show respectively the impulse response and the frequency response of the continuous-time fractional-order filters for $\int_0^{0.5}\frac{2}{Ts^{0.95}+1}\mathrm{d}T$ and the approximated discrete-time IIR filter. The approximate discrete-time IIR filters can accurately portray a time domain characteristic of continuous-time fractional-order for $T_s=0.001$, $a=0$, $b=0.5$ and $\alpha=0.95$. For frequency responses, the impulse

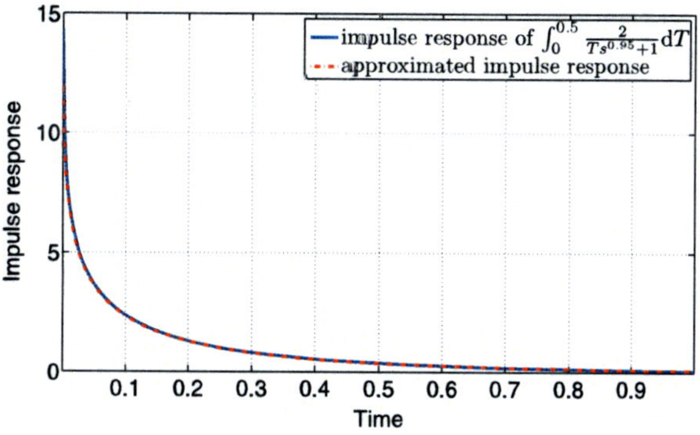

Fig. 7.7 The impulse responses for $\int_0^{0.5} \frac{2}{Ts^{0.95}+1}\mathrm{d}T$ and the approximated discrete-time filter

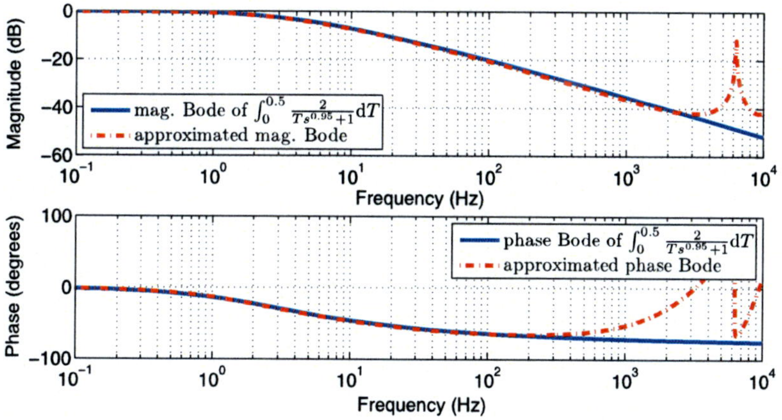

Fig. 7.8 The frequency responses for $\int_0^{0.5} \frac{2}{Ts^{0.95}+1}\mathrm{d}T$ and the approximated discrete-time filter

response invariant discretization method approximates well. Moreover, using Theorem 7.8 and the "*stmcb*" function in MATLAB, the approximated transfer function is

$$\frac{0.01436z^5 - 0.05217z^4 + 0.07106z^3 - 0.04335z^2 + 0.01039z - 0.0002926}{z^5 - 4.477z^4 + 7.966z^3 - 7.036z^2 + 3.08z - 0.5339}. \tag{7.33}$$

7.4 Chapter Summary

In this chapter, we derived the impulse response functions of the distributed-order integrator/differentiator, fractional-order distributed low-pass filter, and the fractional-order distributed parameter low-pass filter from the complex path integral expressed in the definite integral form. Based on these results, we obtained some asymptotic properties, and we can accurately compute the integrals on the whole time domain. Moreover, for practical applications, we presented a technique known as "impulse-response-invariant discretization" to perform the discretization of the above three distributed-order filters. Lastly, it was shown that the distributed-order fractional filters had some unique features compared with the classical integer-order or constant-order fractional filters.

Part IV
Applications of Fractional-Order Signal Processing Techniques

Chapter 8
Fractional Autoregressive Integrated Moving Average with Stable Innovations Model of Great Salt Lake Elevation Time Series

8.1 Introduction

Great Salt Lake (GSL), located in the northern part of the U.S. State of Utah, is the largest salt lake in the western hemisphere, the fourth largest terminal lake in the world. In an average year, the lake covers an area of around 1,700 square miles (4,400 km^2), but the size of the lake fluctuates substantially due to its elevation (shallowness). GSL is located on a shallow playa, so small changes in the water-surface elevation result in large changes in the surface area of the lake. For instance, in 1963 it reached its lowest recorded level at 950 square miles (2,460 km^2), but in 1987 the surface area was at the historic high of 3,300 square miles (8,547 km^2) [306]. The variations of the GSL elevation have an enormous impact on the people who live nearby. The rise in 1987 had caused 285 million U.S. dollars worth of damage to lakeside industries, roads, railroads, wildfowl management areas, recreational facilities and farming that had been established on the exposed lake bed [6].

GSL is divided into a north and a south part by a rock-fill causeway. Because of the importance of the GSL elevation, the United States Geological Survey (USGS) has been collecting water surface elevation data from the south part of GSL since 1875 and continuously since Oct. 1902. The north part of the lake has been monitored since 1960 [7]. The USGS operates gauges that collect water surface elevation data on the south part of the lake at the Boat Harbor gage, and on the north part of the lake at the Saline gage [160]. We found that the distribution of the data from north part is evidently heavy-tailed, so the north part water surface elevation data of the lake was analyzed in this FOSP application. Several studies have been performed to build a precise model of the GSL elevation time series and a variety of techniques have been used to estimate historical GSL elevation time series, including geological and archaeological methods [158, 159, 300]. Despite these preliminary efforts, all the conventional methods and models were found to be insufficient to characterize the lake levels and predict its future. One reason for such inadequacy might be the existence of long-range dependence in the GSL elevation time series [293]. Another reason might be the non-convergence of the second-order moment of the GSL elevation time series. So, FOSP techniques are probably the better techniques to model and predict it [292]. Fractional-order signal processing is, in recent

H. Sheng et al., *Fractional Processes and Fractional-Order Signal Processing*, Signals and Communication Technology,
DOI 10.1007/978-1-4471-2233-3_8,

years, becoming a very active research area. FOSP provides many powerful techniques to analyze fractional processes which have both short and long-term memories or time series with heavy-tailed distribution [56]. FOSP is based on the knowledge of α-stable distribution, FrFT and fractional calculus (FC). FC is a generalization of the conventional integer-order differential and integral operators [181]. It is the mathematical basis of fractional-order systems described by fractional-order differential equations. The simplest fractional-order dynamic systems include the fractional-order integrators and fractional-order differentiators. The FARIMA with stable innovations model is a typical fractional-order system, which combines both features: infinite variance and long-range dependence [143]. FARIMA with stable innovations model is based on linear fractional stable noise which is a stationary, self-similar and heavy-tailed process. "Model accepts physical interpretation, since it explains how the observed data appear as a superposition of independent effects" [219]. The traditional models, such as AR, MA, ARMA and ARIMA (Autoregressive Integrated Moving Average) processes, can only capture the short-range dependence [31]. FARIMA and FIGARCH models give a good fit for LRD time series, but it cannot characterize the time series with the heavy-tailed property precisely. Therefore, we propose to use the FARIMA with the stable innovations model to characterize the GSL elevation time series.

8.2 Great Salt Lake Elevation Data Analysis

The levels of the north GSL are measured twice a month including the 1st day and 15th day of the month. The observed one-dimensional Jan. 1960–Feb. 2009 GSL north-level time series is shown in Fig. 8.1. The figure shows dramatic rise and falls of GSL levels at different times throughout the measurement period. From the peak of the plot, the flood in 1987 can be seen clearly.

We propose to build an FARIMA with stable innovations model for GSL elevation time series. This model can characterize both short/long-range dependence and heavy-tailed properties. It provides a more powerful technique to analyze long-memory time series with heavy-tailed distribution processes. The precondition is that the data indeed has LRD and stable properties. Therefore, we analyze the LRD and stable properties of the GSL elevation time series first.

Time series is said to be stationary when the mean, variance and autocorrelations can be well approximated by sufficiently long time averages based on the single set of realizations. For a stationary process, the effects of the shocks are temporary and the series reverts to its long-run level. Under that condition, any long-term forecast of a stationary series will converge to the unconditional mean of the series. For a non-stationary process, time-dependence exists and matters. So a long-run mean may not exist and the variance diverges to infinity.

Standard stable ARMA analysis requires the assumption that the considered time series is stationary, which can be analyzed using the autocorrelation function (ACF) and the partial autocorrelation function (PACF). ACF determines the correlation between a time series value and the lag of that value. The PACF specifies the additional

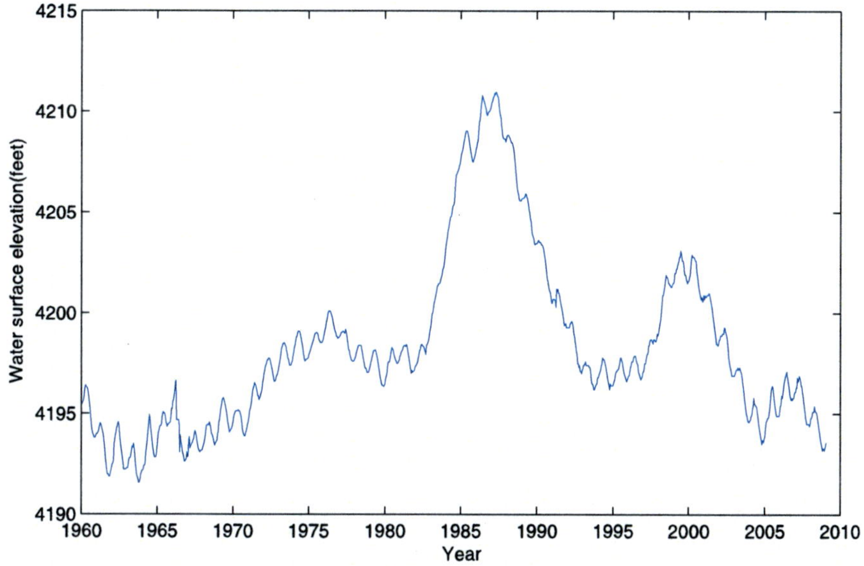

Fig. 8.1 Water surface elevation graph of the GSL, Jan. 1960–Feb. 2009

correlation between a time series value and a special lag value, withdrawing the influence of other lag values. When the ACF of a time series decreases slowly, the time series is likely to be non-stationary. Figure 8.2 shows the ACF and PACF of surface water levels and the difference of surface water levels. From Fig. 8.2 we can observe that the GSL elevation time series is non-stationary. In this study, the Aggregated Variance method and the Absolute Values of the Aggregated method are adopted to estimate the Hurst parameter of the GSL elevation data. The outputs of these two estimators are 0.9970 and 0.9880 respectively, which indicates that the GSL elevation time series is an LRD process which is predictable [165].

It is also important to know whether the distribution is Gaussian ($\alpha = 2$) or non-Gaussian ($\alpha < 2$) in time series analysis. To decide whether to use the FARIMA model or FARIMA with stable innovations model to characterize the variability of GSL elevation, we only need to know whether the distribution of elevation levels of north part of GSL is Gaussian or non-Gaussian. A property that differentiates the Gaussian and non-Gaussian stable distributions is that the non-Gaussian stable distribution does not have a finite variance. For simplicity, we analyze the sample variance of GSL elevation time series. Specifically, let $X_k, k = 1, \dots, N$, be samples from the same stable distribution. For each $1 \le n \le N$, form the sample variance based on the first n observations as follows

$$S_n^2 = \frac{1}{n}\sum_{k=1}^{n}(X_k - \overline{X}_n)^2, \tag{8.1}$$

Fig. 8.2 ACF and PACF for GSL water surface elevation time series

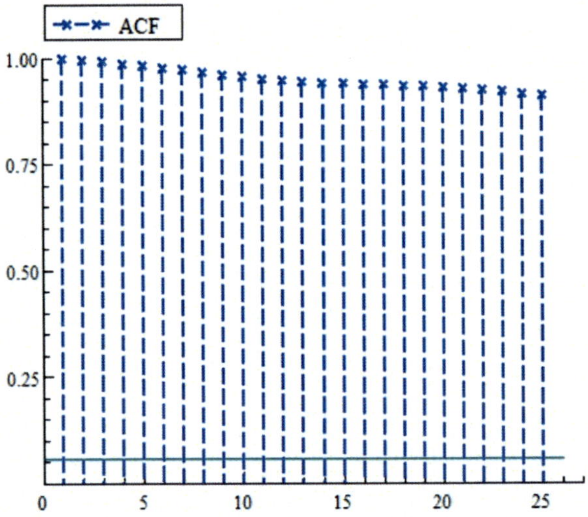

(a) ACF for GSL water surface elevation time series

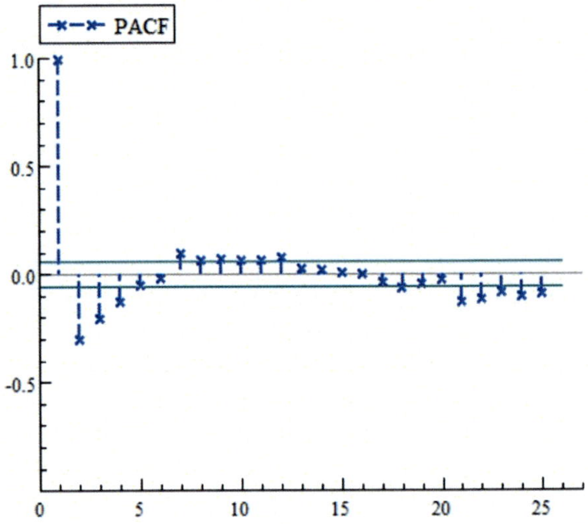

(b) PACF for GSL water surface elevation time series

where

$$\overline{X}_n = \frac{1}{n}\sum_{k=1}^{n} X_k, \tag{8.2}$$

and plot the sample variance estimate S_n^2 against n. If the population distribution $F(x)$ has a finite variance, S_n^2 should converge to a finite constant value as n increases. Otherwise, S_n^2 will diverge [103]. Figure 8.3 shows the sample variance

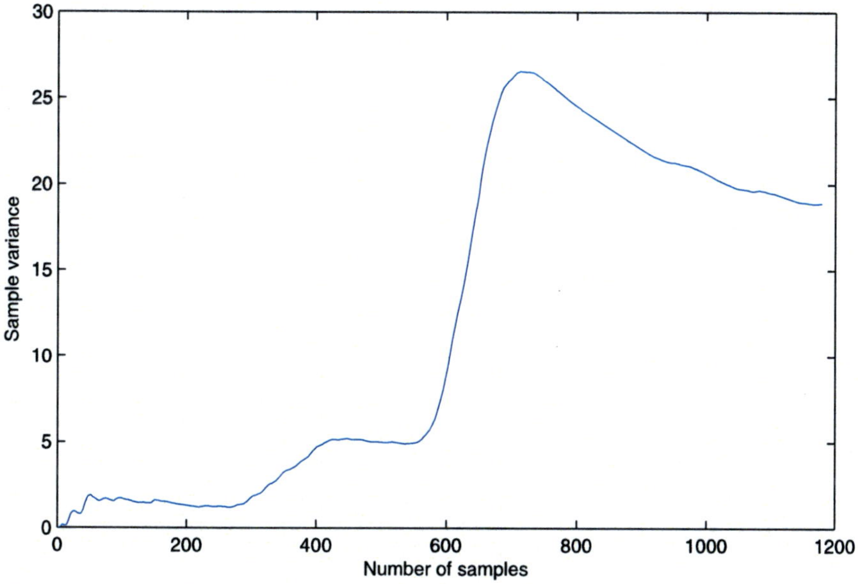

Fig. 8.3 Running variance of the GSL elevation time series

trend of the north levels of the GSL elevation measurements. From Fig. 8.3 we can see that the variance does not converge to a finite constant value.

On the other hand, unlike the Gaussian density which has exponential tails, the stable densities have algebraic tails. The $S\alpha S$ densities have heavier tails than that of the Gaussian processes. The smaller α is, the heavier the tails are [99]. This is a desirable feature for many applications in signal processing since many non-Gaussian phenomena are similar to the Gaussian phenomenon, but with heavier tails [186]. Figure 8.4 shows the estimated probability density and histogram of the GSL elevation time series. The green curve indicates the probability density of the normal distribution. The red curve indicates the estimated probability density of the GSL elevation time series. From the comparison between estimated probability density of GSL elevation and the normal distribution, we can see that the estimated probability density of GSL elevation has heavier tails than that of the normal distribution. The characteristic exponent α for the distribution of GSL elevation time series was calculated using McCullochs method [254], where $\alpha = 1.4584$. Therefore, the GSL elevation data is a non-Gaussian process with infinite variance. From the above analysis we can conclude that the FARIMA with stable innovations model is valid to characterize the variation of GSL elevation time series.

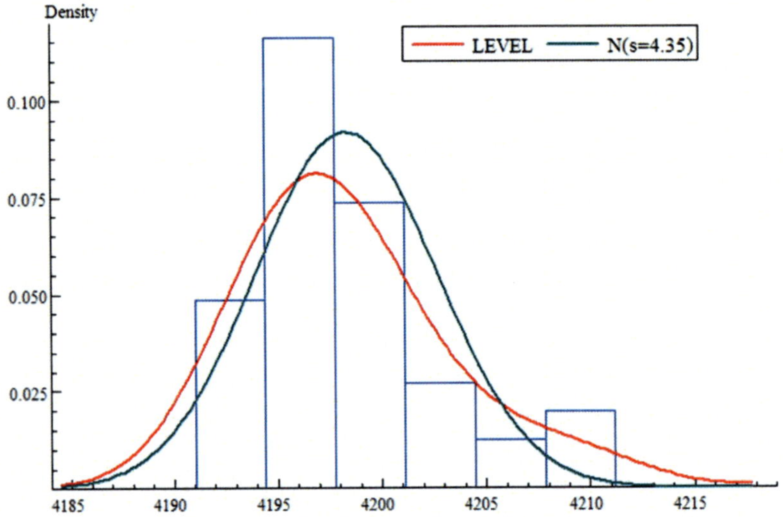

Fig. 8.4 Estimated density and histogram of GSL

8.3 FARIMA and FIGARCH Models of Great Salt Lake Elevation Time Series

Several models have been adopted to estimate and predict the GSL elevation time series, such as the ARMA, ARIMA, FARIMA, GARCH and FIGARCH models. Among these models, the FARIMA and FIGARCH have been considered as the better models for LRD processes, because they not only can characterize the short-range dependence property but also can capture the long-range dependence property of the processes [166]. In order to evaluate the FARIMA with stable innovations model, we estimate the effectiveness of FARIMA and FIGARCH models using G@RCH 4.2, which is a OxMetrics software package for estimating and forecasting GARCH model [161]. OxMetrics is a single product that includes and integrates all the important components for theoretical and empirical research in econometrics, time series analysis and forecasting [82]. The FARIMA and FIGARCH models were introduced in Chap. 3.

The north part level values of the GSL from Jan. 1960 to Feb. 2009 are used to evaluate FARIMA and FIGARCH models. The elevation data is split in two parts: the estimation part (Jan. 1960–Dec. 2007) and the forecasting part (Jan. 2008–Feb. 2009). The efficiency of each model will be estimated by the performance of the forecasting accuracy, which will be measured using mean squared error. FARIMA and FIGARCH forecast results are illustrated in Fig. 8.5. The blue line is the actual value of GSL elevations; the black line with dot marks is the forecasted data using FARIMA model; and the red one with star marks is the forecasted data using FIGARCH model. The mean squared errors of FARIMA and FIGARCH are 0.3864 and 0.2884, respectively.

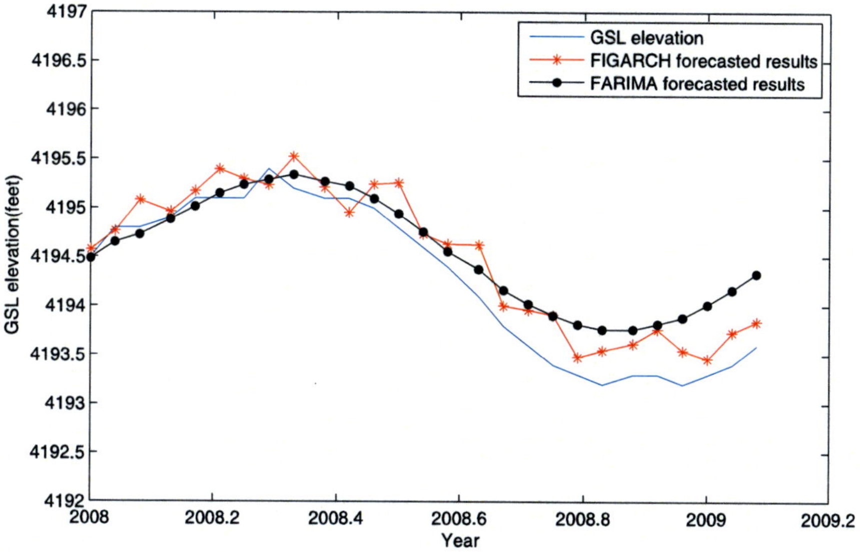

Fig. 8.5 Jan. 2008–Feb. 2009 GSL elevation forecast using FARIMA and FIGARCH models

8.4 FARIMA with Stable Innovations Model of Great Salt Lake Elevation Time Series

In this section, FARIMA with stable innovations model is adopted to characterize the GSL elevation time series which is shown to be LRD. The definition of the model was introduced in Chap. 3. For the parameter estimation of the model, a three-step parameter estimation method for FARIMA with stable innovations processes based on the given time series $X_n, n = 1, \ldots, N$ was proposed by Harmantzis [110]:

- First step: Estimate the characteristic exponent $\widehat{\alpha}$.
- Second step: Estimate the differencing parameter $\widehat{d}$.
- Third step: Estimate the coefficients of the ARMA part using normalized correlations and cumulates.

The key technique of the parameter estimation algorithm is to remove the LRD by passing the time series through the filter $(1 - z^{-1})^d$. The three-step estimation scheme described in this section is illustrated in Fig. 8.6.

The length of the available north part GSL elevation time series is 1179. The first 1152 (Jan. 1960–Dec. 2007) data samples are used as experimental time series for fitting FARIMA with stable innovations model. Then, the FARIMA with stable innovations model will be used for forecasting the GSL water surface elevation data from Jan. 2008 to Feb. 2009, which are the remaining 27 measurements. In order to analyze data easily, we turn first to draw the mean value of the GSL elevation data.

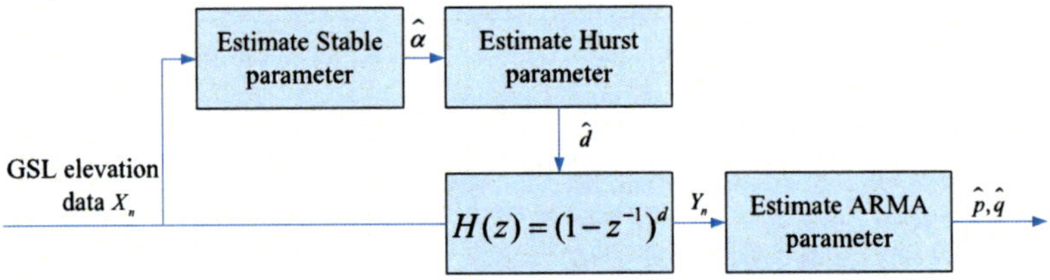

Fig. 8.6 Parameter estimation of FARIMA with stable innovations model

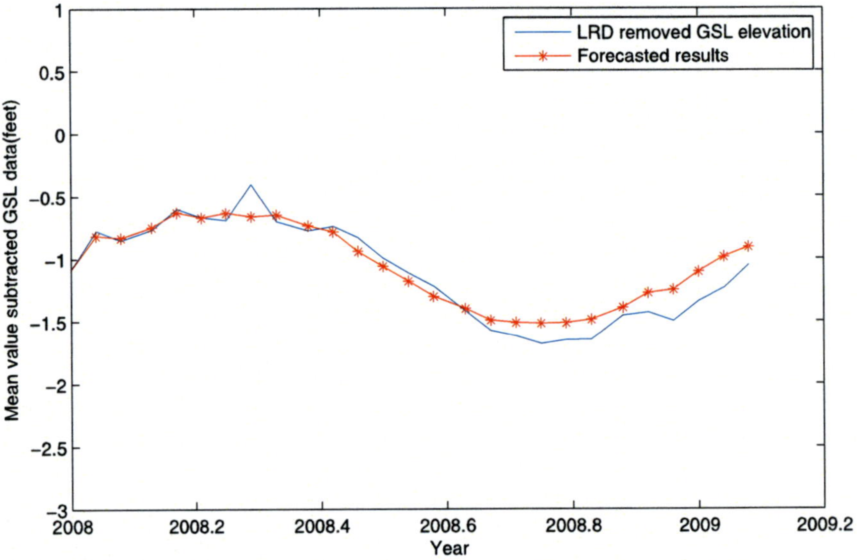

Fig. 8.7 Jan. 2008–Feb. 2009 LRD removed GSL time series forecast

That is

$$Y(n) = X(n) - \frac{1}{N}\sum_{i=1}^{N} x_i. \tag{8.3}$$

The α-stable parameter $\widehat{\alpha} = 1.4584$ is estimated using the method which was introduced in Chap. 2. Then, the Hurst parameter $\widehat{H} = 0.9970$ is calculated using the Aggregated Variance method. According to the estimated $\widehat{\alpha}$ and $\widehat{H}$, the differencing parameter $\widehat{d}$ of the fractional-order system can be computed. The LRD removed GSL data sequence can be obtained by passing the GSL elevation time series through the discrete fractional-order system. Here, the autoregressive moving $S\alpha S$ model is used to model the LRD removed GSL data sequence because any ARMA or MA processes can be approximately represented by a high order AR process.

After the above analytic processing, we can forecast the GSL elevation data easily. At first the LRD removed GSL data sequence are forecasted using AR model. The forecast is compared with the true LRD removed GSL data as shown in Fig. 8.7.

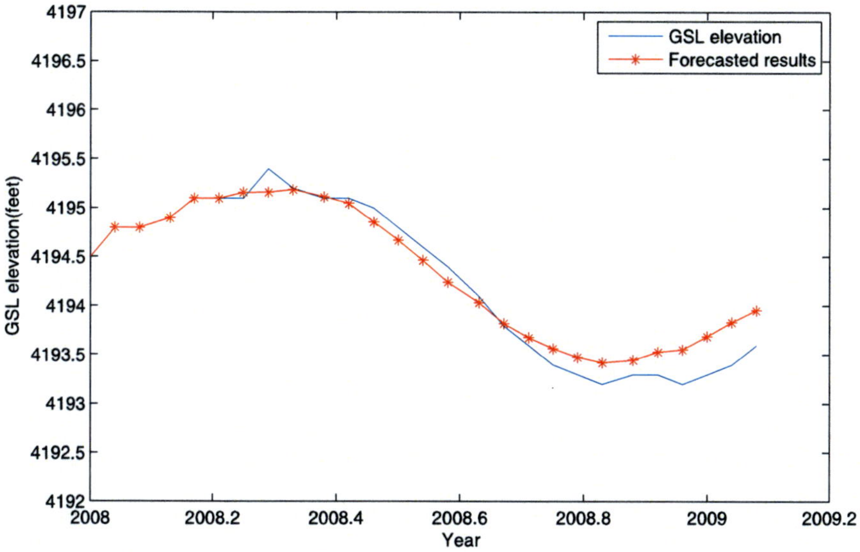

Fig. 8.8 Jan. 2008–Feb. 2009 GSL elevation forecast using FARIMA with stable innovations model

The blue line is the actual value of LRD removed GSL elevations from Jan. 2008–Feb. 2009, while the red one with star marks is the forecasted data.

Figure 8.8 shows the forecast result for the GSL elevation time series. The blue line is the actual value of GSL elevations from Jan. 2008 to Feb. 2009, while the red one with star marks is the forecasted data. As shown in Fig. 8.8, the forecasted time series fits the actual time much better than the FARIMA and FIGARCH model. The mean squared error of the GSL elevation forecasting is only 0.0044. FARIMA with stable innovations model successfully characterizes the variation of the GSL elevation.

From the above results we can see that the FARIMA with stable innovations model can characterize the GSL elevation time series more accurately than FARIMA and FIGARCH models. The GSL levels forecasted by FARIMA with stable innovations model successfully predict the rise and fall of GSL elevation with higher precision. In conclusion, the FARIMA with stable innovations model is capable of characterizing and forecasting the GSL elevation time series very well.

8.5 Chapter Summary

This chapter presented an application example of FOSP techniques in hydrology. The FOSP techniques presented in Chap. 5 were used to study the north part of GSL water-surface elevation time series, which possess long-range dependence and infinite variance properties. The LRD property and infinite variance property of GSL elevation time series explain the insufficiency of conventional methods on modeling and prediction and suggest the necessity of implementing FOSP techniques. In this application example we also show that FARIMA with stable innovations model can successfully characterize the GSL historical water levels and predict its future rise and fall with much better accuracy. Therefore, we can observe that FOSP techniques provide more powerful tools for forecasting the GSL elevation time series with LRD and infinite variance properties.

Chapter 9
Analysis of Biocorrosion Electrochemical Noise Using Fractional Order Signal Processing Techniques

9.1 Introduction

There are several electrochemical techniques used in determining of corrosion behaviors, such as linear polarization (LP), open circuit potential, electrochemical impedance spectroscopy (EIS) and the electrochemical noise (ECN) which monitors the corrosion process on two similar electrodes coupled through a zero resistance ammetry (ZRA) [80, 195]. ECN is considered a nondestructive technique and is extensively used for assessment of biocorrosion processes [316]. The use of ZRA in ECN also provides useful measurements:

- of Ece, the potential of counter electrode (CE) versus the reference electrode (RE);
- of Icoupling, the current between the working electrode (WE) and CE; and
- of Ewe, the corrosion potential, that is the potential of working electrode (WE) versus the reference electrode (RE).

Electrochemical corrosion processes can be analyzed by several conventional time domain or frequency domain analysis techniques [234]. In time domain, statistical methods are often applied to quantify ECN data, in terms of statistical parameters such as moments, skewness and kurtosis. In particular, the noise resistance, the standard deviation of potential divided by the standard deviation of current, has been used considerably to evaluate the biocorrosion behavior [52]. For frequency domain techniques, Fourier transforms are often used to transform ECN data from time domain into the frequency domain. Spectral noise resistance or spectral noise impedance, the square root of the potential noise PSD (power spectral density) to that of current noise PSD, has been demonstrated to be a valid method to assess biocorrosion behaviors [324]. In this chapter, we explore the application potentials of some FOSP techniques in the analysis of ECN signal.

In this study, four FOSP techniques: FrFT, FPSD (fractional power spectral density), self-similarity, and local self-similarity are used to analyze ECN data generated by TiO_2 nano-particle coated Ti-6Al-4V and the bare Ti-6Al-4V corrosion process in three different simulated biofluid solutions.

H. Sheng et al., *Fractional Processes and Fractional-Order Signal Processing*, Signals and Communication Technology,
DOI 10.1007/978-1-4471-2233-3_9,

Table 9.1 Compositions of the three simulated biofluid solutions

No.	Composition	Solution A	Solution C	Solution B
1	NaCl	10.00 g/L	8.75 g/L	8.00 g/L
2	$NaHCO_3$	–	0.35 g/L	0.35 g/L
3	NaH_2PO_4	–	0.06 g/L	0.10 g/L
4	$Na_2HPO_4 \cdot 2H_2O$	–	0.06 g/L	0.06 g/L
5	KCl	–	–	0.40 g/L
6	$CaCl_2$	–	–	0.14 g/L
7	Glucose	–	–	1.00 g/L
8	$MgCl_2 \cdot 6H_2O$	–	–	0.10 g/L
9	$MgSO_4 \cdot 7H_2O$	–	–	0.06 g/L

9.2 Experimental Approach and Data Acquisition

The details of the experimental approaches including the preparation of the Ti-6Al-4V bioimplants, preparation of the simulated biofluids, and the preparation of the TiO2 nano-particle are referred to [61, 328, 329]. The compositions of the three simulated biofluid solutions (named as solutions A, B, C) are listed in Table 9.1.

ECN data typically consist of three sets of noise measurements, the corrosion potential of the working electrode (Ewe), the corrosion potential of the counter electrode (Ece), and the coupling current (Icoupling) between WE and CE. The ECN data between two identical Ti-6Al-4V electrodes (WE) immersed in the above simulated solutions were recorded by using a VMP (versatile multichannel potentiostat) multichannel potentiostat (PAR, TN) with a data acquisition software ECLab-Win version 9.01. The time course fluctuation of the potential of WE versus reference electrode (RE) as well as the coupling current between WE and CE were measured simultaneously by ECLab software. Figure 9.1 presents the examples of the time profiles of the electrochemical noise responses obtained from a TiO_2 nano-particle coated electrode Ti-6Al-4V exposed to the simulated Solution C for 30 minutes. In this study, the potential noise Ewe and the current noise Icoupling are analyzed using conventional and FOSP techniques.

9.3 Conventional Analysis Techniques

9.3.1 Conventional Time Domain Analysis of ECN Signals

In order to draw a comparison with FOSP analysis method, the conventional time domain parameters of two bioimplants in three simulated biofluid solutions are provided in this section. In time domain, statistical analyses are mainly used to characterize the electrochemical response of systems undergoing corrosion behavior. Some typical statistical parameters of ECN data [234] are illustrated in Table 9.2.

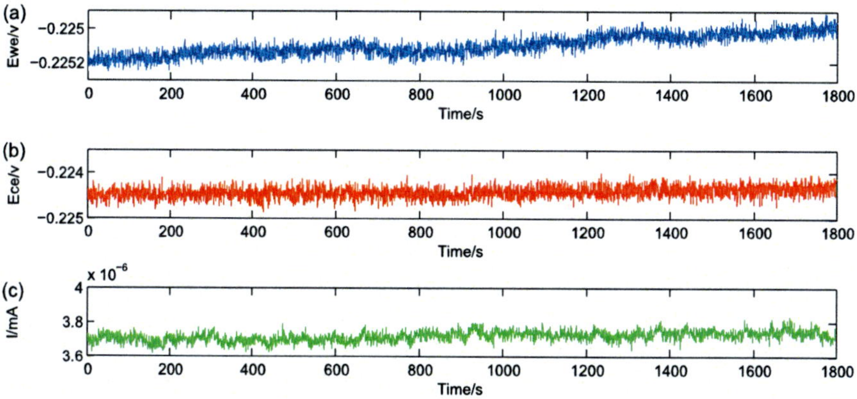

Fig. 9.1 An example of ECN measurement in 30 minutes. (**a**) The potential noise of the WE. (**b**) The potential noise of the CE. (**c**) The corresponding coupling current between WE and CE. Solution used: Cigada solution; electrode materials: TiO_2 nano-particle coated on WE (Ti-6Al-4V)

Table 9.2 The definitions of conventional time-domain parameters

Parameter	Function
Mean	$\bar{x} = \frac{1}{N}\sum_{k=1}^{N} x_i$
Variance	$S = \frac{1}{N}\sum_{k=1}^{N}(x_k - \bar{x})^2$
Third moment	$m_3 = \frac{1}{N}\sum_{k=1}^{N}(x_k - \bar{x})^3$
Fourth moment	$m_4 = \frac{1}{N}\sum_{k=1}^{N}(x_k - \bar{x})^4$
Skewness	$g_1 = \frac{m_3}{S^{2/3}}$
Kurtosis	$g_2 = \frac{m_4}{S^2}$
Standard deviation	$\sigma = S^{1/2}$
Noise resistance	$R_n = \frac{\sigma V}{\sigma I}$
Coefficient of variance	$C_{cf} = \frac{\sigma}{\bar{x}}$
Root mean square	$rms = \sqrt{\frac{1}{N}\sum_{k=1}^{N} x_k^2}$

- The mean value of current or potential values may only be applied to provide a rough estimate of biocorrosion rate;
- the variance of the ECN signal relates to the power in the ECN data;
- the third moment is a measure of the asymmetry of the ECN data around the mean value;
- the fourth moment is used to calculate the kurtosis reflecting the distribution of the ECN signals, and for data which exhibits spontaneous changes in amplitude distribution;
- the skewness of the ECN signals can be used to identify particular biocorrosion mechanisms;
- the standard deviation relates to the broadband alternating current (AC) component of the ECN signal;

Table 9.3 Time-domain parameters of bare Ti-6Al-4V and TiO_2 nano-particle coated Ti-6Al-4V electrode in simulated biofluid solutions A, B and C for 24 hours

	Solution A		Solution B		Solution C	
	Bare	TiO_2	Bare	TiO_2	Bare	TiO_2
Mean	−0.22771	−0.15839	−0.19834	−0.25370	−0.23120	−0.23423
Variance	0.000693	0.001625	0.001729	0.000450	0.001783	0.000392
Third moment	0.000016	0.000045	0.000099	−0.00001	0.000057	−0.00001
Fourth moment	0.000001	0.000006	0.000012	0.000000	0.000008	0.000001
Skewness	−0.89910	−0.67417	−1.38225	−0.76369	−0.75580	−1.52527
Kurtosis	2.872957	2.178911	4.093310	2.459635	2.561102	4.430374
Standard deviation	0.026328	0.040642	0.041580	0.021223	0.042227	0.019799
Coefficient of variance	−0.11562	−0.25658	−0.20963	−0.08365	−0.18263	−0.08452
Root mean square	0.052545	0.026741	0.041068	0.064815	0.055238	0.055256
Noise resistance R_n	83454.17	13637.02	8696.22	3736.31	11426.88	2967.83

- the coefficient of variance is a measure of the distribution of the ECN data around the mean or root mean square value respectively;
- the noise resistance can be used to determine the biocorrosion resistance [85, 118].

Typically, the ECN measurement was collected in a short term, from several minutes to a couple of hours. However, by doing so, the long-range biocorrosion process of bioimplant cannot be obtained. In order to avoid loss of important information, in this study, long-term (24 hours) ECN data for bare Ti-6Al-4V electrode and TiO_2 nano-particle coated Ti-6Al-4V electrodes in three different simulated solutions (A, B and C) for 24 hours were collected, respectively. Table 9.3 lists the time domain statistic parameters of bare Ti-6Al-4V electrode and TiO_2 nano-particle coated Ti-6Al-4V electrode in these three solutions, respectively. Among these statistic parameters, the noise resistance R_n, which has been found to be inversely related to the localized corrosion rate, has been frequently used to indicate the biocorrosion behavior. From Table 9.3 we can see that the corrosion noise resistant value for bare Ti-6Al-4V electrode in these three solution follows: Solution A > Solution C > Solution B, and for TiO_2 nano-particle coated Ti-6Al-4V electrode follows: Solution A > Solution B > Solution C. All the time domain parameters analyzed above can also be studied from the profiles of corrosion potential Ewe data in Fig. 9.2, which represent the corrosion potential of bare Ti-6Al-4V and TiO_2 nano-particle coated Ti-6Al-4V electrodes in solution A, B and C for 24 hours, respectively.

9.3.2 Conventional Frequency Domain Analysis

Frequency domain is a term used to describe the analysis of signals with respect to frequency, rather than time. Fourier transform is often used to transform the ECN

Fig. 9.2 The corrosion potentials (Ewe) of bare Ti-6Al-4V and TiO_2 nano-particle coated Ti-6Al-4V electrode in solutions A, B and C for 24 hours. (**a**) Bare Ti-6Al-4V electrode, (**b**) TiO_2 nano-particle coated Ti-6Al-4V electrode

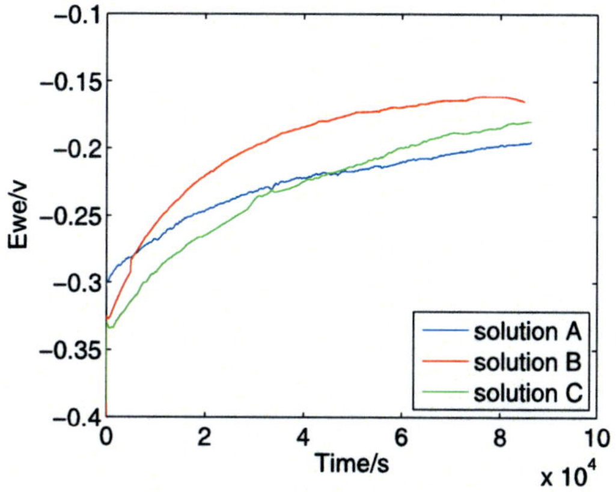

(a) Bare Ti-6Al-4V electrode

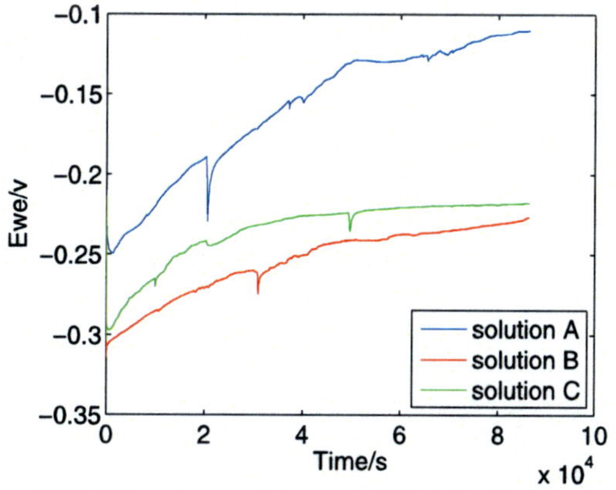

(b) TiO2 nano-particle coated Ti-6Al-4V electrode

noise data from time domain into the frequency domain. For ECN signal analysis, we deal with signals that are discretely sampled at constant intervals, and of finite duration or period. So, the discrete Fourier transform (DFT), which is normally computed using the so-called fast Fourier transform (FFT), is appropriate in the analysis of ECN signal. The DFT of N uniformly sampled ECN data points x_n ($n = 0, \ldots, N-1$) is defined by

$$X_k = \sum_{n=0}^{N-1} x_n e^{\frac{-2\pi ikn}{N}}, \quad k = 0, \ldots, N-1, \tag{9.1}$$

where i is the imaginary unit. Another often used frequency domain technique is PSD, which describes how the power of an ECN signal is distributed with frequency

Fig. 9.3 Magnitudes of the Fourier transform for ECN data of bare Ti-6Al-4V and TiO_2 nano-particle coated Ti-6Al-4V electrode in solutions A, B and C for 24 hours. (**a**) Bare Ti-6Al-4V electrode, (**b**) TiO_2 nano-particle coated Ti-6Al-4V electrode

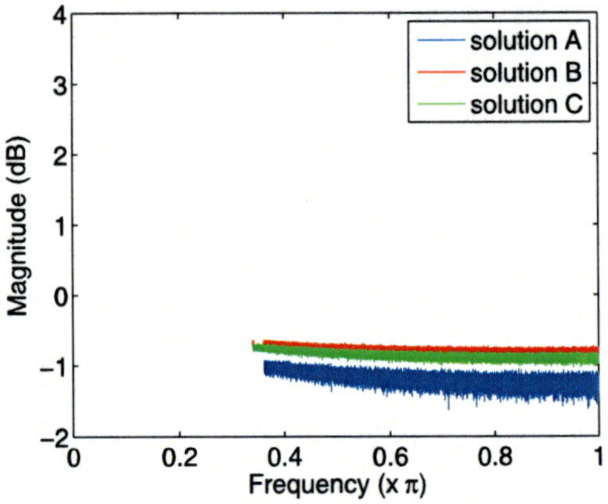

(a) Bare Ti-6Al-4V electrode

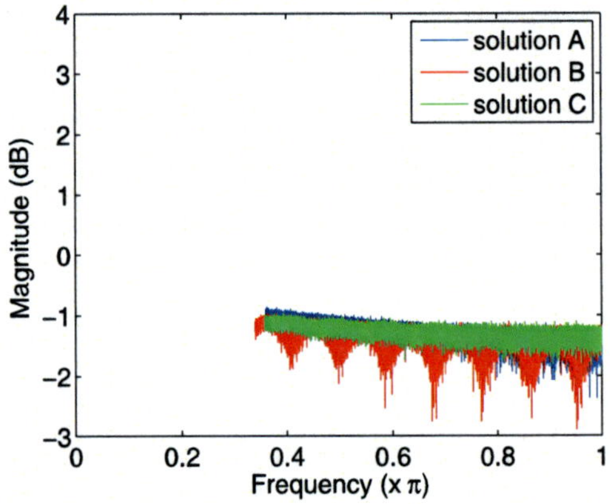

(b) TiO_2 nano-particle coated Ti-6Al-4V electrode

[212]. The PSD of a discrete signal f_n is determined as

$$\Phi(\omega) = \left| \frac{1}{\sqrt{2\pi}} \sum_{n=-\infty}^{\infty} f_n e^{-i\omega n} \right|^2 = \frac{F(\omega)F^*(\omega)}{2\pi}, \tag{9.2}$$

where ω is the angular frequency, $F(\omega)$ is the discrete-time Fourier transform of f_n, and $F^*(\omega)$ is the complex conjugate of $F(\omega)$.

Figure 9.3 shows the Fourier transform analysis of the ECN data obtained by the bare Ti-6Al-4V electrode and the TiO_2 nano-particle coated Ti-6Al-4V electrode in three different simulated solutions for 24 hours. It has been found in our previous study [328] that different measurement durations lead to different spectral characteristics (or noise patterns). And the magnitudes of the Fourier transform obtained in three solutions become more and more indistinguishable with the increase of measurement duration. As shown in Fig. 9.3, it is hard to distinguish the magnitude of the Fourier transform for the bare Ti-6Al-4V electrode and the TiO_2 nano-particle

Fig. 9.4 Spectrum noise impedances of bare Ti-6Al-4V and TiO_2 nano-particle coated Ti-6Al-4V electrode in solutions A, B and C for 24 hours. **(a)** Bare Ti-6Al-4V electrode, **(b)** TiO_2 nano-particle coated Ti-6Al-4V electrode

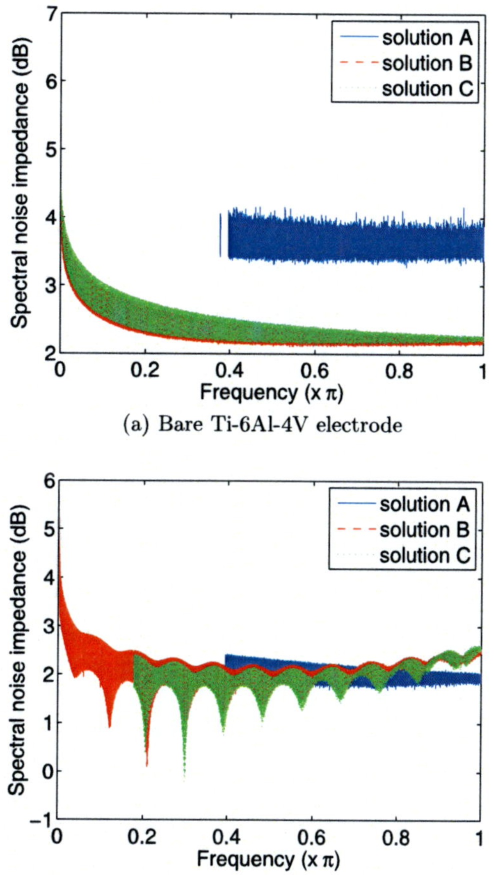

(a) Bare Ti-6Al-4V electrode

(b) TiO_2 nano-particle coated Ti-6Al-4V electrode

coated Ti-6Al-4V electrode in different simulated biofluid solutions when longer measurements are made.

The spectral noise impedance is also a very useful evaluation of biocorrosion in frequency domain. The spectral noise impedance is defined as

$$R_{sn} = \sqrt{\frac{S_v(f)}{S_i(f)}}, \tag{9.3}$$

where $S_v(f)$ is the PSD of the potential noise and $S_i(f)$ is the PSD of the current noise [23].

Figure 9.4 shows the spectral noise impedance comparison of bare Ti-6Al-4V Ewe electrode and the TiO_2 nano-particle coated Ti-6Al-4V electrode in solutions A, B and C for 24 hours, respectively. The spectral noise impedance of bare Ti-6Al-4V electrode in solution A is obviously higher than that in solution B and C. However, the spectral noise impedances of bare Ti-6Al-4V electrode are almost the same in solution B and C. For TiO_2 nano-particle coated Ti-6Al-4V electrode, we also cannot distinguish the spectral noise impedance in the three solutions. Therefore, in the next section we investigate the FOSP methods in order to obtain clear analysis results.

9.4 Fractional-Orders Signal Processing Techniques

In this section, we provide some application examples of FOSP in ECN signal analysis. Four FOSP methods: FrFT, fractional power spectrum, self-similar, and local self-similar analysis techniques are used to analyze ECN signal generated by bare and TiO_2 nano-particle coated Ti-6Al-4V electrodes in three simulated biofluid solutions.

9.4.1 Fractional Fourier Transform Technique

Richer in theory and more flexible in applications, FrFT is well suitable for analyzing time-varying signals for which the conventional Fourier transform may fail to work as desired.

The discrete fractional Fourier transform (DFrFT) has been studied by many researchers. A definition of DFrFT (1.75) has been introduced in Chap. 1. A fast DFT-based DFrFT algorithm [16] is used in this research to estimate the DFrFT of signals from the bare and TiO_2 nano-particle coated Ti-6Al-4V electrodes in solutions A, B and C. Figure 9.5 presents the FrFT analysis results for the same ECN data as in Fig. 9.3, and parameter $a = 2\alpha/\pi$ ranging from 0.1 to 0.7 with step 0.2. It is evident that the magnitude of the FrFT of these ECN signals are much clearer than that of the Fourier transform. The magnitudes of the FrFT for the corrosion potentials of bare Ti-6Al-4V electrode in three simulated biofluid solutions are presented in Fig. 9.5(a). The magnitudes in these three solutions follow: solution A > solution C > solution B. Figure 9.5(b) presents the magnitudes of the FrFT for the corrosion potentials of TiO_2 nano-particle coated Ti-6Al-4V electrode in three solutions: solution B > solution C > solution A, which are different from the results of bare Ti-6Al-4V electrode.

Figure 9.6 shows the FrFT based spectrum noise impedance for the bare Ti-6Al-4V electrode and the TiO_2 nano-particle coated Ti-6Al-4V electrode in solutions A, B and C. Comparing with the traditional FFT based spectrum impedance (Fig. 9.4), we can get clear spectrum noise impedance and better results by using the FrFT techniques. The FrFT based spectrum noise impedance of the corrosion potentials of bare Ti-6Al-4V electrode in three simulated biofluid solutions follow: solution A > solution C > solution B. The FrFT based spectrum noise impedance of the corrosion potentials of TiO_2 nano-particle coated Ti-6Al-4V electrode in three solutions follow: solution A > solution B > solution C.

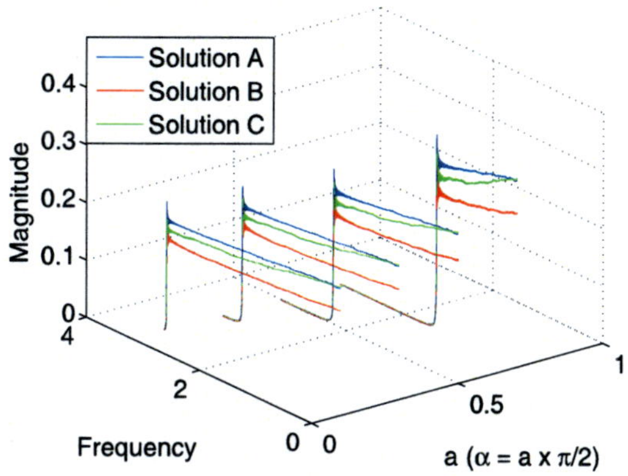

(a) Bare Ti-6Al-4V electrode

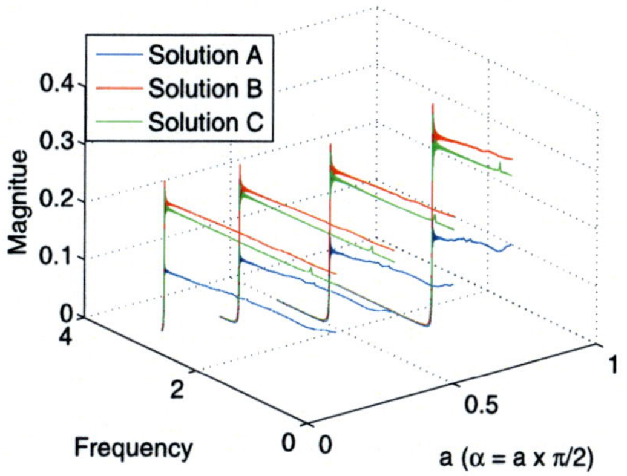

(b) TiO_2 nano-particle coated Ti-6Al-4V electrode

Fig. 9.5 Magnitudes of the FrFT for fGn of bare Ti-6Al-4V and TiO_2 nano-particle coated Ti-6Al-4V electrode in solutions A, B and C for 24 hours. (**a**) Bare Ti-6Al-4V electrode, (**b**) TiO_2 nano-particle coated Ti-6Al-4V electrode

9.4.2 Fractional Power Spectrum Density

Based on the conventional PSD and the FrFT, Tao and Zhang investigated the properties of the FPSD and the fractional correlation function for the random case in detail [295]. FPSD can be regarded as the generalization of the conventional PSD. FPSD is more flexible and suitable for processing non-stationary signals due to the flexibility of FrFT. FPSD is useful in detection and parameter estimation of chirp signals, and system identification in the fractional Fourier domain. Fractional power spectrum is defined as (2.44), and the αth fractional cross-power spectrum of $\varepsilon(t)$ and $\eta(t)$ is determined as

$$P^{\alpha}_{\varepsilon\eta}(\mu) = \lim_{T\to\infty} \frac{E|\xi_{\alpha,T}(\mu)\cdot\zeta^{*}_{\alpha,T}(\mu)|}{2T}, \tag{9.4}$$

where $\zeta^{*}_{\alpha,T}(\mu)$ denotes the complex conjugate of the αth FrFT of $\eta_T(t)$. $\eta_T(t)$ is the truncation function in $[-T, T]$ of the sample function of the random process $\eta(t)$.

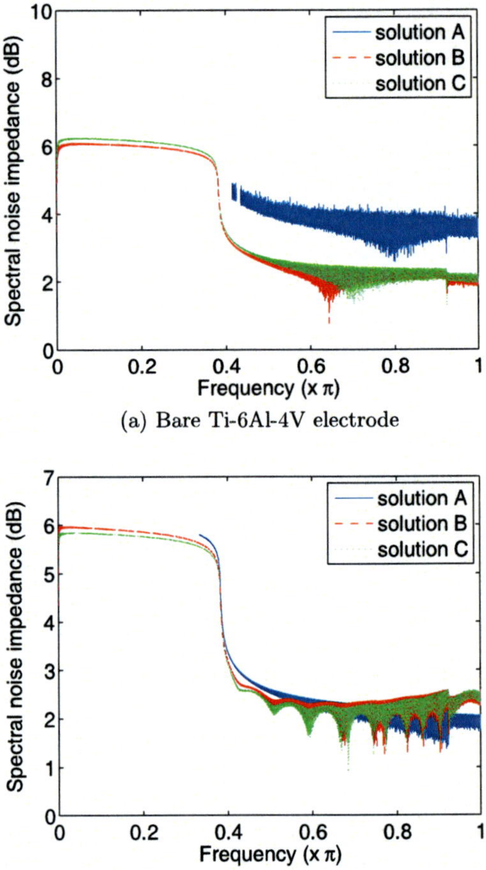

(a) Bare Ti-6Al-4V electrode

(b) TiO_2 nano-particle coated Ti-6Al-4V electrode

Fig. 9.6 FrFT based spectrums noise impedances of bare Ti-6Al-4V and TiO_2 nano-particle coated Ti-6Al-4V electrode in solutions A, B and C for 24 hours. (**a**) Bare Ti-6Al-4V electrode, (**b**) TiO_2 nano-particle coated Ti-6Al-4V electrode

The discrete fractional power spectrum is defined as

$$P^{\alpha}_{\xi\xi}(\omega) = A_{-\alpha} \cdot \tilde{F}^{\alpha}[R^{\alpha,1}_{\xi\xi}](\omega) e^{-j(\omega^2/2T^2)\cot\alpha}, \tag{9.5}$$

where $\tilde{F}^{\alpha}[\cdot]$ denotes the DFrFT, and

$$R^{\alpha,1}_{\xi\xi}[m] = \lim_{M\to\infty} \frac{1}{2M+1} \sum_{n_2=-M}^{M} R_{\xi\xi}[n_2+m, n_2] e^{jn_2 m T^2 \cot\alpha}, \tag{9.6}$$

$$R_{\xi\xi}[n_2+m, n_2] = E\{\xi[n_1]\xi^*[n_2]\}. \tag{9.7}$$

Figure 9.7 presents the discrete FPSD ($\alpha = \pi/4$) of the bare and TiO_2 nano-particle coated Ti-6Al-4V electrodes in three solutions. It can be seen from Fig. 9.7 that the analysis results of FPSD are very distinct. For bare Ti-6Al-4V electrode, the FPSD is the highest in solution A, followed by solution C and Solution B. However,

Fig. 9.7 Fractional power spectra comparison of bare Ti-6Al-4V and TiO_2 nano-particle coated Ti-6Al-4V electrode in solutions A, B and C for 24 hours. (**a**) Bare Ti-6Al-4V electrode, (**b**) TiO_2 nano-particle coated Ti-6Al-4V electrode

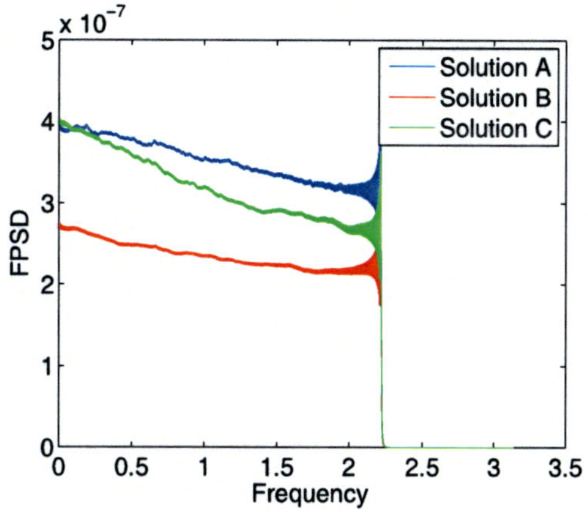

(a) Bare Ti-6Al-4V electrode

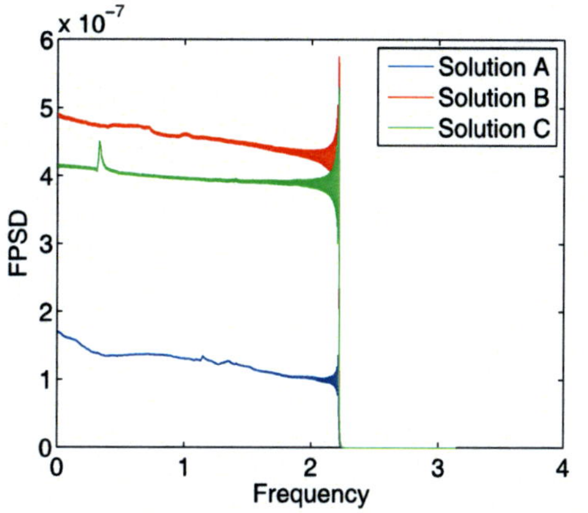

(b) TiO_2 nano-particle coated Ti-6Al-4V electrode

for TiO_2 nano-particle coated Ti-6Al-4V electrode, the FPSD is opposite: solution B > solution C > solution A. In solution A, the FPSD of bare Ti-6Al-4V electrode is much higher than that of nano-particle coated one. But in solution B, the FPSD of bare Ti-6Al-4V electrode is much lower than that of nano-particle coated one. It is obviously that the analysis results of FPSD have the same clarity and flexibility as FrFT.

9.4.3 Self-similarity Analysis

Self-similar processes are increasingly concerned in areas ranging from financial science to computer networking. Strong coupling between values at different times can be found in self-similar processes [152]. This indicates that the decay of the autocorrelation function is hyperbolic, decaying slower than exponential decay, and

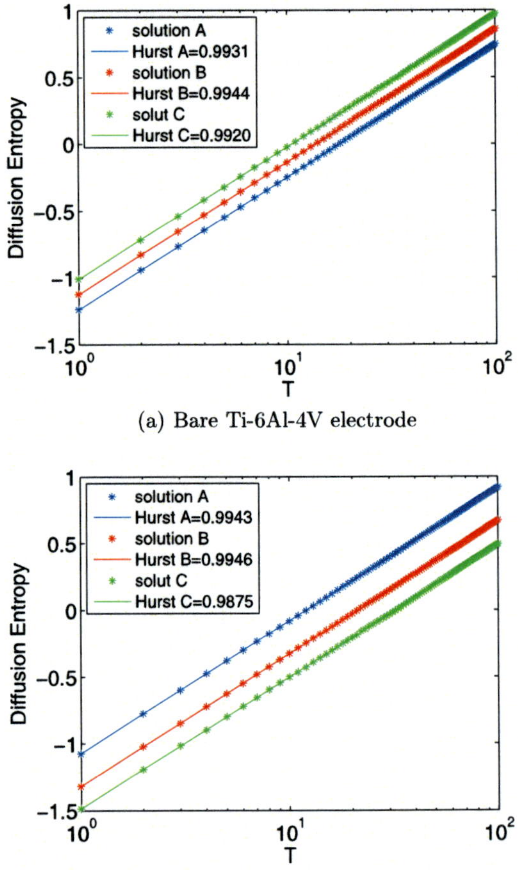

(a) Bare Ti-6Al-4V electrode

(b) TiO_2 nano-particle coated Ti-6Al-4V electrode

Fig. 9.8 Diffusion Entropy algorithm (DEA) comparison of bare Ti-6Al-4V and TiO_2 nano-particle coated Ti-6Al-4V electrode in solutions A, B and C for 24 hours. (**a**) Bare Ti-6Al-4V electrode, (**b**) TiO_2 nano-particle coated Ti-6Al-4V electrode

that the area under the function curve is infinite. Hurst parameter ($0 < H < 1$) is a simple direct parameter which can characterize the level of self-similarity. Some research efforts have proved that the Hurst parameter is a qualitative guide for the biocorrosion processes [101, 183, 246]. Moon and Skerry proposed three protection levels to corrosion conditions [208]. Up to now, however, there is a shortage of certain accepted criterion for evaluating corrosion rate using the Hurst exponent. In this section, we aim at exploring the relationship between self-similarity and biocorrosion behavior by analyzing the Hurst parameters of ECN signals of two electrodes in three simulated biofluid solutions.

Many Hurst parameter estimators have been proposed to analyze the LRD time series. Among them, the DEA method has good accuracy and robustness [105]. Figure 9.8 shows the DEA analysis results of the bare Ti-6Al-4V electrode and the TiO_2 nano-particle coated Ti-6Al-4V electrode in three solutions for 24 hours. For bare Ti-6Al-4V electrode, the Diffusion Entropy values in three simulated solutions follow: solution C > solution B > solution A. On the contrary, for TiO_2 nano-particle coated Ti-6Al-4V electrode, the Diffusion Entropy values in three simulated solutions follow: solution A > solution B > solution C, which indicates the difference of corrosion behaviors of bare and TiO_2 nano-particle coated Ti-6Al-4V electrode in three solutions. It also can be seen that the Hurst parameters of two electrodes in three solutions are almost the same. That clearly shows that the constant Hurst parameter cannot be used to assess the corrosion behaviors of bare and TiO_2 nano-particle coated Ti-6Al-4V electrodes in three solutions effectively, so the local self-similar analysis is applied to estimate the biocorrosion in the next section.

9.4.4 Local Self-similarity Analysis

Most of the previous study considered the Hurst parameter of ECN data as a constant value. However, the biocorrosion process is a complex dynamic process. The biocorrosion rates could be different for a bioimplant in the same solution (environment) in different time segments. Local self-similar technique can dynamically process ECN signals. So, the local Hurst exponents may provide some valuable local information of biocorrosion behavior, although there is a lack of definitive evaluation criterion. In this section the local self-similarities of bare Ti-6Al-4V electrode and the TiO_2 nano-particle coated Ti-6Al-4V electrode in three simulated biofluid solutions are analyzed. The time series are divided into 24 segments in order to accurately estimate the local Hurst parameter of ECN data in every hour. The local Hurst parameters are estimated using LASS (local analysis of self-similarity) which is a tool for the local analysis of self-similarity [287]. By doing so we can estimate the local Hurst parameters at different time segments. Figure 9.9 shows the local Hurst parameter variation for both electrodes in simulated biofluid solutions A, B and C, respectively, for 24 hours. It can be seen that, in solutions B and C, most of the local Hurst values of TiO_2 nano-particle coated Ti-6Al-4V electrode are lower than those of the bare Ti-6Al-4V electrode. On the contrary, many local Hurst values of TiO_2 nano-particle coated Ti-6Al-4V electrode is higher than that of the bare one in solution A. In contrary, the biocorrosion rate of TiO_2 nano-particle coated Ti-6Al-4V electrode is higher in Solution B. Moreover, the local Hurst value of TiO_2 nano-particle coated Ti-6Al-4V electrode fluctuates drastically in solution B, so the biocorrosion rate of TiO_2 nano-particle coated Ti-6Al-4V electrode changes obviously during the entire corrosion process in Solution B.

In brief, local self-similar analysis method provides some detailed local information of biocorrosion behaviors in these experiments. Nevertheless, further studies are required to uncover the essence of the relationship between the biocorrosion rate and the local Hurst exponent.

9.5 Chapter Summary

This chapter provided an application example of FOSP techniques in biomedical signals. The ECN data of the bare Ti-6Al-4V bioimplant and the TiO_2 nano-particle

Fig. 9.9 Local Hurst values comparison of bare Ti-6Al-4V and TiO_2 nano-particle coated Ti-6Al-4V electrode in solutions A, B and C for 24 hours. (**a**) Bare Ti-6Al-4V electrode, (**b**) TiO_2 nano-particle coated Ti-6Al-4V electrode

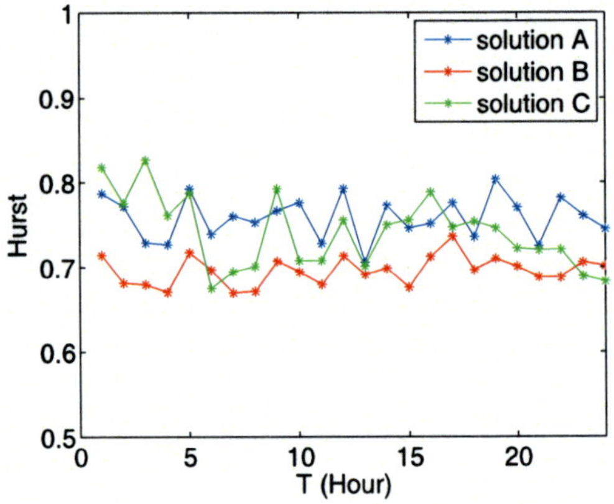

(a) Bare Ti-6Al-4V electrode

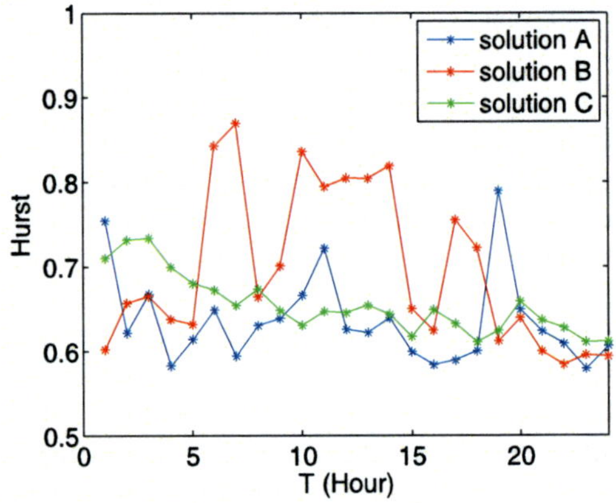

(b) TiO_2 nano-particle coated Ti-6Al-4V electrode

coated Ti-6Al-4V bioimplant in three simulated biofluid solutions were analyzed. To draw a comparison between conventional analysis methods and FOSP techniques, we first characterized the biocorrosion behavior using time domain statistic parameters, magnitudes of the Fourier transform, and spectral noise impedance. Although these techniques provided some valuable biocorrosion information, it is difficult to assess biocorrosion rate in long term ECN measurement. Compared with the conventional time or frequency domain based analysis techniques, FrFT, fractional power spectrum, and the local Hurst parameter analysis of ECN data all provided improved results as observed from the signal processing figures.

Chapter 10
Optimal Fractional-Order Damping Strategies

10.1 Introduction

A damper is a valuable component for reducing the amplitude of dynamic instabilities or resonances in system stabilization [264]. In physics and engineering, the mathematical model of the conventional damping can be represented by

$$f(t) = -cv(t) = -c\frac{\mathrm{d}x(t)}{\mathrm{d}t}, \tag{10.1}$$

where $f(t)$ is the time varying force, c is the viscous damping coefficient, $v(t)$ is the velocity, and $x(t)$ is the displacement [145]. Taking advantage of fractional calculus, fractional-order damping with a viscoelastic damping element provides a better model to describe a damping system [142]. Fractional-order damping is modeled as a force proportional to the fractional-order derivative of the displacement [175]

$$f(t) = c\, {}_0D_t^{\alpha} x(t), \tag{10.2}$$

where ${}_0D_t^{\alpha} x(t)$ is the fractional-order derivative defined by (1.64) [237]. Motivated by potential benefits of fractional damping, many efforts have been made to investigate the modeling of systems with damping materials using fractional-order differential operators [74, 77, 229, 248, 249, 272]. However, up to now, little attention has been paid to time-delayed fractional-order damping, and distributed-order fractional damping. In this chapter, we investigate the potential benefits of a non-delayed fractional-order damping system, a time-delayed fractional-order damping system, and a distributed-order fractional damping system.

In order to design an optimal transfer function form, the performance of a control system should be measured, and the parameters of the system should be adjusted to derive the desirable response. The performance of a system is usually specified by several time response indices for a step input, such as rise time, peak time, overshoot, and so on [83]. Furthermore, the performance index, a scalar, is adequately used to represent the important system specifications instead of a set of indices. The transfer function of a system is considered as an optimal form when the system

H. Sheng et al., *Fractional Processes and Fractional-Order Signal Processing*, Signals and Communication Technology,
DOI 10.1007/978-1-4471-2233-3_10,

parameters are adjusted so that the performance index reaches an extremum value [83]. The well-known integral performance indices are the integral of absolute error (IAE), the integral of squared error (ISE), the integral of time multiplied absolute error (ITAE), the integral of time multiplied squared error (ITSE), and the integral of squared of time multiplied error (ISTE) [83, 299]. Hartley and Lorenzo studied the single term damper that minimizes the time domain ISE and ITSE, and found that the optimal fractional-order damping is more optimal than the optimal integer-order damping [112]. In this chapter, we investigate three types of optimal fractional-order damping systems using frequency-domain and time-domain optimization methods. In frequency domain, the time-delayed fractional-order and distributed-order fractional damping systems are optimized using ISE criterion. In time domain, three types of fractional-order damping systems are optimized using ISE, ITSE, IAE and ITAE criteria. Unlike integer-order differential equations, fractional-order always leads to difficulties in solving differential equations. So, the numerical inverse Laplace transform algorithm is used to calculate the numerical results of the performance indices, and to find the optimum coefficients. The comparisons of an optimal integer-order damping system and three optimal fractional-order damping systems indicate that three types of ISE and ITSE optimal fractional-order damping systems perform better than ISE and ITSE optimal integer-order damping systems. The ITAE optimal fractional-order damping system outperforms the ITAE optimal integer-order damping system, and IAE and ITAE optimal time-delayed fractional-order damping systems turn out IAE and ITAE optimal time-delayed integer-order damping systems. The optimal time-delayed fractional-order damping system performs the best among the optimal integer-order damping system and optimal fractional-order damping systems.

10.2 Distributed-Order Fractional Mass-Spring Viscoelastic Damper System

In this section, we explain the distributed-order fractional mass-spring viscoelastic damper system in detail. At first, we briefly review the mass-spring-damper, mass-spring viscoelastic damper, and time-delayed mass-spring viscoelastic damper. An ideal mass-spring-damper system with mass m, spring constant k, and viscous damper of damping coefficient c can be described by a second-order differential equation

$$f(t) = m\frac{\mathrm{d}^2x(t)}{\mathrm{d}t^2} + c\frac{\mathrm{d}x(t)}{\mathrm{d}t} + kx(t), \tag{10.3}$$

where $f(t)$ is the time varying force on the mass, $x(t)$ is the displacement of the mass relative to a fixed point of reference. The transfer function from force to displacement for the ideal mass-spring-damper system can be expressed as

$$G(s) = \frac{1}{ms^2 + cs + k}. \tag{10.4}$$

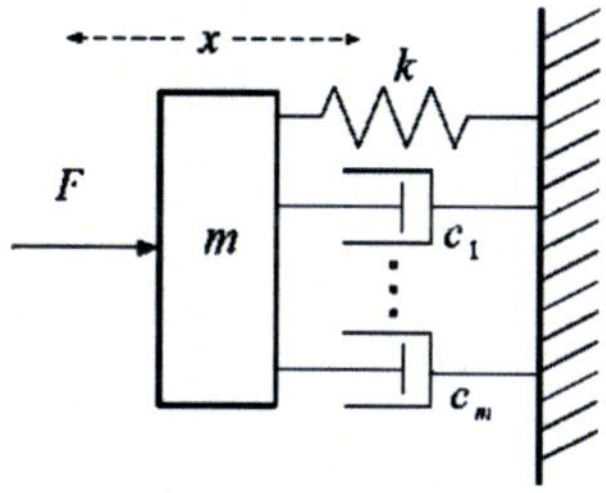

Fig. 10.1 A distributed-order fractional mass-spring viscoelastic damper system

A mass-spring viscoelastic damper system can be described by a fractional-order differential equation

$$f(t) = m\frac{\mathrm{d}^2 x(t)}{\mathrm{d}t^2} + c\,{}_0D_t^{\alpha} x(t) + kx(t), \tag{10.5}$$

where $0 < \alpha < 2$. The transfer function form of a mass-spring viscoelastic damper system can be expressed as

$$G(s) = \frac{1}{ms^2 + cs^{\alpha} + k}. \tag{10.6}$$

Similarly, the transfer function form of a time-delayed mass-spring viscoelastic damper system can be expressed as

$$G(s) = \frac{1}{ms^2 + cs^{\alpha}e^{-\tau s} + k}, \tag{10.7}$$

where $0 < \alpha < 2$.

A distributed-order fractional mass-spring viscoelastic damper system with mass m (in kilograms), spring constant k (in Newton per meter) and an assembly of viscoelastic dampers of damping coefficient c_i $(1 < i < n)$ is subject to an oscillatory force

$$f_s(t) = -kx(t), \tag{10.8}$$

and damping force

$$f_d(t) = -\sum_{i=1}^{n} c_i \cdot {}_0D_t^{\alpha_i} x(t), \tag{10.9}$$

where c_i is the viscoelastic damping coefficient. Figure 10.1 illustrates a distributed-order fractional mass-spring viscoelastic damper system. According to the Newton's second law, the total force $f_{tot}(t)$ on the body is

$$f_{tot}(t) = ma = m\frac{\mathrm{d}^2 x(t)}{\mathrm{d}t^2}, \tag{10.10}$$

where a is the acceleration (in meters per second squared) of the mass, and $x(t)$ is the displacement (in meters) of the mass relative to a fixed point of reference. The time varying force on the mass can be represented by

$$\begin{aligned} f(t) &= f_{tot}(t) - f_d(t) - f_s(t) \\ &= m\frac{\mathrm{d}^2 x(t)}{\mathrm{d}t^2} + \sum_{i=1}^{n} c_i \, {}_0D_t^{\alpha_i} x(t) + kx(t). \end{aligned} \tag{10.11}$$

Assuming elements with orders that vary from a to b, the above mass-spring viscoelastic damper system of (10.11) can be replaced by an integral over the system order,

$$f(t) = m\frac{\mathrm{d}^2 x(t)}{\mathrm{d}t^2} + \int_a^b c(\alpha)\, {}_0D_t^{\alpha} x(t)\mathrm{d}\alpha + kx(t), \tag{10.12}$$

where $0 < a < b < 2$. The transfer function from the force to displacement x for the spring-mass-viscoelastic damper system of (10.12) can be expressed as

$$G(s) = \frac{X(s)}{F(s)} = \frac{1}{ms^2 + \int_a^b c(\alpha)s^{\alpha}\mathrm{d}\alpha + k}. \tag{10.13}$$

What we are concentrating on in this study is the normalized transfer functions of above three types of the spring-mass-viscoelastic damper systems. They are normalized transfer function of the spring-mass-viscoelastic damper system

$$G(s) = \frac{1}{s^2 + cs^{\alpha} + 1}, \quad 0 < \alpha < 2, \tag{10.14}$$

normalized transfer function of the time-delayed spring-mass-viscoelastic damper system

$$G(s) = \frac{1}{s^2 + cs^{\alpha}e^{-\tau s} + 1}, \quad 0 < \alpha < 2, \tag{10.15}$$

and the normalized transfer function of the constant damper coefficient distributed-order spring-mass-viscoelastic damper system

$$G(s) = \frac{1}{s^2 + c\int_a^b s^{\alpha}\mathrm{d}\alpha + 1}, \quad 0 < a < b < 2. \tag{10.16}$$

10.3 Frequency-Domain Method Based Optimal Fractional-Order Damping Systems

The ISE optimal integer-order damping system with transfer function

$$G(s) = \frac{1}{s^2 + s + 1} \tag{10.17}$$

has been investigated in [217], and the ISE optimal fractional-order damping with transfer function

$$G(s) = \frac{1}{s^2 + 0.8791 s^{0.8459} + 1} \tag{10.18}$$

has been found in [180] using a frequency-domain method. In this section, ISE optimal time-delayed and distributed-order fractional mass-spring viscoelastic damper systems are studied in frequency-domain. The ISE performance measure is the integral of the squared error of the step response $e(t) = u(t) - x(t)$

$$J_{\mathrm{ISE}} = \int_0^\infty e^2(t)\mathrm{d}t, \tag{10.19}$$

where $x(t)$ is the output of the system [76]. Using Parseval's identity

$$J_{\mathrm{ISE}} = \int_0^\infty e^2(t)\mathrm{d}t = \frac{1}{2\pi}\int_{-\infty}^\infty |E(j\omega)|^2 \mathrm{d}\omega, \tag{10.20}$$

where $E(j\omega)$ is the Fourier transform of the error $e(t)$. For a system with transfer function $G(s)$, the Laplace transform of the error can be written as

$$E(s) = \frac{1}{s} - \frac{1}{s}G(s). \tag{10.21}$$

In frequency domain, (10.21) is represented by

$$E(j\omega) = \frac{1}{j\omega} - \frac{1}{j\omega}G(j\omega). \tag{10.22}$$

For a time-delayed spring-mass-viscoelastic damper system with the normalized transfer function (10.15), the Laplace transform of the step response error is

$$E(s) = \frac{1}{s} - \frac{1}{s}\left(\frac{1}{s^2 + cs^\alpha e^{-\tau s} + 1}\right) = \frac{1}{s}\left(\frac{s^2 + cs^\alpha e^{-\tau s}}{s^2 + cs^\alpha e^{-\tau s} + 1}\right). \tag{10.23}$$

The frequency response of the error is

$$E(j\omega) = \frac{1}{j\omega}\left(\frac{(j\omega)^2 + c(j\omega)^\alpha e^{-\tau(j\omega)}}{(j\omega)^2 + c(j\omega)^\alpha e^{-\tau(j\omega)} + 1}\right). \tag{10.24}$$

Using the frequency-domain method in [112], the minimum $J_{\mathrm{ISE}} = 0.8102$ was obtained when $\tau = 0.635$, $c = 1.12$ and $\alpha = 1.05$. The step response using optimum coefficients for the ISE criterion is given in Fig. 10.2.

For a mass-spring viscoelastic damper model with the normalized distributed-order fractional transfer function (10.16), the Laplace transform of the step response error is

$$\begin{aligned} E(s) &= \frac{1}{s} - \frac{1}{s}\left(\frac{1}{s^2 + c\int_a^b s^\alpha \mathrm{d}\alpha + 1}\right) \\ &= \frac{1}{s}\left(\frac{\ln(s)s^2 + c(s^b - s^a)}{\ln(s)s^2 + c(s^b - s^a) + \ln(s)}\right). \end{aligned} \tag{10.25}$$

Fig. 10.2 Step responses of ISE optimal damping systems based on frequency-domain method

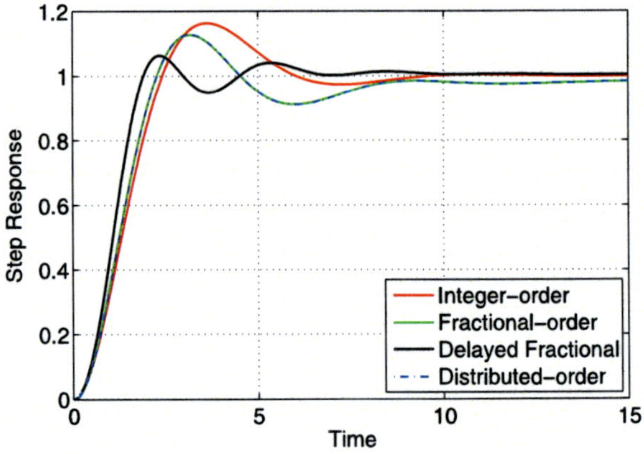

Table 10.1 ISE optimum coefficients and minimum ISE performance indexes using frequency-domain method

	Optimal form	J_{ISE}
Integer	$G_{\text{ISE}}(s) = 1/(s^2 + s + 1)$	1.0000
Fractional	$G_{\text{ISE}}(s) = 1/(s^2 + 0.8791s^{0.8459} + 1)$	0.9494
Delayed	$G_{\text{ISE}}(s) = 1/(s^2 + 1.12s^{1.05}e^{-0.635s} + 1)$	0.8102
Distributed	$G_{\text{ISE}}(s) = 1/(s^2 + 10\int_{0.8015}^{0.8893} s^{\alpha}\mathrm{d}\alpha + 1)$	0.9494

The frequency response of the error is

$$E(j\omega) = \frac{1}{j\omega}\left(\frac{\ln(j\omega)(j\omega)^2 + c[(j\omega)^b - (j\omega)^a]}{\ln(j\omega)(j\omega)^2 + c[(j\omega)^b - (j\omega)^a] + \ln(j\omega)}\right). \tag{10.26}$$

Then, we can search the optimum coefficients of the distributed-order fractional damping system. The optimum coefficients are $a = 0.8015$, $b = 0.8893$ and $c = 10$, which can minimize the ISE performance measure to $J_{\text{ISE}} = 0.9494$. Figure 10.2 shows the step responses of integer-order, non-delayed fractional-order, time-delayed fractional-order, and distributed-order fractional damping systems using optimum coefficients for ISE. It can be seen that the step responses of optimal distributed-order fractional damping system with transfer function (10.16) are almost as good as that of the optimal non-delayed fractional-order damping system. The optimal time-delayed fractional-order damping system performs the best among these four types of damping systems. The ISE optimal forms and ISE performance indexes of integer-order, non-delayed fractional-order, time-delayed fractional-order, and distributed-order fractional damping systems are summarized in Table 10.1.

10.4 Time-Domain Method Based Optimal Fractional-Order Damping Systems

The optimal integer-order damping systems based on ISE and ITSE criteria have been analytically studied in [217]. The ITAE optimal integer-order damping system with optimum coefficient $c = 1.4$ and performance index $J_{\mathrm{ITAE}} = 1.97357$ was provided in [83]. But the optimum coefficient and performance index for ITAE were corrected, and the new coefficient $c = 1.505$ and performance index $J_{\mathrm{ITAE}} = 1.93556$ were found in [42]. In this section, three types of fractional-order damping systems are numerically optimized using ISE, ITSE, IAE and ITAE criteria in time domain. Unlike ISE which can be manipulated easily in frequency domain, ITSE, IAE and ITAE should be calculated numerically in time domain. In order to provide a clear comparison, the ISE performance index is also numerically calculated using the time-domain method in this section. A numerical inverse Laplace transform algorithm NILT was used to calculate the numerical results of performance measures. The NILT fast numerical inversion of Laplace transforms method was provided in [33]. The method is based on the application of fast Fourier transformation followed by so-called ε-algorithm to speed up the convergence of infinite complex Fourier series. The accuracy of the NILT method for fractional-order differential equations has been studied in [269]. The NILT method performs very well for fractional-order and distributed-order fractional differential equations. The optimization method used in this study is introduced as follows. The step responses of transfer function (10.14), (10.15) and (10.16) were numerically calculated using a NILT inversion of the Laplace transforms method. The error of the step response to a unit step input was sampled from $t = 0$ to 50 in constant steps of $\mathrm{d}t = 0.001$. The integral was then computed numerically with those limits and the step size. The different integration upper limits 50, 100 and 1000 have been tested to calculate the performance index, but there was almost no difference in the numerical results. Therefore, the upper limit 50 is chosen to speed up the computation. Then, the parameters of the system might be adjusted to minimize the performance measure, and the transfer function form of the damping system is optimized based on the performance criterion. To test the accuracy of the numerical optimization method, the integer-order damping system is optimized based on ISE, ITSE, IAE and ITAE criteria. The calculated optimum coefficients and the minimum performance indexes are provided in Table 10.2. It can be seen that the calculated optimum coefficients and the minimum performance indexes for ISE and ITSE criteria are the same as the analytical results which were provided in [217]. The calculated coefficients and the performance index for ITAE criterion are almost the same as the corrected results provided in [42]. Therefore, the numerical optimization method used in this research is valid and reliable.

Using the above numerical optimization method, the optimum coefficients of non-delayed fractional-order damping system are $c = 0.8745$ and $\alpha = 0.8367$, which can minimize the ISE performance measure to $J_{\mathrm{ISE}} = 0.9485$. The minimum ISE performance index $J_{\mathrm{ISE}} = 0.8111$ of the time-delayed fractional-order damping system is obtained when $c = 1.1157$, $\alpha = 1.0604$ and $\tau = 0.6435$. The optimum

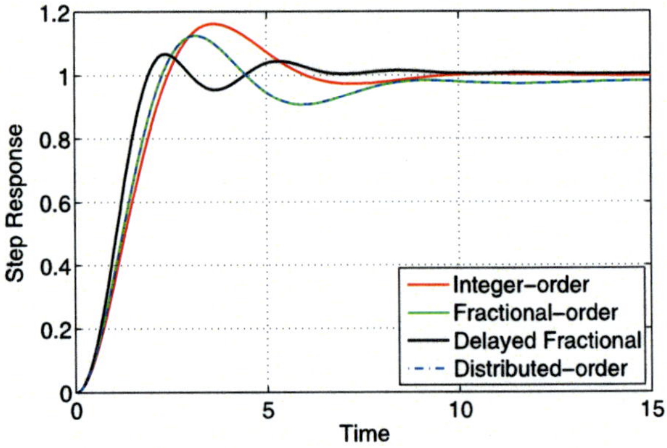

Fig. 10.3 Step responses of ISE optimal damping systems

coefficients of the distributed-order fractional damping system are $c = 40.1367$, $a = 0.8260$ and $b = 0.8478$, which can minimize the ISE performance measure to $J_{\text{ISE}} = 0.9485$. The step responses of the ISE optimal integer-order and three types of fractional-order damping systems are provided in Fig. 10.3. From Fig. 10.2 and Fig. 10.3 we can see that the step responses of ISE optimal damping systems using both time-domain and frequency-domain methods are almost the same, although the optimum coefficients are slightly different. The ISE optimal time-delayed fractional-order damping system performs better than other ISE optimal integer-order and fractional-order damping systems.

The ITSE performance measure is the integral of the squared error of the step response $e(t)$ multiplied by time [76]

$$J_{\text{ITSE}} = \int_0^\infty te^2(t)\mathrm{d}t. \tag{10.27}$$

Based on the above numerical optimization method, the optimum coefficients of the non-delayed fractional-order damping system are $c = 1.0855$ and $\alpha = 0.9372$, which can minimize the ITSE performance measure to $J_{\text{ITSE}} = 0.6593$. The optimum coefficients of the time-delayed fractional-order damping system are $c = 1.1019$, $\alpha = 1.0212$ and $\tau = 0.5662$, which can minimize the ITSE performance measure to $J_{\text{ITSE}} = 0.4115$. Based on ITSE criterion, the distributed-order fractional damping system can be optimized with $c = 40.1368$, $a = 0.9237$ and $b = 0.9508$, and the minimum ITSE performance index is $J_{\text{ITSE}} = 0.6595$. The step responses of these optimal damping systems for ITSE criterion are provided in Fig. 10.4. Similar to ISE optimal damping systems, the optimal time-delayed fractional-order damping system performs the best, and the optimal distributed-order fractional and non-delayed fractional-order damping systems have almost the same step responses. All the ITSE optimal damping systems perform a little bit better than ISE optimal damping systems.

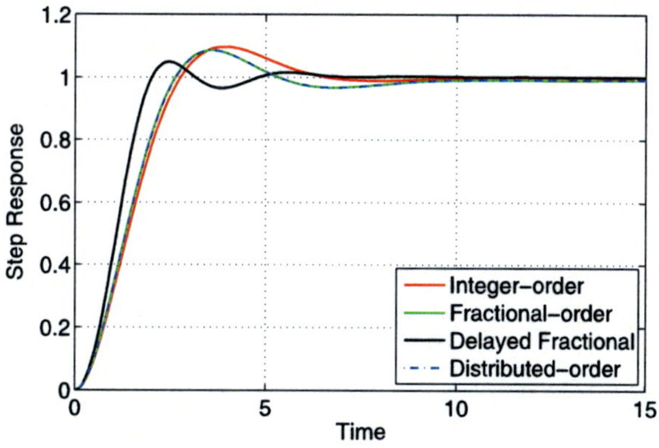

Fig. 10.4 Step responses of ITSE optimal damping systems

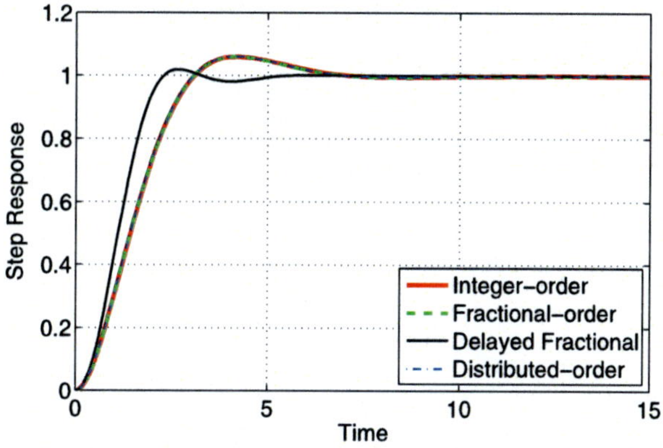

Fig. 10.5 Step responses of IAE optimal damping systems

The IAE performance measure is the integral of the absolute magnitude of the error [76]

$$J_{\mathrm{IAE}} = \int_0^\infty |e(t)|\mathrm{d}t. \tag{10.28}$$

Using the same numerical optimization method, the minimum IAE performance measure $J_{\mathrm{IAE}} = 1.6051$ of non-delayed fractional-order damping system is obtained with $c = 1.3204$ and $\alpha = 0.9985$. The optimum coefficients of the time-delayed fractional-order damping system are $c = 1.1288$, $\alpha = 1.0000$ and $\tau = 0.4801$, which can minimize the IAE performance measure to $J_{\mathrm{IAE}} = 1.1567$. Based on IAE criterion, the distributed-order damping system can also be optimized with $c = 40.3136$, $a = 0.9824$ and $b = 1.0151$, and the minimum IAE performance index is $J_{\mathrm{IAE}} = 1.6058$. The step responses of these optimal damping systems for IAE criterion are provided in Fig. 10.5. The optimal time-delayed fractional-order damping system is the best among these IAE optimal damping systems. The Optimal IAE integer-order and other two types of fractional-order damping systems perform almost the same.

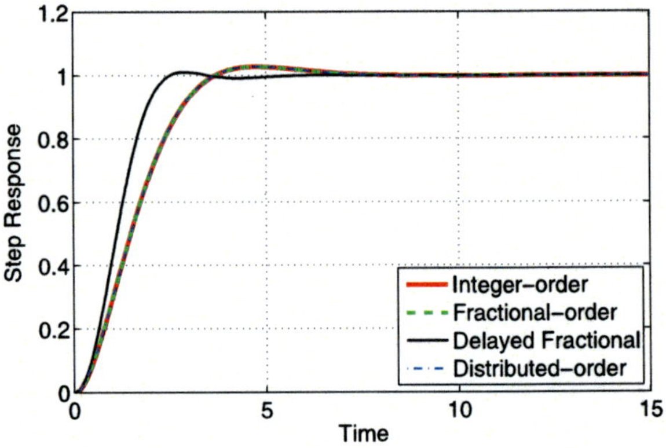

Fig. 10.6 Step responses of ITAE optimal damping systems

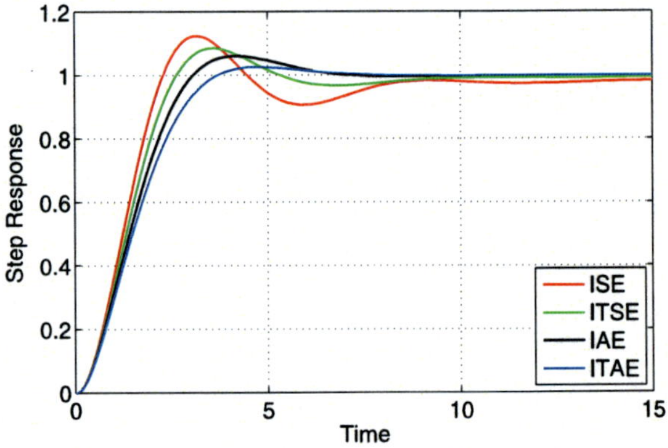

Fig. 10.7 Step responses of optimal non-delayed fractional-order damping system

The ITAE performance measure is the integral of time multiplied by absolute error [76]

$$J_{\text{ITAE}} = \int_0^\infty t|e(t)|\mathrm{d}t. \tag{10.29}$$

The optimum coefficients of non-delayed fractional-order damping system are $c = 1.5047$ and $\alpha = 1.0000$, which can minimize the ITAE performance measure to $J_{\text{ITAE}} = 1.9518$. The ITAE performance measure $J_{\text{ITAE}} = 0.8755$ of time-delayed fractional-order damping system is obtained with $c = 1.1504$, $\alpha = 1.0000$ and $\tau = 0.4393$. The optimum coefficients of the distributed-order fractional damping system are $c = 40.3140$, $a = 0.9818$ and $b = 1.0191$, which can minimize the ITAE performance measure to $J_{\text{ITAE}} = 1.9581$. The step responses of these optimal damping systems for ITAE criterion are provided in Fig. 10.6. The optimal time-delayed fractional-order damping system is the best among these four ITAE optimal damping systems.

Furthermore, in order to clearly compare the performances of ISE, ITSE, IAE and ITAE criteria to each optimal damping system, the step responses of four performance indexes based fractional-order damping systems are presented in Figs. 10.7,

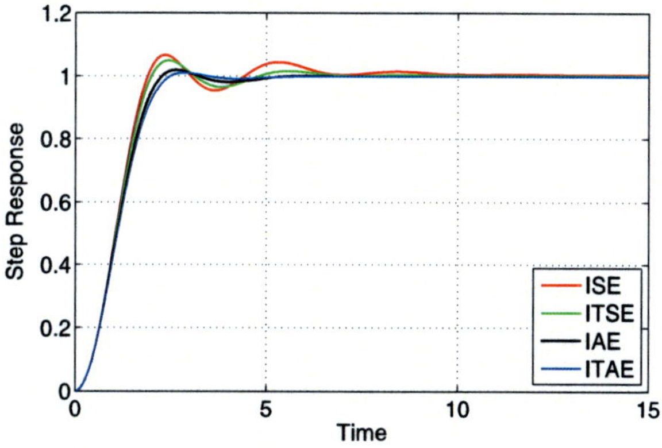

Fig. 10.8 Step responses of optimal time-delayed fractional-order damping system

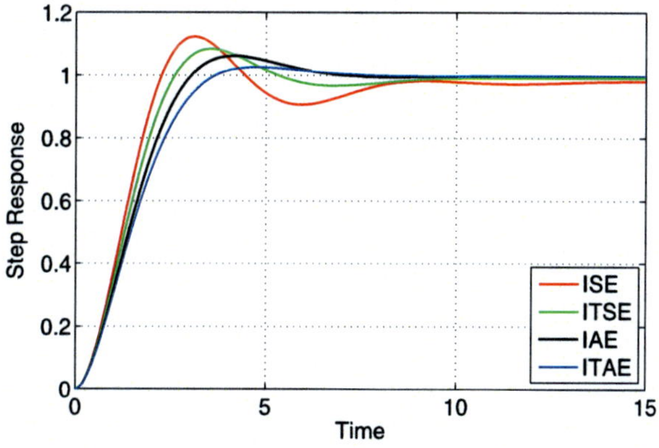

Fig. 10.9 Step responses of optimal distributed-order fractional damping system

10.8 and 10.9, respectively. Obviously, for all the optimal integer-order and three types of optimal fractional-order damping systems, the ITAE performance index produces smaller overshoot and oscillations than the ISE, ITSE and IAE indices, but the ISE performance index results in faster rise time.

Table 10.2 provides the ISE, ITSE, IAE and ITAE criteria based optimal transfer functions and minimum performance indexes of the optimal integer-order and three types of fractional-order damping systems. From Table 10.2 we can see that, based on ISE, ITSE, IAE and ITAE performance measures, the optimal distributed-order damping system performs as well as optimal non-delayed fractional-order damping system. The step responses of the non-delayed fractional-order and the distributed-order fractional damping systems using optimum coefficients for four performance criteria are almost the same. The ITAE optimal non-delayed fractional-order damping outperforms the ITAE optimal non-delayed integer-order damping. Similarly, the IAE and ITAE optimal time-delayed fractional-order damping outperforms IAE and ITAE optimal time-delayed integer-order damping. Based on ISE, ITSE, IAE and ITAE criteria, the optimal time-delayed fractional-order damping system performs better than the integer-order and the other two types of optimal fractional-order damping systems.

Table 10.2 Optimal coefficients and minimum performance indexes using time-domain method

	ISE	
	Optimal form	J_{ISE}
Integer	$1/(s^2+s+1)$	1.005
Fractional	$1/(s^2+0.8745s^{0.8367}+1)$	0.9485
Delayed	$1/(s^2+1.1157s^{1.0604}e^{-0.6435s}+1)$	0.8111
Distributed	$1/(s^2+40.1367\int_{0.8260}^{0.8478}s^{\alpha}\mathrm{d}\alpha+1)$	0.9485
	ITSE	
	Optimal form	J_{ITSE}
Integer	$1/(s^2+1.1892s+1)$	0.7071
Fractional	$1/(s^2+1.0855s^{0.9372}+1)$	0.6593
Delayed	$1/(s^2+1.1019s^{1.0212}e^{-0.5662s}+1)$	0.4115
Distributed	$1/(s^2+40.1368\int_{0.9237}^{0.9508}s^{\alpha}\mathrm{d}\alpha+1)$	0.6595
	IAE	
	Optimal form	J_{IAE}
Integer	$1/(s^2+1.3247s+1)$	1.6056
Fractional	$1/(s^2+1.3204s^{0.9985}+1)$	1.6051
Delayed	$1/(s^2+1.1288s^{1.0000}e^{-0.4801s}+1)$	1.1567
Distributed	$1/(s^2+40.3136\int_{0.9824}^{1.0151}s^{\alpha}\mathrm{d}\alpha+1)$	1.6058
	ITAE	
	Optimal form	J_{ITAE}
Integer	$1/(s^2+1.5049s+1)$	1.9518
Fractional	$1/(s^2+1.5047s^{1.0000}+1)$	1.9518
Delayed	$1/(s^2+1.1504s^{1.0000}e^{-0.4393s}+1)$	0.8755
Distributed	$1/(s^2+40.3140\int_{0.9818}^{1.0191}s^{\alpha}\mathrm{d}\alpha+1)$	1.9581

10.5 Chapter Summary

In this chapter, we tried to determine the optimal non-delayed fractional-order damping, time-delayed fractional-order damping, and optimal distributed order fractional damping based on ISE, ITSE, IAE and ITAE performance criteria. The comparisons of the step responses of the integer-order and the three types of fractional-order damping systems indicate that the optimal fractional-order damping systems achieve much better step responses than optimal integer-order systems in some instances, but sometimes the integer-order damping systems performs as well as fractional-order ones. Furthermore, time delay can sometimes be used to gain benefit in control systems, and, especially, the fractional-order damping plus properly chosen delay can bring outstanding performance. Time-delayed fractional-order damping systems can produce a faster rise time and less overshoot than others. Besides, although the distributed-order fractional damping system does not perform better than non-delayed and time-delayed fractional-order damping systems, it has much potential to improve the damping system by choosing an appropriate viscoelastic damping coefficient weighting function.

Chapter 11
Heavy-Tailed Distribution and Local Memory in Time Series of Molecular Motion on the Cell Membrane

11.1 Introduction

Surface protein tracking technique has become an important method for characterizing the mechanisms underlying cell membrane organization [44, 86, 327]. Single-particle tracking (SPT) provides a powerful tool for observing the motion of a single particle to study the behaviors that may go undetected in measurements of a large population of particles [4, 132, 133, 156]. The information extracted from the measurement of particle trajectories provides some essential insights into the regulation mechanisms and forces that drive and constrain the particles motion, so it has been used in various fields of cell biology. The motion modes of membrane dynamics were studied using SPT technique in [258]. SPT technique was also used to study nuclear trafficking of viral genes and applied to confined diffusion of the cystic fibrosis transmembrane conductance regulator in [14]. Based on the SPT technique, this chapter studies some statistical analyses of experimental biological data which track the motion of 40 nm gold particles bound to Class I major histocompatibility complex (MHCI) molecules on the membranes of mouse hepatoma cells similar to [327].

The materials and methods used in tracking and recording the MHCI molecules, which are studied in this chapter, were introduced in [44, 327]. The MHCI molecules were labeled with 40 nm gold particles and tracked by differential interference contrast microscopy with a 63×NA 1.4 objective on a Zeiss Axiovert Microscope. Video sequences were captured with a fast charge-coupled device (CCD) camera (CCD72S model, DAGE-MTI, Michigan City, IN) with a time resolution of thirty three millisecond and were recorded to tape on a SONY EVO-9650 Hi8 VCR. Then, the ISee software was used to track the centroid of a given particle through the sequence of images, and outputs the $x-y$ coordinates of the particle in successive image frames [44]. The MHCI molecular trajectories data consist of thirty three gold particles' paths having between 623 and 2,117 points in a single path. The coordinates of the positions are in nanometers and the time step is given by $1/30$ second. Detailed information of these biological data can be obtained from [44, 327]. Based on these SPT data, Capps et al. found that short cytoplasmic tails can influence

H. Sheng et al., *Fractional Processes and Fractional-Order Signal Processing*, Signals and Communication Technology,
DOI 10.1007/978-1-4471-2233-3_11,

markedly class I MHC mobility, and that cytoplasmic tail length and sequence affect the molecule's diffusion in the membrane [44]. Ying et al. found that the jump data have significant autocorrelations and fitted the data using four statistical models [327]. In this chapter, we focus on the heavy-tailed distribution and local memory characteristics for ten jump time series of these MHCI molecules.

The phenomena of heavy-tailed distribution and long memory have been observed in many branches of sciences, such as insurance and economics [43, 270]. Heavy-tailed data frequently exhibit large extremes and may even have infinite variance, while long memory data exhibit a slow decay of correlations. The joint presence of heavy-tailed distribution and long memory has been founded in many data sets, such as teletraffic data, financial data and biomedical data [43, 70, 270]. A typical heavy-tailed distribution is α-stable distribution with an adjustable tail thickness parameter α. α-stable distribution has been successfully applied in modeling underwater signals, low-frequency atmospheric noise and many types of man-made noises [215]. Long memory process can be characterized by the Hurst parameter $H \in (0, 1)$. However, the constant Hurst parameter cannot well capture the local scaling characteristic of the stochastic processes. So, the long memory process with a time-varying long-memory parameter is investigated to explain the complex physical phenomena [232]. In this chapter, the α-stable model was used to characterize the MHCI molecular jump time series with infinite second-order statistics and heavy-tailed distribution. The long memory and local long memory characteristics were detected using the Diffusion Entropy Hurst estimator and the sliding-windowed Koutsoyiannis' local Hölder exponent estimator, respectively.

11.2 Heavy-Tailed Distribution

Heavy-tailed distributions have been observed in many natural phenomena, such as physics, hydrology, biology, financial and network traffic [43, 70, 253, 270]. Heavy-tailed distributions are probability distributions whose tails are heavier than the exponential distribution. The tails of heavy-tailed distributions cannot be cut off, and the large-scale and rare events cannot be neglected. Because the heavier than those of Gaussian distribution, this kind of time series always exhibits more sharp spikes or occasional bursts of outlying observations than one would expect from normally distributed signals. So, heavy-tailed distributions generally have high variabilities or infinite second-order moment. Infinite second-order moment makes it meaningless to discuss the variance and correlation function. Similarly, many standard signal processing tools, which are based on the assumption of finite second-order moment, such as least-squares techniques and spectral analysis, may give misleading results [215].

The distribution of a random variable X with distribution function $F(x)$ is said to have a heavy tail if its complementary cumulative distribution $F_c(x) = 1 - F(x)$ decays more slowly than exponentially [8], that is, for all $\gamma > 0$,

$$\lim_{x\to\infty} \exp(\gamma x) F_c(x) \to \infty.$$

A typical heavy-tailed distribution is power-tailed if $F_c(x) \sim \alpha x^{-\beta}$ as $x \to \infty$ for constants $\alpha > 0$ and $\beta > 0$, where '$\sim$' means the ratio of the left hand and the right hand sides converges to 1. The classic examples of one-sided distribution exhibiting heavy-tailed behavior are Pareto distribution, Log-normal distribution, Lévy distribution, Weibull distribution, and so on. The typical two-sided heavy-tailed distributions are α-stable distribution, Cauchy distribution, skew lognormal cascade distribution, and so on. Among them, the α-stable distribution is a good heavy-tailed distribution model for signals and noises of impulsive nature, because of the generalized central limit theorem as shown, if the sum of independent and identically distributed (i.i.d.) random variables with or without finite variance converges to a distribution by increasing the number of variables, the limit distribution must belong to the family of stable laws [36, 94]. In addition, the α-stable distribution has a characteristic exponent parameter α $(0 < \alpha \leq 2)$, which controls the heaviness of its tails. So, it can exactly model heavy-tailed distribution non-Gaussian processes with different tail thickness [215]. In the next section, we will use the α-stable distribution to characterize the jump time series of MHCI molecular motion data [327].

11.3 Time Series of Molecular Motion

In this section, the same jump time series of MHCI molecular trajectories data are analyzed as used in [327]. The collected MHCI molecular trajectories data consist of thirty three gold particles trajectories with between 623 and 2,117 points in a path. In our study, ten of these thirty three were re-analyzed: Experiments 1, 3, 4, 7, 16, 19, 24, 27, 28 and 32. Some further analysis results using the new methods are provided in Table 11.2 and Figs. 11.13–11.18.

It has been shown in [327] that time series analysis is a useful analytic tool for analyzing the motion of membrane proteins. Similarly, in our study, the MHCI molecular motion data is also viewed as a time series. The jump time series L_n $(1 \leq n \leq N)$ is defined as

$$L_n = \sqrt{\Delta X_n^2 + \Delta Y_n^2}, \tag{11.1}$$

where ΔX_n and ΔY_n are the displacement changes in x-axis and y-axis, respectively. Figures 11.1, 11.2, 11.3 and 11.4 show the trajectories and jump time series for Experiments 1, 7, 16 and 27. It can be seen from Fig. 11.1(b), Fig. 11.2(b), Fig. 11.3(b) and Fig. 11.4(b) that, different from Gaussian processes, all these four jump time series exhibit sharp spikes or occasional bursts of outlying observations. To characterize this kind of processes in the next subsection, the variance trend and the histogram of these jump time series are plotted, and the α-stable distribution model is employed to fit them. The characteristic exponent parameter α $(0 < \alpha \leq 2)$ can be used to evaluate the tail thickness of the distribution. Furthermore, the long memory and local memory characteristics are studied by estimating the Hurst parameter H and the local Hölder exponent $H(t)$.

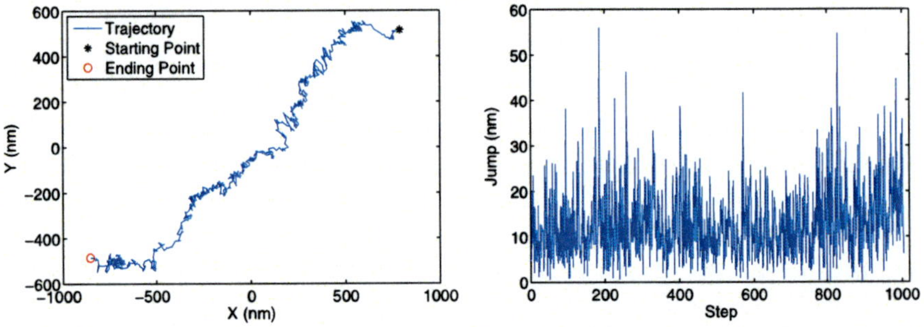

(a) Trajectory (b) Jump time series

Fig. 11.1 Trajectory and jump time series for Experiment 1

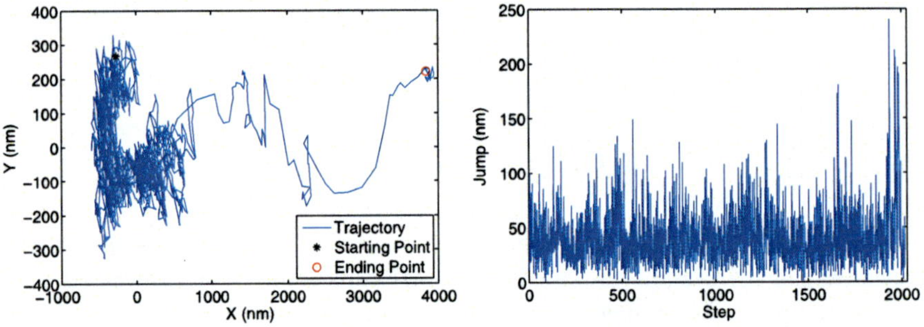

(a) Trajectory (b) Jump time series

Fig. 11.2 Trajectory and jump time series for Experiment 7

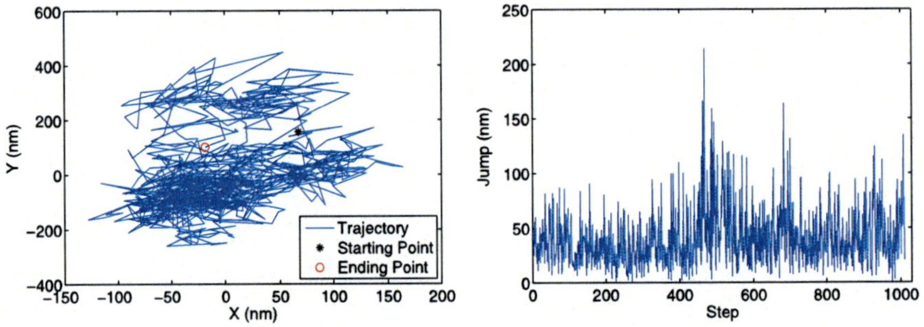

(a) Trajectory (b) Jump time series

Fig. 11.3 Trajectory and jump time series for Experiment 16

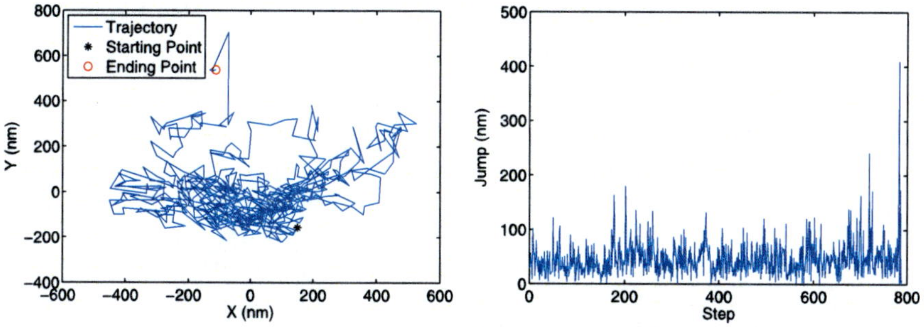

(a) Trajectory

(b) Jump time series

Fig. 11.4 Trajectory and jump time series for Experiment 27

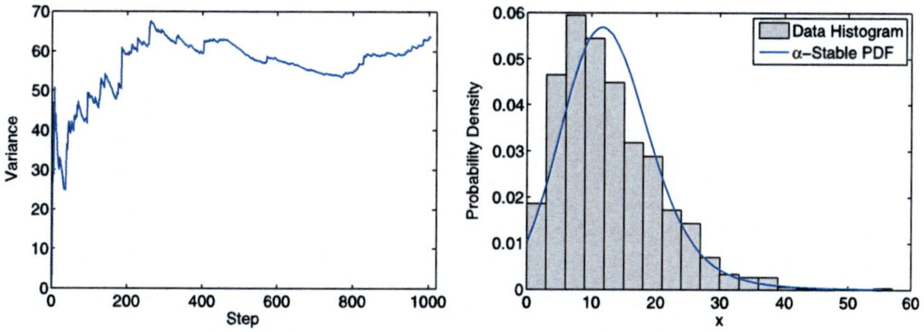

(a) Variance trend

(b) Histogram and fitted α-stable PDF ($\alpha = 1.8026$)

Fig. 11.5 Variance trend and fitted α-stable PDF for Experiment 1

11.4 Infinite Second-Order and Heavy-Tailed Distribution in Jump Time Series

It has been introduced in Sect. 11.2 that time series with heavy-tailed distribution generally have high variation or infinite second-order moment. α-stable distribution is a basic statistical modeling tool to model heavy-tailed distributions with infinite variance non-Gaussian signals. In this subsection, the α-stable distribution model is used to fit the histogram distributions for these MHCI molecular jump time series, and to quantify the tail thickness of the distributions.

It has been introduced in Chap. 1 that the α stable characteristic function (or distribution) is determined by the parameters α, a, β and γ, where α controls the tail heaviness of distribution, and γ is a scale parameter. These two parameters influence the PDF of an α-stable distribution.

Figures 11.5, 11.6, 11.7 and 11.8 show the variance trends and the fitted α-stable densities for Experiments 1, 7, 16 and 27, respectively. It can be seen from Fig. 11.8(a) that the variance of Experiment 27 is obviously divergent. The variance of the other three time series are also not converging. Figures 11.5(b)–11.8(b) indi-

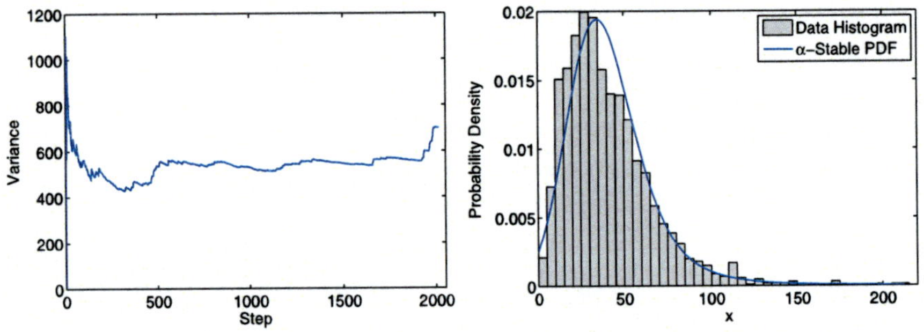

(a) Variance trend

(b) Histogram and fitted α-stable PDF ($\alpha = 1.6912$)

Fig. 11.6 Variance trend and fitted α-stable PDF for Experiment 7

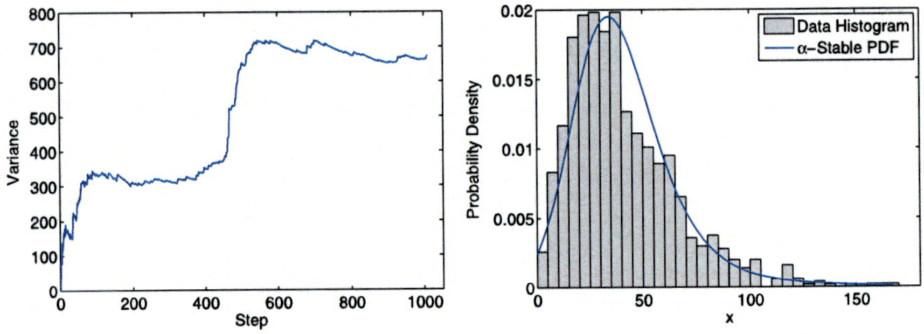

(a) Variance trend

(b) Histogram and fitted α-stable PDF ($\alpha = 1.6602$)

Fig. 11.7 Variance trend and fitted α-stable PDF for Experiment 16

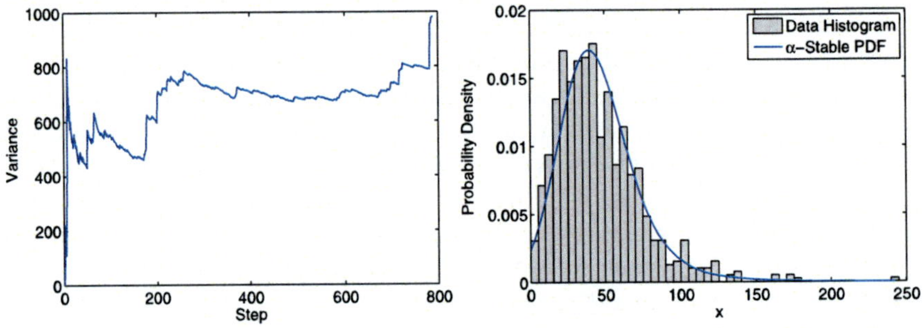

(a) Variance trend

(b) Histogram and fitted α-stable PDF ($\alpha = 1.7229$)

Fig. 11.8 Variance trend and fitted α-stable PDF for Experiment 27

Table 11.1 α-stable parameters

	α	β	γ	a
Experiment 01	1.8026	1.0000	04.9491	13.4427
Experiment 03	1.7905	1.0000	11.2125	32.6742
Experiment 04	1.5740	1.0000	12.1260	37.1829
Experiment 07	1.6912	1.0000	14.4688	43.4335
Experiment 16	1.6602	1.0000	14.3755	43.9307
Experiment 19	1.0608	0.9599	08.9666	105.7616
Experiment 24	1.5961	1.0000	16.0943	48.3427
Experiment 27	1.7229	1.0000	16.4552	48.1364
Experiment 28	1.7623	1.0000	11.2344	31.5433
Experiment 32	1.6983	1.0000	07.2060	21.2042

cate that the histograms of these four MHCI molecular jump time series all have heavy tails, and these histograms can be fitted well using α-stable distributions. The plots of the variance trend and the fitted α-stable distributions of the MHCI molecular jump time series for Experiments 3, 4, 19, 24, 28 and 32 are displayed in Figs. 11.13–11.18, and the parameters for the fitted α-stable distributions of all ten experiment time series are summarized in Table 11.1. From Table 11.1 we can see that all the parameters $\alpha \in (1, 2)$, which indicates that the distributions of these ten time series all have heavier tails than Gaussian processes ($\alpha = 2$). The tail thickness of the distribution for these ten MHCI molecular jump time series can be quantified by the characteristic exponent α. The smaller the characteristic exponent parameter α, the heavier the tail of the distribution.

11.5 Long Memory and Local Memory in Jump Time Series

Long memory and local memory processes have been introduced in Chaps. 3 and 4. Figures 11.9, 11.10, 11.11 and 11.12 present the estimated Hurst parameter $\hat{H}$ and the estimated local Hölder exponent $\hat{H}(t)$ for Experiments 1, 7, 16 and 27, respectively. The long memory and local long memory analysis results for Experiments 3, 4, 19, 24, 28 and 32 are shown in Figs. 11.13–11.18. The summary of the estimated Hurst parameters for all ten time series is presented in Table 11.2. It can be seen that all these ten jump time series have long memory parameter $H \in (0.5, 1)$, which means that all ten MHCI molecular jump time series have long memory characteristics. Furthermore, the local memory analysis results illustrate that the variable local memory parameter can characterize the local scaling property well for these MHCI molecular jump time series. From Fig. 11.2(a) and the Fig. 11.12(b) we can observe that the variation of $H(t)$ $(55 < t < 60)$ clearly reflects the abnormal motion of the last steps of the molecular jump time series, which indicates that the time-varying local long memory parameter can reflect the essential changes which underlie the surface phenomenon.

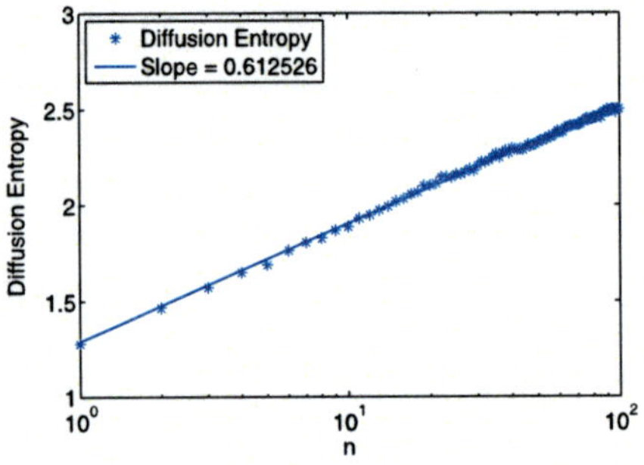

(a) Diffusion Entropy method estimation result

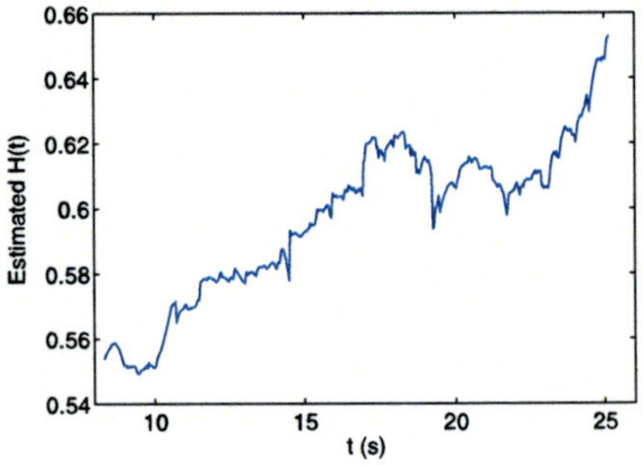

(b) Estimated $\hat{H}(t)$

Fig. 11.9 Diffusion Entropy method estimated Hurst and estimated $\hat{H}(t)$ for Experiment 1

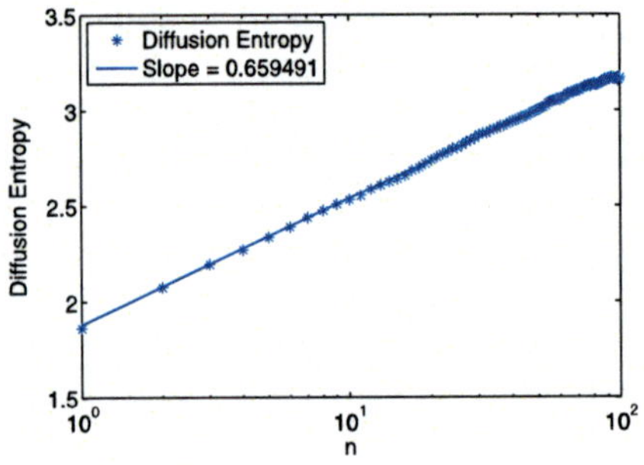

(a) Diffusion Entropy method estimation result

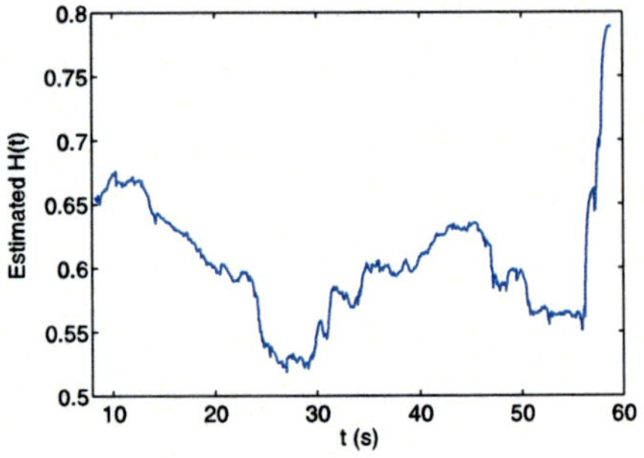

(b) Estimated $\hat{H}(t)$

Fig. 11.10 Diffusion Entropy method estimated Hurst and estimated $\hat{H}(t)$ for Experiment 7

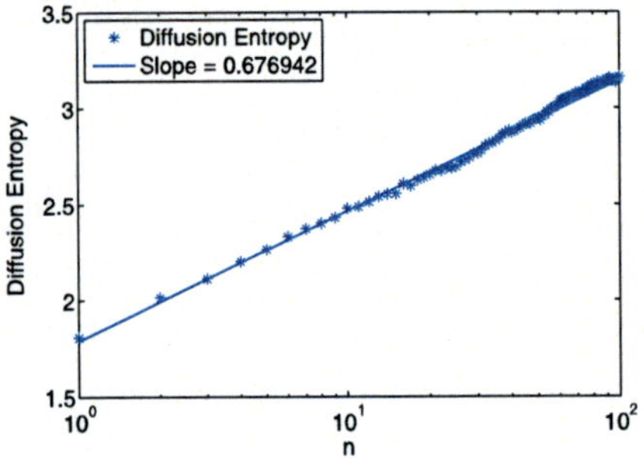

(a) Diffusion Entropy method estimation result

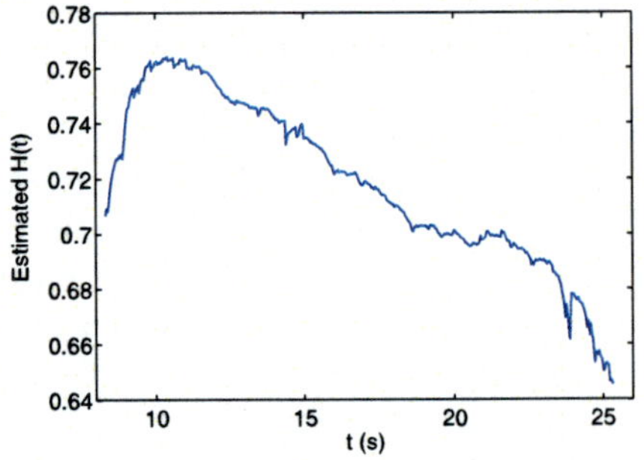

(b) Estimated $\hat{H}(t)$

Fig. 11.11 Diffusion Entropy method estimated Hurst and estimated $\hat{H}(t)$ for Experiment 16

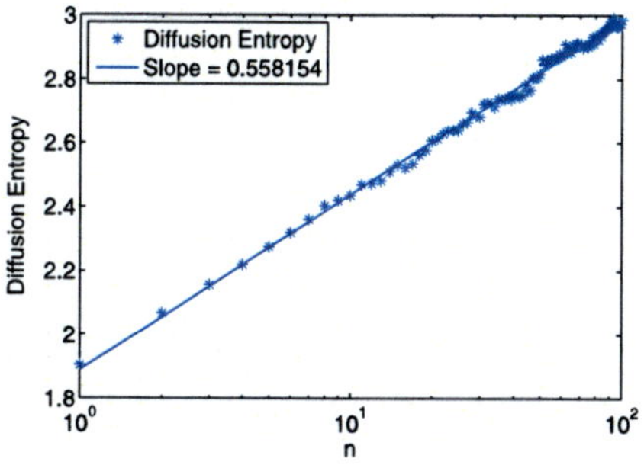

(a) Diffusion Entropy method estimation result

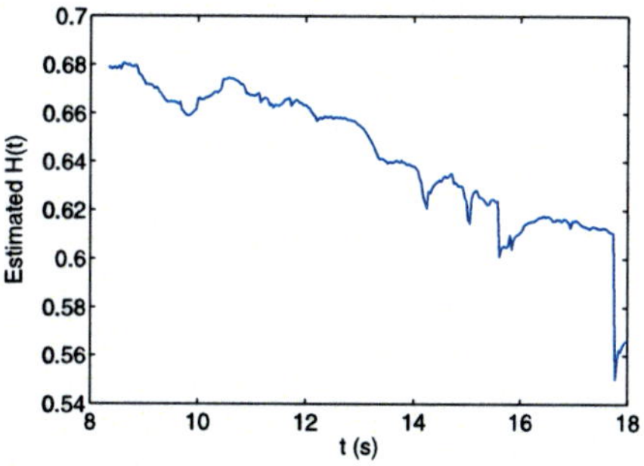

(b) Estimated $\hat{H}(t)$

Fig. 11.12 Diffusion Entropy method estimated Hurst and estimated $\hat{H}(t)$ for Experiment 27

Table 11.2 Long memory parameters

	Long memory parameter				
	Exp. 01	Exp. 03	Exp. 04	Exp. 07	Exp. 16
Diffusion Entropy	0.6125	0.6357	0.7882	0.6595	0.6769
Koutsoyiannis	0.6208	0.6493	0.7960	0.6928	0.7156
Kettani and Gubner	0.5896	0.6130	0.6766	0.6524	0.6477
R/S	0.5969	0.6128	0.7201	0.7135	0.7224
Aggregated variance	0.6561	0.7058	0.8980	0.7111	0.7928
Absolute value	0.6728	0.7454	0.9167	0.7299	0.8003
Variance of Residuals	0.5300	0.5591	0.5559	0.6538	0.6562
Periodogram	0.6358	0.6323	0.7605	0.7587	0.7818
Modified Periodogram	0.6943	0.5526	0.7727	0.8029	0.8888
Whittle	0.7825	0.7189	0.9329	0.7959	0.8493
Abry and Veitch	0.3533	0.6353	0.3867	0.6230	0.8688
	Long memory parameter				
	Exp. 19	Exp. 24	Exp. 27	Exp. 28	Exp. 32
Diffusion Entropy	0.7329	0.7407	0.5582	0.6635	0.6474
Koutsoyiannis	0.6873	0.7950	0.6141	0.6393	0.6726
Kettani and Gubner	0.6967	0.7160	0.5595	0.5852	0.6194
R/S	0.6898	0.7543	0.7226	0.6982	0.6083
Aggregated variance	0.6717	0.8356	0.6430	0.7223	0.7466
Absolute value	0.6633	0.8574	0.6901	0.7247	0.7910
Variance of Residuals	0.6072	0.8538	0.6939	0.5879	0.5626
Periodogram	0.7464	0.8226	0.7096	0.7099	0.6285
Modified Periodogram	0.6585	0.9971	0.7211	0.7315	0.5479
Whittle	0.6820	1.0158	0.7013	0.7113	0.7695
Abry and Veitch	0.5331	0.6589	0.3606	0.7518	0.5416

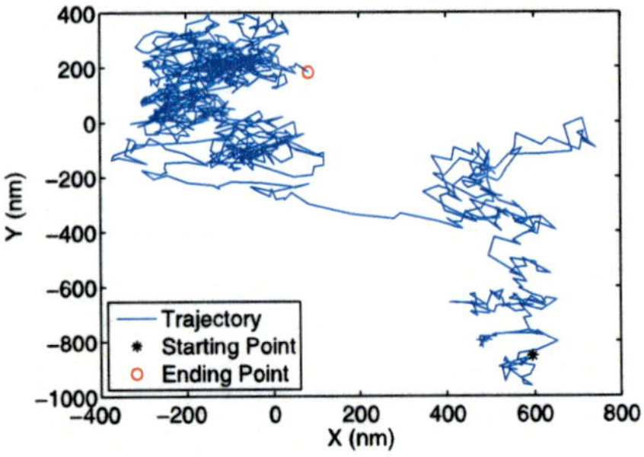

(a) Trajectory

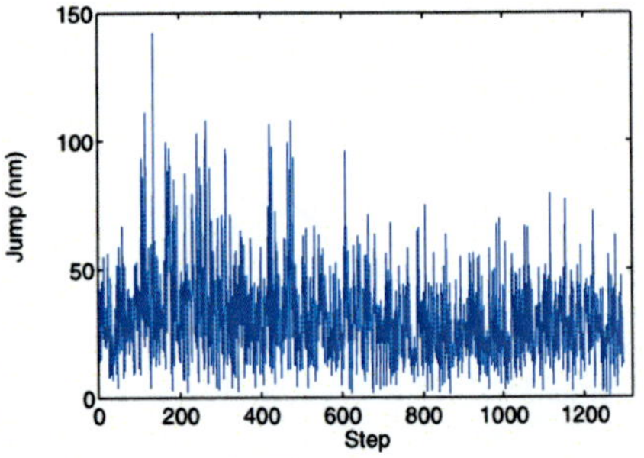

(b) Jump time series

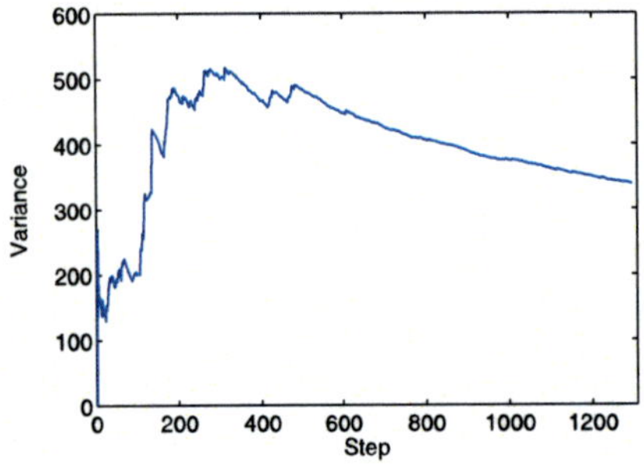

(c) Variance trend

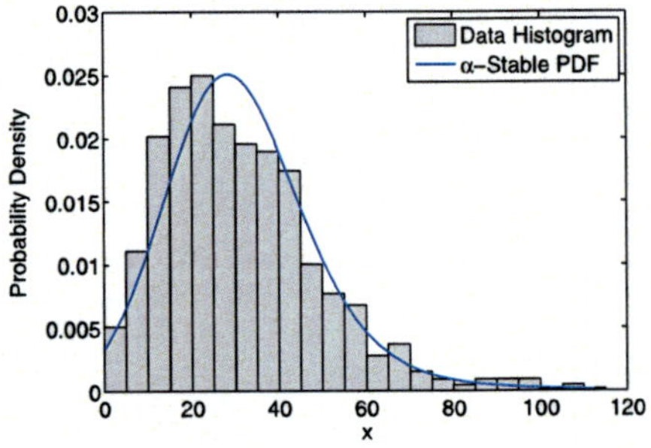

(d) Histogram and fitted α-stable PDF

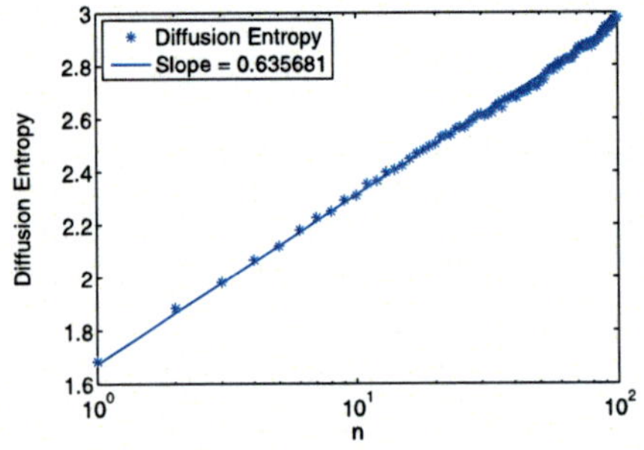

(e) Diffusion Entropy method estimation result

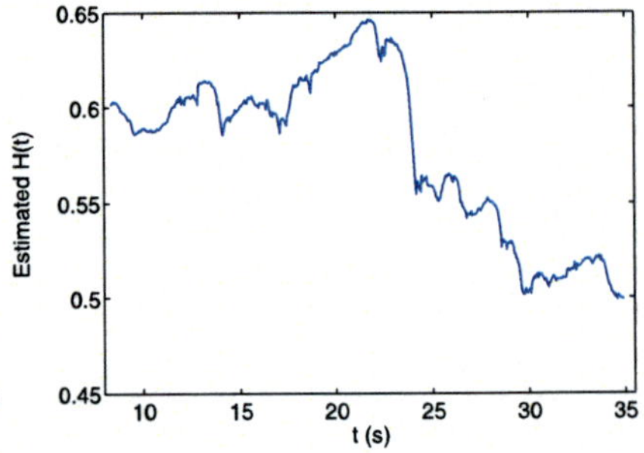

(f) Estimated $\hat{H}(t)$

Fig. 11.13 Analysis results for Experiment 3

11.6 Chapter Summary

In this chapter, the heavy-tailed distribution and local memory characteristics of ten MHCI molecular jump time series were analyzed. The histograms of these ten jump time series were fitted by α-stable distributions, and the tail thickness of the distributions was quantified using the characteristic exponent parameter α. The long memory and local memory characteristics were tested using the Diffusion Entropy method and the sliding-windowed Koutsoyiannis' method, respectively. The levels

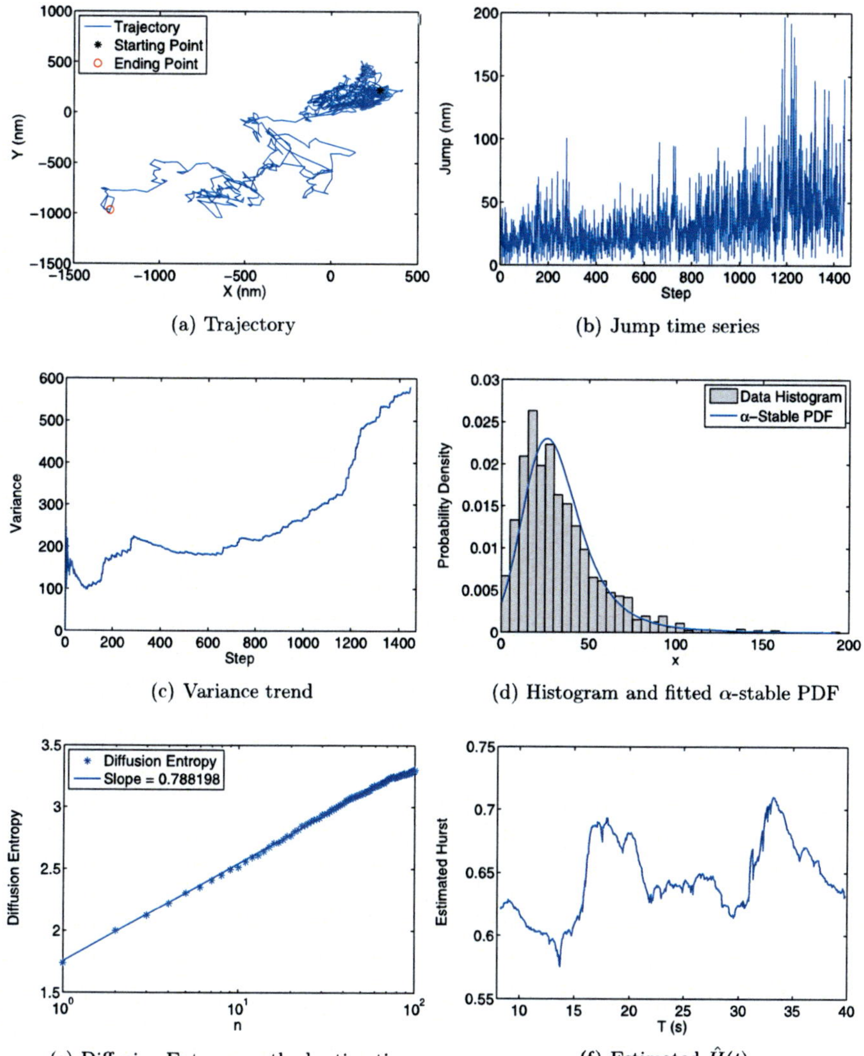

(a) Trajectory

(b) Jump time series

(c) Variance trend

(d) Histogram and fitted α-stable PDF

(e) Diffusion Entropy method estimation result

(f) Estimated $\hat{H}(t)$

Fig. 11.14 Analysis results for Experiment 4

of long memory and local memory are quantified by Hurst parameter H and local Hölder exponent $H(t)$. The analysis results show that the MHCI molecular jump time series obviously have heavy-tailed distribution and local memory characteristics. The local Hölder exponent can reflect the essential changes of these MHCI molecular motions. The analysis results of heavy-tailed distribution, long memory and local long memory for these MHCI molecules provide some additional yet essential insights into the regulation mechanisms underlying cell membrane organization. Besides, the analysis results provide some useful information for understanding

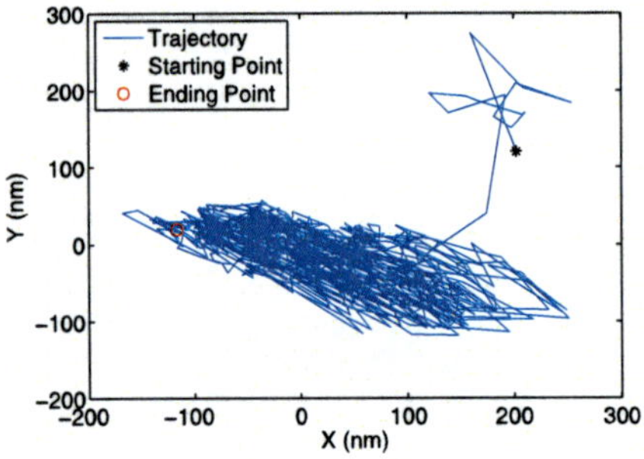

(a) Trajectory

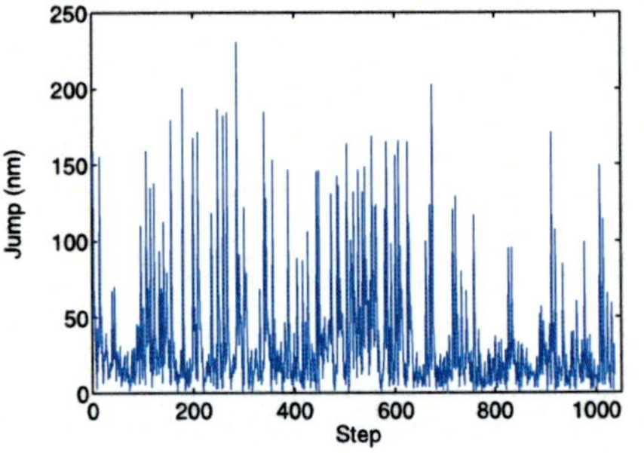

(b) Jump time series

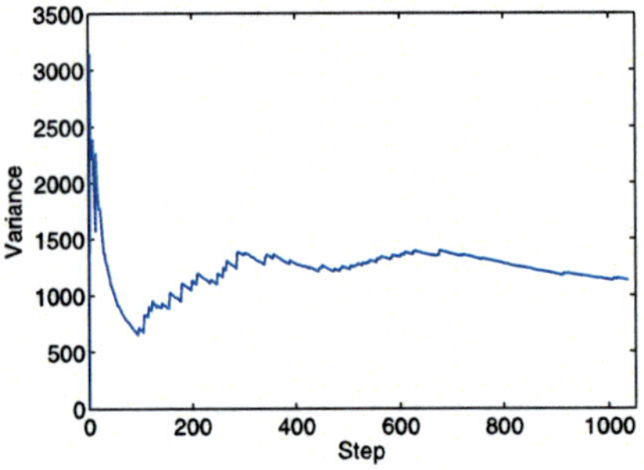

(c) Variance trend

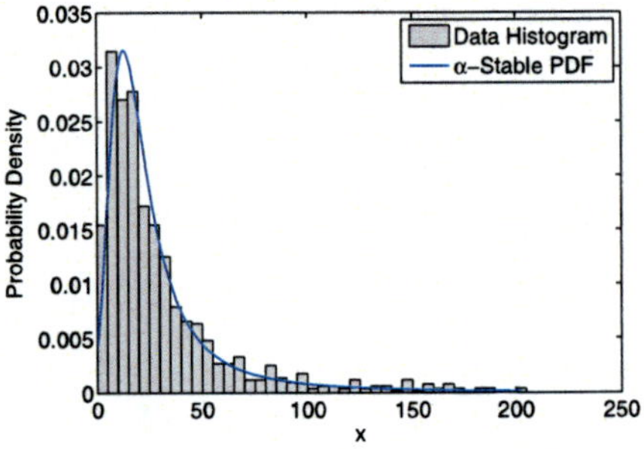

(d) Histogram and fitted α-stable PDF

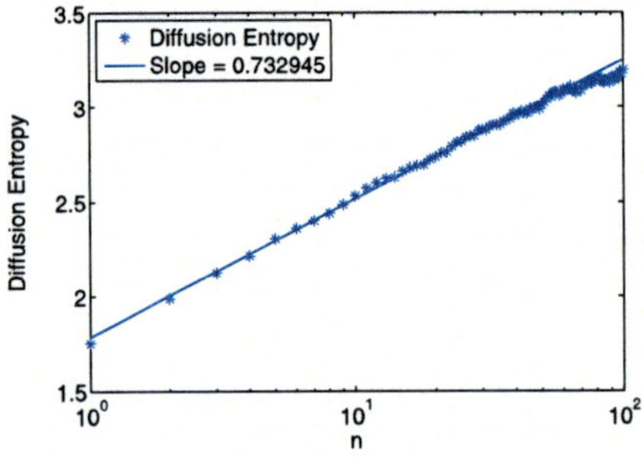

(e) Diffusion Entropy method estimation result

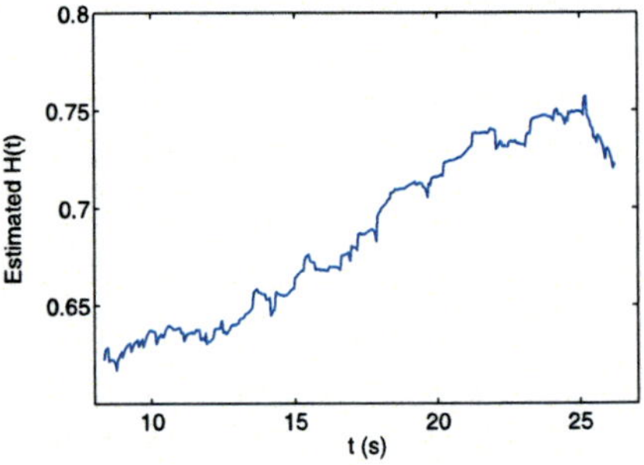

(f) Estimated $\hat{H}(t)$

Fig. 11.15 Analysis results for Experiment 19

the motion of the individual molecule. According to the above analysis, the motion of a single MHCI molecule can be well modeled neither by fractional Brownian motion [24], which has long memory with a constant Hurst parameter but has no heavy-tailed distribution, nor by Lévy motion [19], which has heavy-tailed distribution but has no long memory characteristic. The most appropriate model is one which can capture both the heavy-tailed distribution and local memory characteristics of the motion of a single MHCI molecule. We believe the data processing method in this chapter can find wide applications in processing other bioengineering signals such

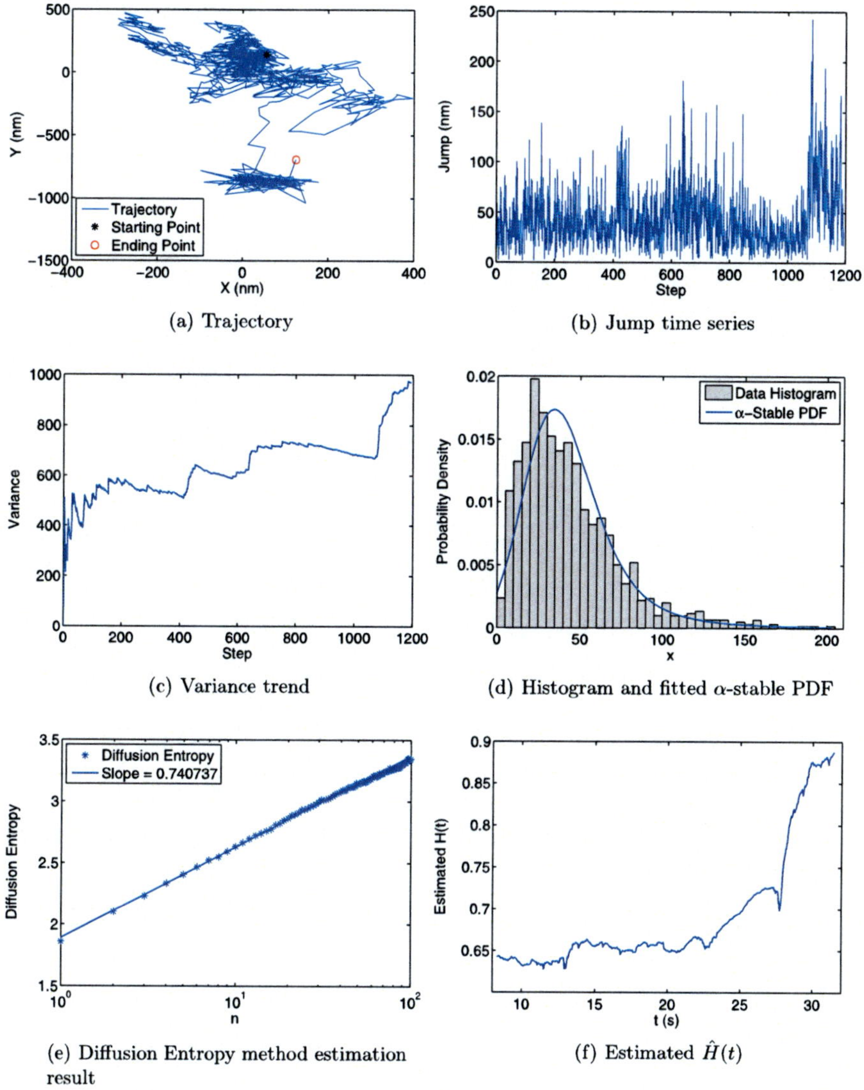

(a) Trajectory

(b) Jump time series

(c) Variance trend

(d) Histogram and fitted α-stable PDF

(e) Diffusion Entropy method estimation result

(f) Estimated $\hat{H}(t)$

Fig. 11.16 Analysis results for Experiment 24

as bacteria chemotaxis behavior quantification and other cases involving random motions.

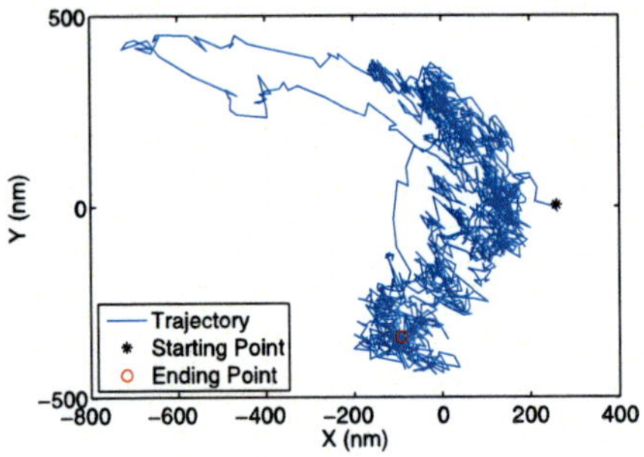

(a) Trajectory

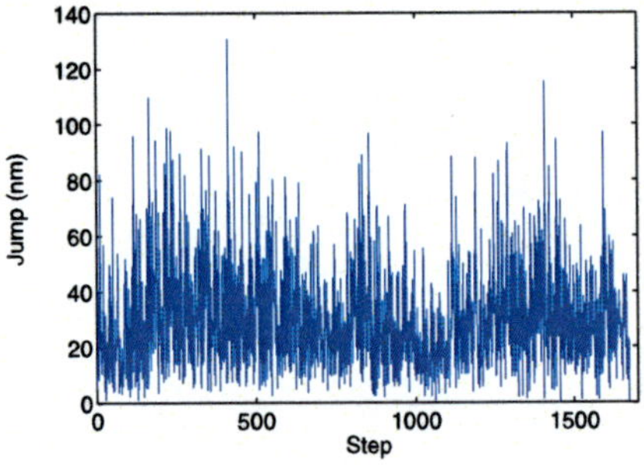

(b) Jump time series

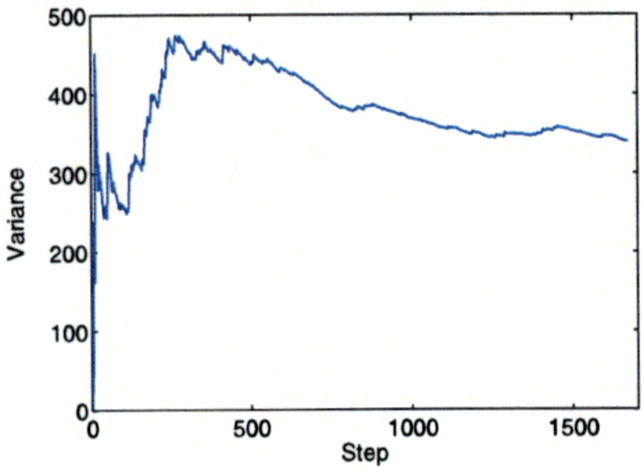

(c) Variance trend

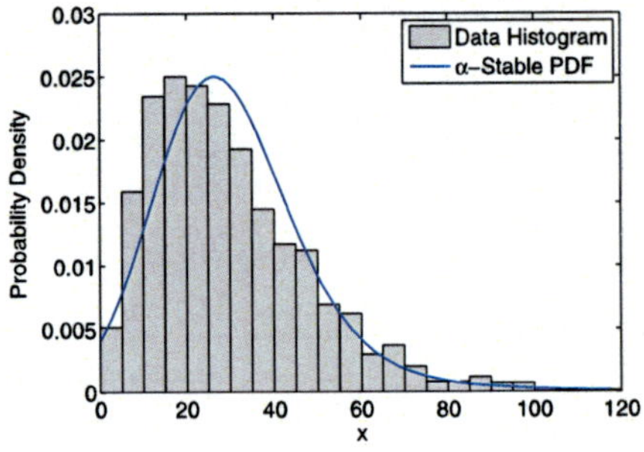

(d) Histogram and fitted α-stable PDF

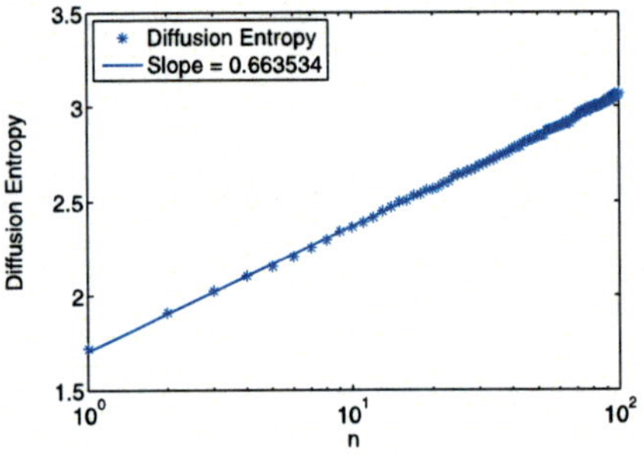

(e) Diffusion Entropy method estimation result

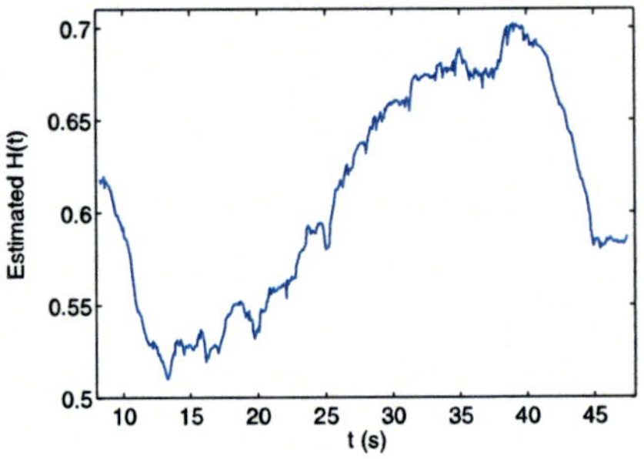

(f) Estimated $\hat{H}(t)$

Fig. 11.17 Analysis results for Experiment 28

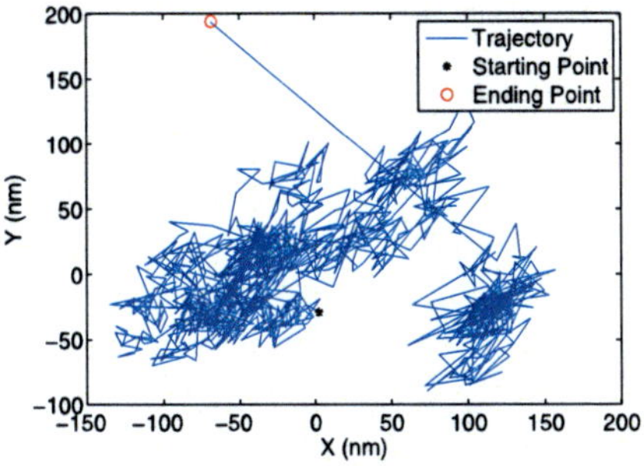

(a) Trajectory

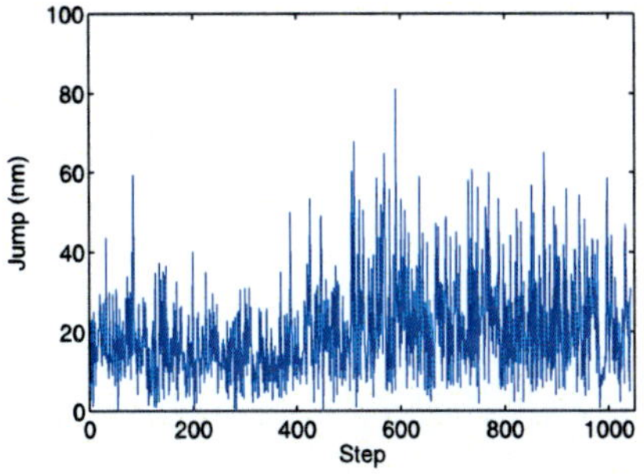

(b) Jump time series

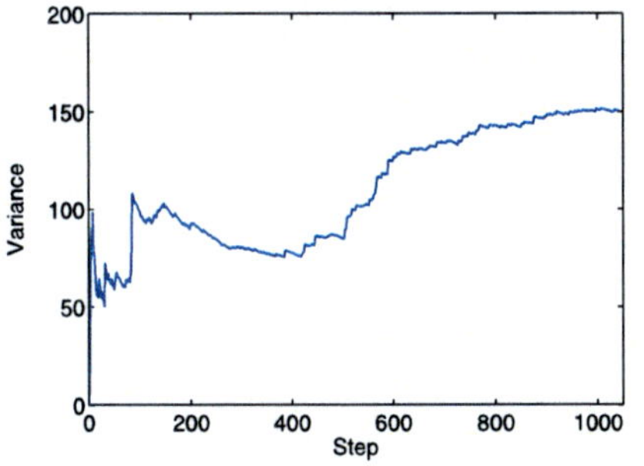

(c) Variance trend

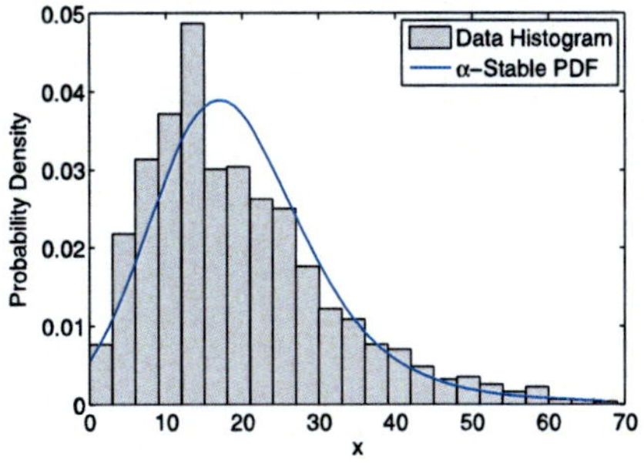

(d) Histogram and fitted α-stable PDF

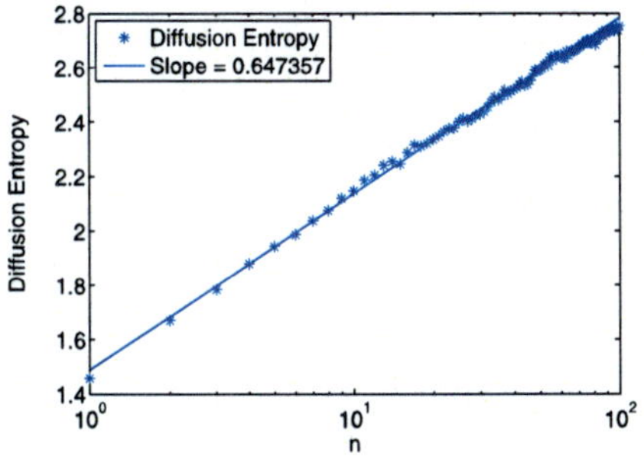

(e) Diffusion Entropy method estimation result

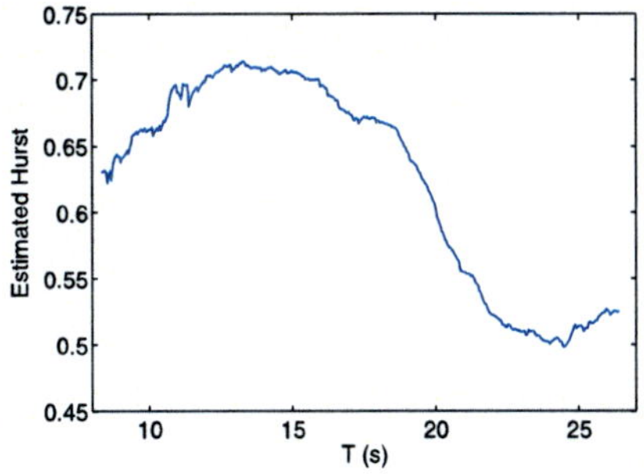

(f) Estimated $\hat{H}(t)$

Fig. 11.18 Analysis results for Experiment 32

Chapter 12
Non-linear Transform Based Robust Adaptive Latency Change Estimation of Evoked Potentials

12.1 Introduction

It is very important to monitor and detect latency changes of evoked potentials (EP) during an operation, so as to find and diagnose the possible disease or injury in the central nervous system of the patient [147, 149, 310]. Compared to the received noises, such as electroencephalogram (EEG), the EP signal obtained from the detector is very weak. The SNR is usually -10 dB or even lower [108]. Thus the principal issue for detecting the latency changes in the noise contaminated EP is to minimize the impact of these noises.

Traditionally, noises in EP signals are considered to be i.i.d. Gaussian random processes [200]. This assumption is reasonable in many situations. It is also convenient to analyze and to process the EP signals under this assumption. However, the EEG signals were found to be non-Gaussian in some studies [115, 146], and the measurement noise in the EP signals obtained in an operating room or other hostile environment may contain artifacts with characteristics far from being Gaussian. In particular, the measurement noise in the impact acceleration experiment, one of the applications we are interested in, is very impulsive and thick or heavy tailed in its distribution function [146, 198], both of which are distinctive features of the non-Gaussian lower order α-stable process [184, 215]. Due to the thick tails, the lower order α-stable processes do not have finite second or higher order moments. This feature may cause all second-order moment based algorithms to degenerate in their performances or to function sub-optimally [215]. A fractional lower order moment based algorithm referred to as the DLMP (direct least mean p-norm) was proposed in [146], sound to be robust under both Gaussian and the lower order α-stable noise conditions.

An α-stable distribution was used to describe the noise contaminated EP signal and its additive noises. The latency change estimation results of the DLMP are more robust than that of the DLMS (direct least mean square) algorithm [88, 149] under both Gaussian and lower order α-stable noise conditions. However, the performance advantage of the DLMP under lower order α-stable noise condition depends on an accurate estimation of the α parameter. Such an accurate estimation of α value of the

H. Sheng et al., *Fractional Processes and Fractional-Order Signal Processing*,
Signals and Communication Technology,
DOI 10.1007/978-1-4471-2233-3_12,

noisy process is not easily achieved in practice, especially in real time applications. In order to solve this problem, $p = 1$ is fixed in the DLMP, and a signed adaptive algorithm (SDA) is formed [148, 239]. A new problem is introduced by the SDA: the estimation error increases because of the sign transform to the adaptive error. This chapter proposes a new nonlinear transform based adaptive latency change estimation algorithm (referred to as NLST) that creates better features than those of the DLMP, without the need to estimate the α value.

12.2 DLMS and DLMP Algorithms

12.2.1 Signal and Noise Model

The signal model for latency change estimation of EP signal is defined as

$$\begin{aligned} x_{1n}(k) &= s_n(k) + v_{1n}(k), \qquad && k = 0, 1, \ldots, K-1; n = 1, 2, \ldots, N, \\ x_{2n}(k) &= s_n(k - D_n) + v_{2n}(k), && k = 0, 1, \ldots, K-1; n = 1, 2, \ldots, N, \end{aligned} \tag{12.1}$$

where $x_{1n}(k)$ and $x_{2n}(k)$ denote the reference and ongoing EP signals, $s_n(k)$ and $s_n(k - D_n)$ are the noise free EP signals, $v_{1n}(k)$ and $v_{2n}(k)$ are the background EEG and other noises in EP signals; D_n is the latency change to be estimated in the nth sweep; k is the discrete time variable. In EP study, $x_{1n}(k)$ is normally obtained by averaging many sweeps of EP so the noise $v_{1n}(k)$ is not significant and can even be negligible (if the sweep number participating in the average is large enough) [146, 147]. When the central nervous system (CNS) condition remains the same, the latency should remain constant or fluctuate only minimally, so the latency change D_n should be close to zero. However, the latency change D_n can be significant when the CNS condition varies, especially with possible injury.

The noise contaminated EP signal and its additive noises are described using α-stable distribution here. The α values of EP signals obtained from the impact acceleration experiments were estimated with the sample fractile method [215]. The results show that the α values of both noise contaminated EP signals and the noises themselves are between 1.06 and 1.94 [146], indicating that the noises in EP signals are lower order α-stable noises. Since the second order moment of a lower order α-stable process tends to be infinity, the properties of second order moment based processors, such as the DLMS, degenerate significantly under such noise conditions.

12.2.2 DLMS and Its Degradation

The DLMS [88] is a widely used time delay estimation algorithm proposed by Etter *et al.* Kong *et al.* [149] applied it in the latency change estimation and analyzed it theoretically. The adaptive iteration equation of this algorithm is given as

$$\widehat{D}_n(k+1) = \widehat{D}_n(k) + \mu e_n(k)[x_{1n}(k - \widehat{D}_n - 1) - x_{1n}(k - \widehat{D}_n + 1)], \tag{12.2}$$

where $e_n(k) = x_{2n}(k) - x_{1n}(k - \widehat{D}_n)$ is the error function, $\widehat{D}_n$ is the estimation of D_n. If the additive noise $v_{2n}(k)$ is an α-stable process, only moments of order less than α of $e_n(k)$ are finite according to the fractional lower order moment theory [184, 215]. This will have the variance of $\Delta\widehat{D}_n = \widehat{D}_n(k+1) - \widehat{D}_n(k)$, and tend to be infinity. It means that the DLMS algorithm degenerates significantly under the lower order α-stable noise conditions.

12.2.3 DLMP and Its Improvement

In the DLMP algorithm, the αth norm of the error function $J = \|e_n(k)\|_\alpha$ is used as the cost function of the adaptive system, by which the degeneration caused by the second order moment is avoided. Based on the fractional lower order moment, the αth order norm of a $S\alpha S$ process is proportional to its pth order moment, if $1 < p \leq \alpha$ is met. Thus, the cost function of the adaptive system can be written as:

$$J = E[|e_n(k)|^p]]. \tag{12.3}$$

By using the gradient technique and identical equation $A^{\langle p-1\rangle} = |A|^{p-1}\operatorname{sgn}(A)$, we get the iteration equation and the limit condition as

$$\begin{aligned}\widehat{D}_n(k+1) = \widehat{D}_n(k) &+ \frac{\mu p}{2}|e_n(k)|^{p-1}\operatorname{sgn}[e_n(k)] \\ &\times [x_{1n}(k - \widehat{D}_n - 1) - x_{1n}(k - \widehat{D}_n + 1)], \\ &1 < p \leq \alpha \leq 2.\end{aligned} \tag{12.4}$$

If we take $p = 2$, the DLMP in (12.4) becomes the DLMS in (12.2). It means that the DLMS is a special case of the DLMP. It can be proven that the DLMP maintains robustness under the lower order α-stable noise conditions because the adaptive error $e_n(k)$ with the lower order α-stable distribution is transformed into a second order moment process by $[e_n]^{\langle p-1\rangle} = |e_n|^{p-1}\operatorname{sgn}[e_n]$.

From the above discussion we know that the p value of the DLMP has to be bounded by the limit condition in (12.4). Otherwise, the DLMP may diverge. A proper selection of p value depends on a proper estimation of the characteristic exponent α of the noise contaminated signals. However, it is not easy to estimate the α parameter continually during the adaptive iteration continually. On the other hand, if we choose $p \to 1$, the limit condition in (12.4) can be met definitely, and the estimation of α parameter becomes unnecessary. Thus, the cost function of the DLMP in (12.3) becomes $J = E[|e_n(k)|]$, and the iteration equation in (12.4) becomes

$$\widehat{D}_n(k+1) = \widehat{D}_n(k) + \frac{\mu}{2}\operatorname{sgn}[e_n(k)][x_{1n}(k - \widehat{D}_n - 1) - x_{1n}(k - \widehat{D}_n + 1)]. \tag{12.5}$$

Equation (12.5) is referred to as the SDA algorithm [148, 239]. The SDA solves the problem of selecting the p parameter by a sign function in its adaptive iteration equation. It also introduces the same such errors in the latency change estimation results from the same kind of sign or binary transform. Actually, the amplitude information of $e_n(k)$ is lost in the transform.

12.3 NLST Algorithm

12.3.1 NLST Algorithm

We have just mentioned that, in essence, the SDA algorithm changes $e_n(k)$ to a binary sequence from a lower order α-stable process with a sign function. Such transform results not only in a suppression of the lower order α-stable noises, but also in a loss of the amplitude information in $e_n(k)$, which causes a significant increase of the latency change estimation error.

In fact, many nonlinear functions can be used to suppress the lower order α-stable noise. The ideal nonlinear transform function should have the following features: it can eliminate the impact of the lower order α-stable noises, but it does not cause a severe distortion to the normal EP signal. The Sigmoid function, widely used in the artificial neural network, is a very good nonlinear function for both purposes. By using the Sigmoid function, this chapter proposes a nonlinear transform based adaptive latency change estimation algorithm (NLST). Our goals are:

- To guarantee the algorithm converges smoothly under the lower order α-stable noise conditions;
- To compensate for the lost amplitude information of $e_n(k)$;
- Never to estimate the α parameter during the process of the adaptive iteration.

The iteration equation of the NLST is given in (12.6) as follows:

$$\widehat{D}_n(k+1) = \widehat{D}_n(k) + \mu \left[\frac{2}{1+\exp[-\lambda e_n(k)]} - 1 \right] \times [x_{1n}(k - \widehat{D}_n - 1) - x_{1n}(k - \widehat{D}_n + 1)], \tag{12.6}$$

where $\frac{2}{1+\exp[-\lambda e_n(k)]} - 1$ is a bipolar Sigmoid function, and $\frac{1}{\lambda} > 0$ is a constant proportional to the power of $e_n(k)$. λ is used as a scale factor to fit various signals and noises.

12.3.2 Robustness Analysis of the NLST

According to the fractional lower order moment theory and the properties of the lower order α-stable process, we analyze the robustness of the NLST algorithm under the lower order α-stable noise conditions. Three results are obtained from the analysis of the transform of the error function with the Sigmoid function in the NLST. We temporarily omit the subscript n and denote $e_n(k)$ with $\mu(k)$ for convenience. The following expression

$$\omega(k) = \frac{2}{1+\exp(-\lambda\mu(k))} - 1 \tag{12.7}$$

is then used to denote the nonlinear transform of the error function $e_n(k)$ in (12.6).

Result 12.1 *If $\mu(k)$ is a $S\alpha S$ process ($\beta = 0$), and $a = 0$, then $\omega(k)$ is symmetric distributed with zero mean in its probability density function when $1 < \alpha \le 2$.*

Proof Since $\mu(k)$ is an $S\alpha S$ process with $\beta = 0$ and $a = 0$, then its probability density function $f(x)$ is an even function, and symmetric with $a = 0$, that is

$$f(\mu) = f(-\mu). \tag{12.8}$$

Omit the discrete time variable k temporally and define

$$\omega = \Psi(\mu) = \frac{2}{1+\exp(-\lambda\mu)} - 1. \tag{12.9}$$

Since $\Psi(-\mu) = \frac{2}{1+\exp(\lambda\mu)} - 1 = -\Psi(\mu)$, we know that $\Psi(\mu)$ is an odd function of μ. Since the first order derivative of (12.9) is

$$\frac{\mathrm{d}\omega}{\mathrm{d}\mu} = \Psi'(\mu) = \frac{2\lambda\exp(-\lambda\mu)}{1+\exp(-\lambda\mu))^2} > 0, \tag{12.10}$$

we know that $\omega = \Psi(\mu)$ is a monotonic incremental function of μ. Suppose its inverse function is

$$\mu = \Psi^{-1}(\mu) = \xi(\omega). \tag{12.11}$$

Then $\mu = \xi(\omega)$ is a monotonic odd function of ω, since $\omega = \Psi(\mu)$ is a monotonic odd function of μ. Thus we have

$$\xi(-\omega) = -\xi(\omega). \tag{12.12}$$

Assume that the probability density function of ω is $g(\omega)$, then from

$$g(\omega) = f(\mu)\left|\frac{\mathrm{d}\mu}{\mathrm{d}\omega}\right| = f[\xi(\omega)]\left|\frac{\mathrm{d}\mu}{\mathrm{d}\omega}\right|, \tag{12.13}$$

and (12.8) and (12.11), we have

$$g(-\omega) = f(-\mu)\left|\frac{\mathrm{d}\mu}{\mathrm{d}\omega}\right| = f[-\xi(\omega)]\left|\frac{\mathrm{d}\mu}{\mathrm{d}\omega}\right| = f[\xi(\omega)]\left|\frac{\mathrm{d}\mu}{\mathrm{d}\omega}\right| = g(\omega). \tag{12.14}$$

Obviously $g(\omega)$ is an even function of ω. On the other hand, we know that

$$E[\omega] = \int_{-\infty}^{\infty} \omega g(\omega)\mathrm{d}\omega = 0. \tag{12.15}$$

So the probability density function of $\omega(k)$ has symmetric distribution and with zero mean. □

Result 12.2 *If $\mu(k)$ is an $S\alpha S$ process with $\gamma > 0$ and $a = 0$, then we have $\|\omega(k)\|_\alpha > 0$, and the mean value of $\omega(k)$ is zero when $1 < \alpha \le 2$.*

Proof We know from the proof of Result 1 that the mean value of $\omega(k)$ is zero. Suppose that the αth order norm of $\mu(k)$ is

$$\|\mu(k)\|_\alpha = \left[|\mu(1)|^\alpha + |\mu(2)|^\alpha + \cdots + |\mu(N)|^\alpha\right]^{1/\alpha} = \Upsilon_\mu^{1/\alpha}, \tag{12.16}$$

where N is the sample number of $\mu(k)$. Since $\Upsilon > 0$, we have $\Upsilon^{1/\alpha} > 0$, then at least one $i \in [1, N]$ can be found to meet $\mu(i) \ne 0$. Thus, at least one $i \in [1, N]$ exists to guarantee $\omega(i) \ne 0$. Then we have

$$\|\omega(k)\|_\alpha = \left[|\omega(1)|^\alpha + |\omega(2)|^\alpha + \cdots + |\omega(N)|^\alpha\right]^{1/\alpha} > 0. \tag{12.17}$$

□

Result 12.3 *If $\mu(k)$ is an $S\alpha S$ process with $1 < \alpha \le 2$ and $a = 0$, then $\omega(k)$ has the finite second order moment with zero mean (referred to as second order moment process) when $1 < p \le \alpha$.*

Proof We know from Result 1 that the mean value of $\omega(k)$ is zero, and $\Psi(\mu)$ is a monotonic incremental function of μ. From (12.9) we have:

$$\max[\omega] = \lim_{\mu\to\infty} \omega = 1, \tag{12.18}$$

and

$$\min[\omega] = \lim_{\mu\to\infty} \omega = -1. \tag{12.19}$$

Then we have:

$$\max[\omega^2] = 1. \tag{12.20}$$

Since $g(\omega)$ is the probability density function of ω and it meets $g(\omega) \ge 0$, we get

$$E[\omega^2] = \int_{-\infty}^{\infty} \omega^2 g(\omega) \mathrm{d}\omega \le \int_{-\infty}^{\infty} \max[\omega^2] g(\omega) \mathrm{d}\omega = 1. \tag{12.21}$$

We know from (12.21) that the second order moment of $\omega = \Psi(\mu) = 2/[1 + \exp(-\lambda\mu)] - 1$ exists and is bounded by 1. Considering the zero mean of $g(\omega)$, we conclude that the $\omega(k)$ is a finite second order moment process. □

We know from Result 12.1 and Result 12.2 that the mean value of $\omega(k) = 2/[1 + \exp(-\lambda\mu)] - 1$ is zero, its probability density function is symmetric, and its αth order norm is positive, if $x(k)$ is a $S\alpha S$ process with $\beta = 0$, $a = 0$, $\Upsilon > 0$ and $1 < \alpha \leq 2$. We also know from Result 3 that $y(k)$ has the finite second order moment. Summarizing the above three results we conclude that $y(k)$ is a second order moment process.

We get from the three results that the direct reason for the robustness of the NLST under the lower order α-stable noise condition is that the iteration equation (12.6) transforms the lower order α-stable process $e_n(k)$ into a second order moment process by the nonlinear transform $\Psi(e_n(k)) = 2/[1 + \exp(-\lambda e_n(k))] - 1$. As a result, the same performance analysis method employed in [310] for the DLMS can be used to analyze the whole performance of the new algorithm.

12.4 Simulation Results and Discussion

Computer simulation is conducted to verify the robustness of the NLST under the lower order α-stable noise conditions. Signals and noises are constructed as (12.1), in which the noise free signal obtained from the average of 1000 pre-impact EP sweeps is used as $s_n(k)$, and the lower order α-stable noises ($\alpha = 1.2, 1.5, 1.8$) are generated to simulate the additive background noises $\nu_{2n(k)}$, respectively. The latency changes are set as follows:

$$D_n = \begin{cases} 0, & 1 \leq n \leq 100, \\ 10T_s, & 101 \leq n \leq 200, \\ 10T_s(400 - n)/200, & 201 \leq n \leq 400, \\ 0, & 401 \leq n \leq 500, \end{cases} \tag{12.22}$$

where T_s is the sampling interval of the EP signals. The mixed Signal to noise ratio (MSNR) [146, 148] defined in (12.23) for the lower order α-stable noises are set to be -10 dB, -15 dB and -20 dB respectively.

$$\mathrm{MSNR} = 10\log_{10}\left(\sigma_s^2/\Upsilon_\nu\right), \tag{12.23}$$

where Υ_ν is the dispersion parameter of the noise $\nu_{2n}(k)$.

Four adaptive EP latency change estimation algorithms (DLMS, DLMP, SDA and NLST) are checked and compared in the simulation. The adaptive gains for different algorithms are adjusted so that the estimation error powers of the first 100 sweeps for the four algorithms are the same, with the DLMS in its critical divergence condition. With these equivalent adaptive gains, the EP latency changes under different MSNRs and α conditions are estimated or detected. The performance of different algorithms is compared by the tracking speed to the step change of the latency between the 100th and 101st sweeps with the average of 20 independent runs of each algorithm. The simulation results are shown in Fig. 12.1. The left part in Fig. 12.1 shows a tracking speed comparison among the four algorithms. The curve with open circle is obtained by the DLMS. It is clear that the tracking speed of the

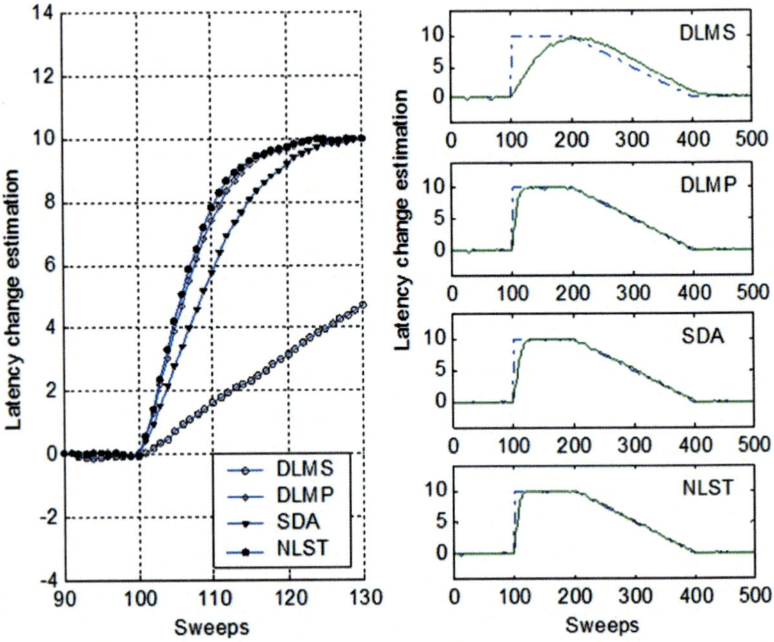

Fig. 12.1 The results of EP latency change estimation (MSNR = −15 dB, $\alpha = 1.5$). The *left part* of the figure shows the results for tracking the step change of the EP latency with the DLMS, DLMP, SDA, and NLST algorithms. The tracking speed of the NLST is the fastest. The *right part* of the figure shows the estimation results obtained from the four algorithms for the whole 500 sweeps. The estimation error powers of the four algorithms under the given condition are 4.8936 (DLMS), 0.7546 (DLMP), 1.0666 (SDA), and 0.6942 (NLST) respectively

DLMS is very slow, caused by a very small adaptive gain so as to ensure its convergence. The tracking speed of the DLMP (the curve with open diamond), the SDA (the curve with solid triangle) and the NLST (the curve with solid circle) are much better than that of the DLMS, with the NLST the fastest. The basic reason for the NLST to have the best performance under the lower order α-stable noise conditions is that the NLST suppresses the impulsive spikes in $e_n(k)$ by the nonlinear transform with the Sigmoid function, and maintains the normal amplitude information in it. The right part in Fig. 12.1 shows the EP latency change in estimation results of the four algorithms to the whole 500 sweeps. We see from the figures that the proposed algorithm has a better estimation result than others.

The Error powers of the EP latency change estimation for the four algorithms under various MSNRs and α values are shown in Fig. 12.2. The results are all the average of 20 independent runs. Obviously, the estimation accuracy of the NLST is best.

As we mentioned above, it is very important to accurately monitor and detect the latency changes of EPs during an operation. With the DLMP algorithm, we can get both high accuracy and convergence speed under both Gaussian and the lower order α-stable noise conditions if the α value of the signal and noise is known.

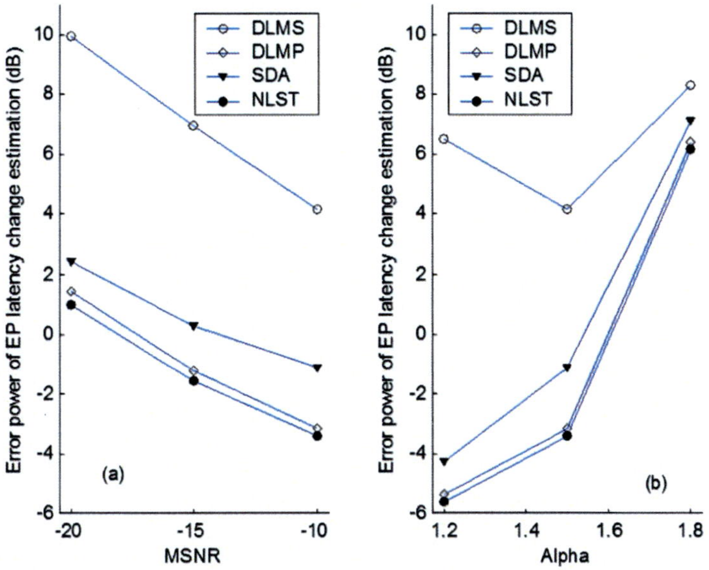

Fig. 12.2 The error powers of the EP latency change estimation of the four algorithms. **(a)** The estimation error power when $\alpha = 1.5$. **(b)** The estimation error power when MSNR $= -10$ dB

Table 12.1 The runtimes of the DLMS, DLMP, SDA and NLST algorithms

Algorithms	DLMS	DLMP	SDA	NLST
Runtimes (s)	3.57	4.88	3.68	4.51

However, the DLMP may not converge if the α value changes during the latency change detection, since the DLMP is not able to estimate the α value dynamically. The SDA algorithm introduces an extra error by using a nonlinear transform with the sign function, although it eliminates the dependence of the DLMP on the estimation of the α value. Such an extra error is not acceptable in a high accuracy detection. The NLST algorithm proposed in this chapter compensates the error caused by the nonlinear transform, and does not depend on the knowledge of the α value since the adoption of the continuous nonlinear transform with the Sigmoid function. As a result, a high accurate and robust latency change detection can be realized with the new algorithm.

On the other hand, the computational complexity is also studied by comparing the runtime of the NLST with the DLMS, DLMP and SDA on a Pentium III computer under the MATLAB® environment (with 500 sweeps and totally 64000 data samples). The runtimes of the four algorithms are shown in Table 12.1.

From Table 12.1 we see that the runtimes of the four algorithms are all less than 5 seconds, which is fast enough for the clinical application. The runtime of the proposed algorithm is about $1/4$ longer than those of the DLMS and SDA, and is a little bit shorter than the DLMP's. It can be said that the improvement on the performance of the NLST does not cost much in term of the computational complexity.

12.5 Chapter Summary

This chapter proposes a new adaptive EP latency change estimation algorithm (NLST) based on the fractional lower order moment and the nonlinear transform of the error function. The computer simulation shows that this new algorithm is robust under the lower order α-stable noise conditions, and it also achieves a better performance than the DLMS, DLMP and SDA algorithms without the need to estimate the α value of the EP signals and noises.

Chapter 13
Multifractional Property Analysis of Human Sleep Electroencephalogram Signals

13.1 Introduction

An electroencephalogram (EEG) provides very useful information on the exploration of brain activity, diagnosis of cerebral disease, causes identification of brain disorders, and so on. The analysis of an EEG signal is complicated due to its nonlinear, irregular and non-stationary properties. In order to investigate the in-depth characteristics of complex EEG signals, many nonlinear techniques were proposed and studied. The scaling property of fluctuations in the human electroencephalogram was studied in [124]. The Hurst parameter estimation method was used for epileptic seizure detection in [224]. The nonlinear properties of EEG signals of normal persons and epileptic patients were investigated in [216]. Fractal analysis of EEG signals in the brain of epileptic rats was investigated in [178], and the dynamics of EEG entropy was analyzed using the Diffusion Entropy method in [127]. In this chapter, the multifractional property of human sleep EEG signals is investigated.

The earliest detailed description of various stages of sleep was provided in [177]. Sleep is generally divided into two broad types: rapid eye movement (REM) and non-rapid eye movement (NREM) sleep. Based on the "Rechtschaffen and Kales sleep scoring manual" [245], NREM is divided further into four stages: Stage I, Stage II, Stage III and Stage IV. The Stages III and IV were combined into Stage III in "The AASM Manual for the Scoring of Sleep and Associated Events" [125]. Sleep progresses in a cycle from Stage I to REM sleep, then the cycle starts over again. Stage I sleep is referred to as the first or earliest stage of sleep. This stage is characterized by low voltage, mixed frequency EEG with the highest amplitude in 2–7 Hz range [314]. Stage II is the principal sleep stage, which occupies 45% to 55% of total sleep in adults during a normal night's sleep. Stage II sleep is characterized by K complexes, which are characterized by a sharp negative wave followed by a slower positive one, and then the absence of slow waves. Stage III sleep is referred to as deep sleep which exhibits slow brain waves. This stage contains waves with 2 Hz or slower and with the amplitudes above 75 μV. REM stage sleep contains low voltage brain waves and rapid eye movements. Most memorable dreaming occurs in the REM stage. The function of sleep includes conservation of energy,

H. Sheng et al., *Fractional Processes and Fractional-Order Signal Processing*, Signals and Communication Technology,
DOI 10.1007/978-1-4471-2233-3_13,

restoration of tissues and growth, thermoregulation, neural maturation and memory and learning, regulation of emotions, and so on. Many efforts have been made to explain the function of sleep using sleep EEG signals. However, the studies and the understanding of sleep EEG signals are incomplete. Motivated by improving the understanding and quantification of sleep EEG signals, the multifractional property of various sleep stages is investigated in this chapter.

It has been found that an EEG signal is a fractional process, which can be characterized by the Hurst parameter $H \in (0, 1)$ [124, 174, 224, 318]. The fractional property of the EEG signals has been studied by many researchers [214, 216, 305]. Fractional process is a generic term for a class of processes which exhibit autocorrelations with hyperbolic decay, such as long memory process, long range dependent process and self similar process [22]. A fractional process with a constant Hurst parameter H can be used to more accurately characterize the long memory process than the traditional short-range dependent stochastic processes, such as Markov, Poisson or ARMA processes [73, 123]. However, the constant Hurst parameter cannot capture the local scaling characteristic of the stochastic processes and cannot describe the time-varying nature of the non-stationary process. So, the multifractional process with a time-varying long-memory parameter was investigated to explain the complex physical phenomena [232]. A multifractional process is the natural extension of a fractional process by generalizing the constant Hurst parameter H to the case where H is indexed by a time-dependent local Hölder exponent $H(t)$ [232]. The extension of a fractional process leads to a set of non-stationary and non-self-similar new stochastic processes, which can describe the complex behavior of non-stationary, nonlinear dynamic systems. The typical examples of local memory time processes are mGn and mBm [67]. In this chapter, the multifractional property of the sleep EEG signals in different sleep stages is studied using local Hölder exponent $H(t)$, which is estimated using sliding-windowed Kettani and Gubner's Hurst estimator [138].

13.2 Data Description and Methods

13.2.1 Data Description

The sleep EEG data for analysis were obtained from the MIT-BIH Polysomnographic Database, a collection of recordings of multiple physiologic signals during sleep, providing a research resource for complex physiologic signals-PhysioBank [100]. In the MIT-BIH Polysomnographic Database, all 16 subjects are male, aged from 32 to 56 (mean age 43), with weights ranging from 89 to 152 kg (mean weight 119 kg). The recording time is between 2 and 7 hours. The sleep EEG signals were digitized at a sampling frequency of 250 Hz and 12 bits/sample [126]. The sleep stage was determined according to the criteria of Rechtschaffen and Kales [245]. In all the recordings from the MIT-BIH Polysomnographic Database, various sleep disorders might manifest themselves through sleep disturbances in different sleep stages, since sleep is an active process involving characteristic physiological changes in the organs of the body. Almost everyone occasionally suffers from short-term sleep disorders. The most common sleep disorders include insomnia, narcolepsy and sleep apnoea. So it is not easy to find the long term sleep EEG recording without sleep disorders in the MIT-BIH Polysomnographic Database. In order to clearly analyze various sleep states, the recording SLP03 was selected. This recording contains long term (30 minutes) continuous EEG signals without any sleep disorders, one channel (C3-O1) of EEG signal in REM and NREM sleep stages, and annotations with sleep staging and apnea information, each applying to thirty-second intervals of the record.

13.2.2 Methods

In our study, the fractional property and the multifractional property of the sleep EEG signals selected from the recording SLP03 are studied using constant Hurst parameters H of short term (1 minutes) sleep EEG signals and local Hölder exponent $H(t)$ of long term (no less than 10 minutes) sleep EEG signals, respectively. The Hurst parameter H is estimated using Kettani and Gubner's Hurst estimator [138], which was evaluated in Chap. 3 and shown to have good robustness and to provide accurate estimation results for fractional processes [267]. The local Hölder exponent $H(t)$ is computed using sliding-windowed Kettani and Gubner's estimator, where the time series is truncated by a sliding window with constant width and the Hurst parameter of each truncated time series is estimated [268]. In the sleep EEG signals analysis, the window width $W_t = 30$ s was settled, since the sleep stage was signed every 30 s.

Many Hurst parameter estimators, including Kettani and Gubner's Hurst estimator, were designed to be applied to stationary fGn-like signals. If the analyzed signals behave as a non-stationary fBm-like time series, these estimators cannot provide the Hurst parameter but indexes related to H with range outside the unit interval [263]. Sleep EEG signals are a non-stationary time series, so the Kettani and Gubner's Hurst estimator cannot directly be used on the original sleep EEG signals. But this Hurst estimator can be applied to non-stationary sleep EEG signals after differentiation, since fGn-like series represent the increments of fBm-like processes and both the fGn-like and fBm-like signals are characterized by the same Hurst parameter by definition [263]. Therefore, the selected sleep EEG data for analysis were studied after differentiation. The analysis results of fractional and multifractional properties for sleep EEG signals during different sleep stages are provided in the next section.

13.3 Fractional Property of Sleep EEG Signals

The fractional property of EEG signals, such as scaling behavior, long range power-law correlation characteristics, and fractal property, have been investigated in [174,

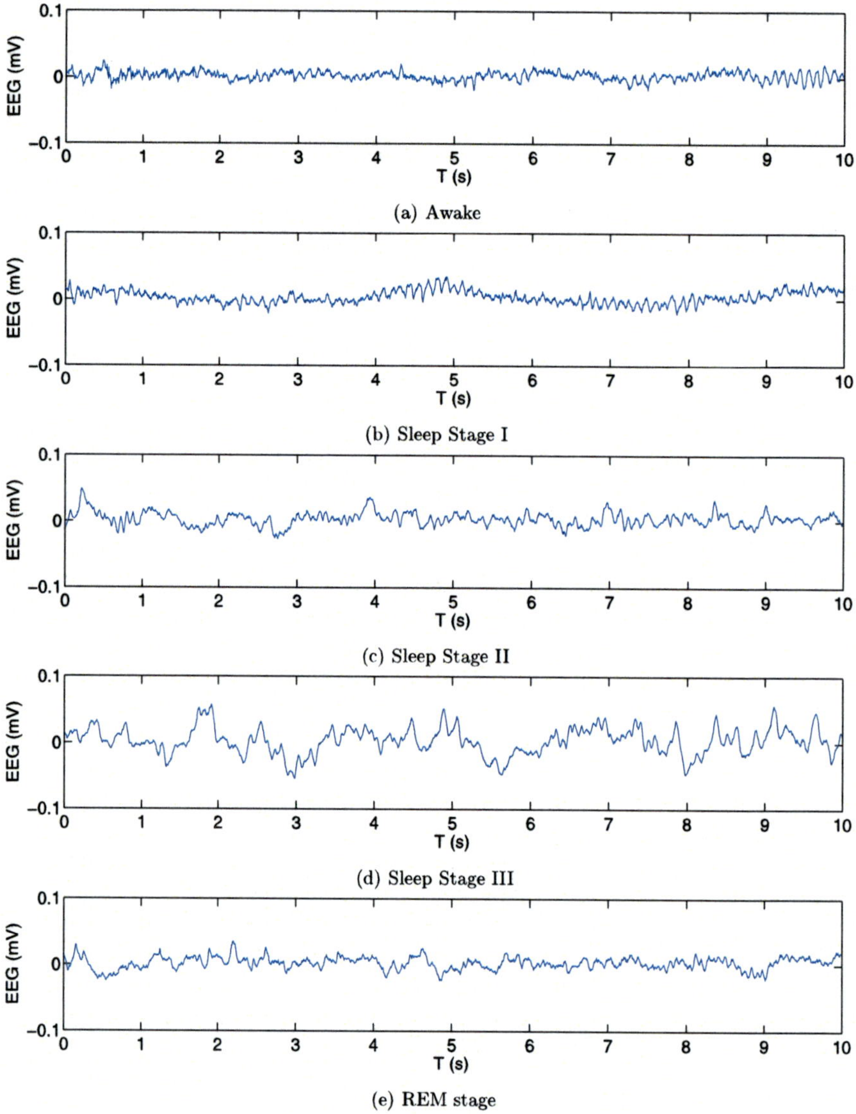

Fig. 13.1 Sleep EEG signals in different stages from recording SLP03

216, 224]. All the properties are based on the fact that the EEG signal is a fractional process, which can be characterized by the Hurst parameter $H \in (0, 1)$. In this subsection the fractional property of sleep EEG signals in REM and NREM sleep stages are analyzed using Kettani and Gubner's Hurst estimator.

Sleep EEG signals in various sleep stages, which were selected from the recording SLP03, are plotted in Fig. 13.1. The length of all the sleep EEG signal segments in Fig. 13.1 is 10 seconds, that is, 2500 data points. The sleep EEG signals in differ-

Table 13.1 Constant Hurst parameter of sleep EEG signal segments

	Awake	Stage I	Stage II	Stage III	REM
Segment 01	0.4588	0.4928	0.5477	0.7509	0.6924
Segment 02	0.5110	0.5530	0.5613	0.7549	0.7350
Segment 03	0.6782	0.6161	0.6015	0.7678	0.6779
Segment 04	0.6382	0.5743	0.6341	0.7451	0.6510
Segment 05	0.6688	0.6021	0.5917	0.7637	0.6444
Segment 06	0.3342	0.5185	0.6809	0.7547	0.6294
Segment 07	0.6239	0.4766	0.6972	0.7843	0.6395
Segment 08	0.7085	0.5688	0.6808	0.7786	0.6929
Segment 09	0.7315	0.4284	0.7055	0.7614	0.6886
Segment 10	0.4771	0.5975	0.6042	0.7729	0.7205
Average	0.5830	0.5428	0.6305	0.7634	0.6772

ent stages exhibit different characteristics. In our study, for each stage, 10 segments of 1-minute sleep EEG signal was analyzed using Kettani and Gubner's Hurst estimator. The estimated constant Hurst parameter $\widehat{H}$ of 10 segments for each sleep stage are presented in Table 13.1. The average of 10 Hurst parameters for each sleep stage is calculated and listed at the bottom of the table. From Table 13.1 we can see that sleep EEG signals are fractional processes, and the fractional properties of sleep EEG signals in various stages are different. During wakefulness, the marked changes of Hurst parameters can be found in EEG signals, because the brain activity is complex during wakefulness and the movement of a limb or eye will obviously influence the EEG signals. In sleep Stage I, a person is in a state of drowsiness with slow rolling eye movements. Most of the Hurst parameters of short term sleep EEG are within the range of 0.45 to 0.60. In sleep Stage II, the eye movements stop and brain waves become slower. Correspondingly, most of the Hurst parameters of short term sleep EEG signals are within the range of 0.6 to 0.7, which is higher than that of sleep Stage I. In sleep Stage III, a person enters a deep or slow-wave sleep. Most of the estimated Hurst parameters $\widehat{H}$ of short term sleep EEG are within the range of 0.75 to 0.80, which is the highest among all the sleep stages. In the REM sleep stage, eye movement rapid and rapid low-voltage EEG appears. Most of the estimated Hurst parameters $\widehat{H}$ of short term sleep EEG are within the range of 0.65 to 0.75, which is higher than that in Stage II, but lower than that in sleep Stage III. The averages of constant Hurst parameters at the bottom of the Table 13.1 also show the difference in Hurst parameters in various sleep stages. However, the constant Hurst parameter cannot capture the dynamic processes of sleep EEG signals. So, in the next subsection, the multifractional property of the sleep EEG signals in different sleep stages are investigated by estimating the local Hölder exponent $H(t)$ of long term sleep EEG signals.

13.4 Multifractional Property of Sleep EEG Signals

In some situations, the assumption that real-world processes exhibit a constant fractional property may not be reasonable [232]. Many recent empirical analyses of complex non-stationary, nonlinear dynamic systems have offered evidence about the limitations of fractional processes with constant Hurst parameter [39, 90]. The main reason is that the constant Hurst parameter cannot capture the local scaling characteristic of the stochastic processes. So, the multifractional processes with time varying local Hölder exponent $H(t)$ were extensively used for non-stationary, nonlinear physical processes. It is known that the sleep is a progressively changing process not a jumping process. Even in the same sleep stage, the EEG signal exhibits different variation. That means the EEG signal is not a stationary fractional process, but a non-stationary, nonlinear dynamic process. So the constant Hurst parameter alone is not sufficient to characterize or quantify the dynamic sleep stages. In this subsection, local Hölder exponent $H(t)$ is used to study the variable property of the sleep EEG signals.

Figure 13.2(a) illustrates the 30-minute sleep EEG signal segment. The 30-minute EEG signal segment includes awake state ($0\text{ s} < t \le 150\text{ s}$), sleep Stage I ($150\text{ s} < t \le 390\text{ s}$), Stage II ($390\text{ s} < t \le 1260\text{ s}$) and sleep Stage III ($1260\text{ s} < t \le 1800\text{ s}$). The difference of the 30-minute sleep EEG signal is illustrated in Fig. 13.2(b), and the estimated $\widehat{H}(t)$ is plotted in Fig. 13.2(c). The red dashed lines in Fig. 13.2(c) are estimated constant Hurst parameter $\widehat{H}$ of EEG signals in different sleep stages. The estimated Hurst parameters $\widehat{H}$ are different in different sleep stages, as follows, $\widehat{H}_{\text{Awake}} < \widehat{H}_{\text{Stage I}} < \widehat{H}_{\text{Stage II}} < \widehat{H}_{\text{Stage III}}$, where $\widehat{H}_{\text{Awake}} = 0.4587$, $\widehat{H}_{\text{Stage I}} = 0.5138$, $\widehat{H}_{\text{Stage II}} = 0.6703$ and $\widehat{H}_{\text{Stage III}} = 0.7612$. The blue line in Fig. 13.2(c) is the estimated local Hölder exponent $\widehat{H}(t)$. During the sleep stages in 30-minute EEG signal segment, the $\widehat{H}(t)$ is increasing gradually from about 0.45 to 0.80. In the awake state, the $\widehat{H}(t)$ of sleep EEG signals displays large fluctuations, which represents the intense brain activities during that stage. Compared with the awake stage, the $\widehat{H}(t)$ is of steadier in the drowsy sleep Stage I. During the sleep Stage II, the $\widehat{H}(t)$ increases gradually from about 0.5 to 0.7, which represents that the sleep proceeds from light sleep to deep sleep status. The $\widehat{H}(t)$ in sleep Stage III is the highest and steadiest, which is in accordance with the unusual eye movements of deep sleep. From local Hölder exponent $\widehat{H}(t)$ during a 30-minute sleep EEG signal segment we can see that the fluctuations of $\widehat{H}(t)$ manifest different characteristics during different sleep stages.

Figure 13.3(a) illustrates the 10-minute sleep EEG signal segment. This segment includes EEG signals in sleep Stage I ($0\text{ s} < t \le 60\text{ s}$), Stage II ($60\text{ s} < t \le 270\text{ s}$), REM sleep stage ($270\text{ s} < t \le 480\text{ s}$), and the stage of alternate REM and Stage I ($480\text{ s} < t \le 600\text{ s}$). The difference of the 10-minute sleep EEG signal segment is illustrated in Fig. 13.3(b), and the estimated constant Hurst parameters and local Hölder exponent $\widehat{H}(t)$ are plotted in Fig. 13.3(c). The red dashed lines in Fig. 13.3(c) are estimated constant Hurst parameters in different sleep stages, and the estimated Hurst parameters are as follows: $\widehat{H}_{\text{Stage I}} < \widehat{H}_{\text{Stage II}} < \widehat{H}_{\text{Stage I\&REM}} < \widehat{H}_{\text{REM}}$, where $\widehat{H}_{\text{Stage I}} = 0.523454$, $\widehat{H}_{\text{Stage II}} = 0.598736$, $\widehat{H}_{\text{Stage I\&REM}} = 0.647535$

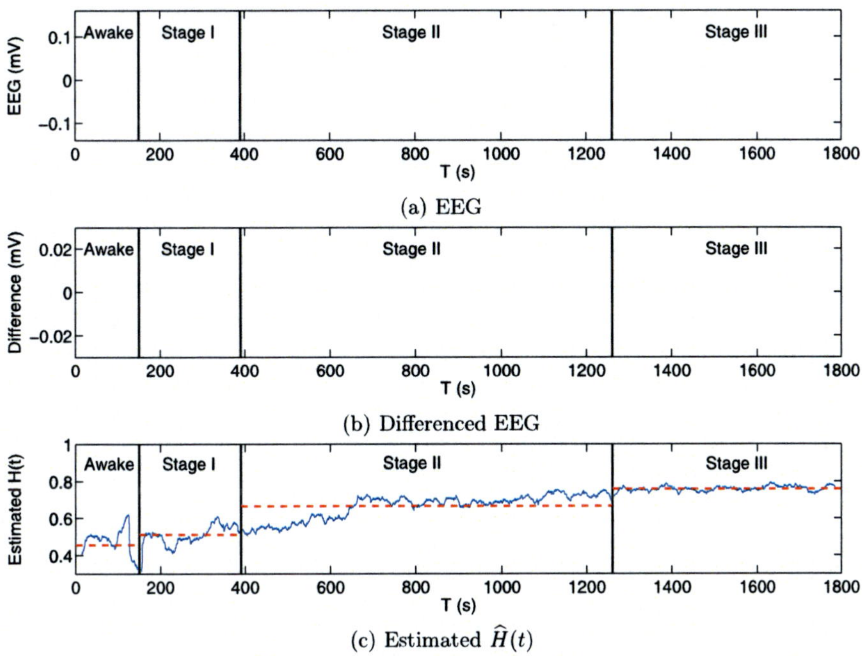

(a) EEG

(b) Differenced EEG

(c) Estimated $\widehat{H}(t)$

Fig. 13.2 Local Hölder exponent of 30-minute sleep EEG signal segment

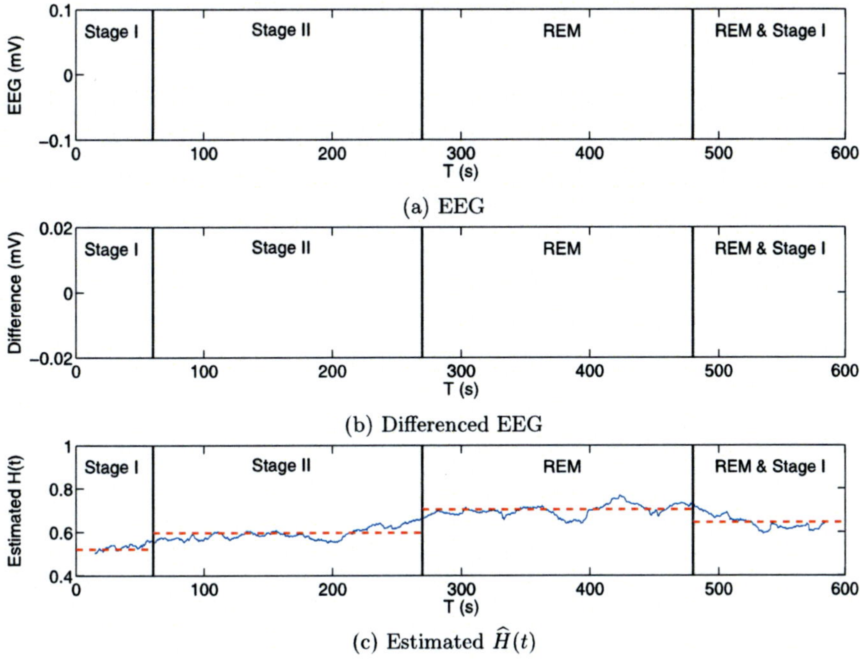

(a) EEG

(b) Differenced EEG

(c) Estimated $\widehat{H}(t)$

Fig. 13.3 Local Hölder exponent of 10-minute sleep EEG signal segment and $\widehat{H}_{\mathrm{REM}} = 0.705560$. The blue line in Fig. 13.3(c) is the estimated local Hölder exponent $\widehat{H}(t)$. During the sleep Stage I, Stage II and REM stage, the $\widehat{H}(t)$ increases gradually from about 0.52 to 0.70. Then, in the stage of alternate REM and Stage I, the $\widehat{H}(t)$ is gradually reduced to about 0.65. In this 10-minute EEG signal segment, the $\widehat{H}(t)$ in sleep Stage I displays almost the same as that in 30-minute EEG signal segment. In sleep Stage II, the $\widehat{H}(t)$ also increases gradually. But in contrast to the 30-minute EEG signal, the $\widehat{H}(t)$ increases from about 0.5 to 0.65, not from 0.5 to 0.7, because the next stage is the REM sleep stage, not deep sleep Stage III. The $\widehat{H}(t)$ of REM stage is also different from sleep Stage III. The $\widehat{H}(t)$ in REM stage is lower, and the fluctuations are bigger than those in Stage III, which might be caused by the rapid eye movements as well as a rapid low-voltage EEG signals. Besides, the most memorable dreaming occurs in REM stage, which can also influence the $\widehat{H}(t)$. The fluctuations of $\widehat{H}(t)$ in the stage of alternate REM and Stage I are different from other stages.

13.5 Chapter Summary

In this chapter, different human sleep stages were investigated by studying the fractional and multifractional properties of sleep EEG signals. From analyzing the results for the fractional property of short term sleep EEG signals in different sleep stages, we can conclude that the average Hurst parameter H is different during different sleep stages. In comparison, the analysis results of multifractional characteristics for long term sleep EEG signals provided more detailed and more valuable information on various sleep stages. In different sleep stages, the fluctuations of local Hölder exponent $H(t)$ exhibit distinctive properties, which are closely related to the distinct characteristics in a specific sleep stage. The emphasis of this study is to provide a novel and more effective analysis technique for dynamic sleep EEG signals, but not yet on performing clinical and cognitive analysis for sleep disorders. A natural next step is to analyze multi-channel EEG signals and evaluate brain disorders by studying the multifractional properties in EEG signals.

Chapter 14
Conclusions

In this monograph, we presented fractional processes and fractional-order signal processing techniques from the perspective of 'fractional signals and fractional-order systems.' Based on the fractional calculus, fractional-order systems are classified into three categories: constant-order fractional systems, variable-order fractional systems, and distributed-order fractional systems. The main characteristics of the fractional-order systems lies in the outstanding abilities in description of memory and hereditary properties. Constant-order fractional systems are characterized by constant memory properties; variable-order fractional systems are characterized by variable memory properties; distributed-order fractional systems can be regarded as the combination of the constant-order fractional systems. Fractional processes, which can be considered as outputs of the fractional-order systems, have significant and complex long-memory properties. According to the classification of the fractional-order systems, fractional processes are categorized as constant-order fractional processes, variable-order fractional processes and distributed-order fractional processes. Constant-order fractional processes are characterized by the constant long-memory parameter $H \in (0, 1)$; variable-order fractional processes are characterized by the variable long-memory parameter $H(t) \in (0, 1)$. To the best of our knowledge, it is not clear how to characterize the distributed-order fractional processes using long-memory parameter or other parameters. In order to best understand the fractional-order systems and extract valuable information from the fractional-order signals, fractional-order signal processing techniques (FOSP) are put forward for different kinds of fractional signals. Constant-order FOSP techniques are used to analyze constant-order fractional signals, including constant-order synthesis of constant-order fractional signals, constant-order fractional system modeling, constant-order fractional filter, and realization of constant-order fractional systems. Variable-order FOSP techniques are suggested to investigate variable-order fractional signals. But the variable-order FOSP techniques are not comprehensively studied because of the difficulties of computation and digital realization. So, only part of the variable-order FOSP techniques are explored in this monograph. Although the distributed-order fractional equations were introduced forty years ago, studies on distributed-order fractional systems and processes have

H. Sheng et al., *Fractional Processes and Fractional-Order Signal Processing*, Signals and Communication Technology,
DOI 10.1007/978-1-4471-2233-3_14,

just started. So the distributed-order fractional signals and analysis techniques were only introduced briefly.

In the previous chapters, we have learned that the essence of fractional processes and FOSP is 'power-law', which externally manifest itself in many distinctive ways, such as heavy-tailed distribution, long-memory, self-similarity, fractal, etc. These distinctive properties are critical in characterizing the intrinsic generating nature of the observed signals or systems. On the other hand, the presence of these distinctive phenomena in random signals or complex systems always leads to certain trouble in correctly analyzing and characterizing them. The purpose of this monograph is to investigate the critical and intrinsic characteristics of some random signals and complex systems with certain extrinsic presentations. It is not surprising that FOSP techniques have been extensively used in econometrics, communication, biomedicine, hydrology, linguistics, and so on. In the third part of the monograph, some application examples were provided. More extensive applications need to be explored in many other areas by readers, and we hope that this monograph will contribute to this end.

There are also some specific research problems to be solved by taking the advantages of FOSP techniques. They are

- Realization of the realtime digital variable-order fractional systems;
- Long memory properties of time-varying Hurst parameter;
- The inherent relationship between fractional Fourier transform and fractional calculus;
- Whitening of various fractional signals;
- Modeling of the distributed-order systems;
- Estimation of the distributed-order parameters;
- Analysis of two dimensional LRD signals with different long-memory parameters in each dimension;
- Physical significance and application of fractional signals and fractional systems.

As stated in the Preface, we hope that the readers will use fractional thinking to understand natural or man-made phenomena, and use fractional techniques to solve the problems and gain additional insights after reading the monograph.

Appendix A
Mittag-Leffler Function

The Mittag-Leffler function plays a very important role in the solution of fractional-order differential equations [255]. The Mittag-Leffler function with form

$$E_\rho(z) = \sum_{k=0}^{\infty} \frac{z^k}{\Gamma(k\rho + 1)},$$

was introduced in [205], where $z \in \mathbb{C}$ and ρ is an arbitrary positive constant. The Laplace transform of the Mittag-Leffler function in one parameter is

$$\mathcal{L}\{E_\rho(-\lambda t^\rho)\} = \frac{s^{\rho-1}}{s^\rho + \lambda}, \tag{A.1}$$

where $\mathrm{Re}\{s\} > |\lambda|^{1/\rho}$.

More general Mittag-Leffler function with two parameters has the form

$$E_{\rho,\mu}(z) = \sum_{k=0}^{\infty} \frac{z^k}{\Gamma(k\rho + \mu)},$$

was introduced in [122], where $z \in \mathbb{C}$, ρ and μ are arbitrary positive constants. When $\mu = 1$, $E_{\rho,1}(z) = E_\rho(z)$. The Laplace transform of the Mittag-Leffler function in two parameters is

$$\mathcal{L}\{t^{\mu-1} E_{\rho,\mu}(-\lambda t^\rho)\} = \frac{s^{\rho-\mu}}{s^\rho + \lambda}, \tag{A.2}$$

where $\mathrm{Re}\{s\} > |\lambda|^{1/\rho}$.

Kilbas et al. studied the generalized Mittag-Leffler function with three parameters [140]

$$E^\gamma_{\rho,\mu}(z) = \sum_{k=0}^{\infty} \frac{(\gamma)_k}{\Gamma(\rho k + \mu)} \frac{z^k}{k!}, \tag{A.3}$$

H. Sheng et al., *Fractional Processes and Fractional-Order Signal Processing*,
Signals and Communication Technology,
DOI 10.1007/978-1-4471-2233-3,

where $z \in \mathbb{C}$, ρ, μ and γ are arbitrary positive constants, and $(\gamma)_k$ is the Pochhammer symbol [256]. When $\gamma = 1$, $E^1_{\rho,\mu}(z) = E_{\rho,\mu}(z)$, and when $\gamma = \mu = 1$, $E^1_{\rho,1}(z) = E_\rho(z)$.

The Laplace transform of Generalized Mittag-Leffler function is

$$\mathcal{L}\{t^{\mu-1}E^{\gamma}_{\rho,\mu}(-\lambda t^{\rho})\} = \frac{s^{\rho\gamma-\mu}}{(s^{\rho}+\lambda)^{\gamma}}, \tag{A.4}$$

where $\mathrm{Re}\{s\} > |\lambda|^{1/\rho}$.

In some applications, generalized Mittag-Leffler function in four parameters has to be used which is defined as follows

$$E^{\gamma,q}_{\alpha,\beta}(z) = \sum_{k=0}^{\infty} \frac{(\gamma)_{kq}}{\Gamma(\alpha k+\beta)} \frac{z^k}{k!}, \tag{A.5}$$

where $\alpha, \beta, \gamma \in \mathbb{C}$, $\mathrm{Re}(\alpha) > 0$, $\mathrm{Re}(\beta) > 0$, $\mathrm{Re}(\gamma) > 0$ and $q \in \mathbb{N}$

$$(\gamma)_0 = 1, \quad \text{and} \quad (\gamma)_k = \gamma(\gamma+1)(\gamma+2)\cdots(\gamma+k-1) = \frac{\Gamma(k+\gamma)}{\Gamma(\gamma)}. \tag{A.6}$$

It is easily seen, by comparing the definitions that

$$E^{\gamma,1}_{\alpha,\beta}(z) = E^{\gamma}_{\alpha,\beta}(z). \tag{A.7}$$

A MATLAB®code `ml_func()` can be used to deal with the generalized Mittag-Leffler function evaluation problems. The syntaxes for the evaluation of generalized Mittag-Leffler functions in three and four parameters are y=`ml_func` `(`$[\alpha, \beta, \gamma]$`,`z`,`n`,`ϵ_0`)` and y=`ml_func(`$[\alpha, \beta, \gamma, q]$`,`$z$`,`$n$`,`$\epsilon_0$`)` respectively. The listing of the function is given in [207].

From [207], the nth-order derivative of the Mittag-Leffler function in four parameters can be evaluated from

$$\frac{\mathrm{d}^n}{\mathrm{d}t^n} E^{\gamma,q}_{\alpha,\beta}(z) = (\gamma)_{qn} E^{\gamma+qn,q}_{\alpha,\beta+n\alpha}(z), \tag{A.8}$$

and in particular, the integer nth-order derivative of the Mittag-Leffler function in two parameters can be evaluated from

$$\frac{\mathrm{d}^n}{\mathrm{d}z^n} E_{\alpha,\beta}(z) = \sum_{j=0}^{\infty} \frac{(j+n)!}{j!\,\Gamma(\alpha j+\alpha n+\beta)} z^j. \tag{A.9}$$

The general form of the Laplace transformation formula is given by [207]

$$L^{-1}\left[\frac{s^{\alpha\gamma-\beta}}{(s^{\alpha}+a)^{\gamma}}\right] = t^{\beta-1}E^{\gamma}_{\alpha,\beta}\left(-at^{\alpha}\right). \tag{A.10}$$

From the above property, many useful formulae can be derived

- When $\gamma = 1$, and $\alpha\gamma = \beta$, the inverse Laplace transform can be interpreted as the analytical response of a fractional-order transfer function $1/(s^\alpha + a)$ driven by an impulse input. In this case, $\beta = \alpha$, and the Laplace transform can be expressed as

$$L^{-1}\left[\frac{1}{s^\alpha + a}\right] = t^{\alpha-1} E_{\alpha,\alpha}\left(-at^\alpha\right). \tag{A.11}$$

- When $\gamma = 1$, and $\alpha\gamma - \beta = -1$, the inverse Laplace transform can be interpreted as the analytical solution of a fractional-order transfer function $1/(s^\alpha + a)$ driven by a step input. In this case, $\beta = \alpha + 1$, and the Laplace transform can be expressed as

$$L^{-1}\left[\frac{1}{s(s^\alpha + a)}\right] = t^{\alpha} E_{\alpha,\alpha+1}\left(-at^\alpha\right). \tag{A.12}$$

  It can also be shown that the inverse Laplace transform of the function can alternatively be written as

$$L^{-1}\left[\frac{1}{s(s^\alpha + a)}\right] = \frac{1}{a}\left[1 - E_{\alpha}\left(-at^\alpha\right)\right]. \tag{A.13}$$

- When $\gamma = k$ is an integer, and $\alpha\gamma = \beta$, the inverse Laplace transform can be interpreted as the analytical solution of a fractional-order transfer function $1/(s^\alpha + a)^k$ driven by an impulse input. In this case, $\beta = \alpha k$, and the Laplace transform can be expressed as

$$L^{-1}\left[\frac{1}{(s^\alpha + a)^k}\right] = t^{\alpha k-1} E^k_{\alpha,\alpha k}\left(-at^\alpha\right). \tag{A.14}$$

- When $\gamma = k$ is an integer, and $\alpha\gamma - \beta = -1$, the inverse Laplace transform can be interpreted as the analytical solution of a fractional-order transfer function $1/(s^\alpha + a)^k$ driven by a step input. In this case, $\beta = \alpha k + 1$, and the Laplace transform can be expressed as

$$L^{-1}\left[\frac{1}{s(s^\alpha + a)^k}\right] = t^{\alpha k} E^k_{\alpha,\alpha k+1}\left(-at^\alpha\right). \tag{A.15}$$

For a two term fractional-order filter with the following transfer function,

$$G(s) = \frac{1}{a_2 s^{\beta_2} + a_1 s^{\beta_1} + a_0}, \tag{A.16}$$

it is interesting to note that we can also obtain its analytical step response using Mittag-Leffler function as follows [207]:

$$y(t) = \frac{1}{a_2}\sum_{k=0}^{\infty} \frac{(-1)^k \hat{a}_0^k t^{-\hat{a}_1+(k+1)\beta_2}}{k!} E^{(k)}_{\beta_2-\beta_1,\beta_2+\beta_1 k+1}\left(-\hat{a}_1 t^{\beta_2-\beta_1}\right), \tag{A.17}$$

where $\hat{a}_0 = a_0/a_2, \hat{a}_1 = a_1/a_2$. A MATLAB function `ml_step` is written to implement the step response of the system. The syntax $\boldsymbol{y}$=`ml_step`$(\boldsymbol{a}, \boldsymbol{b}, \boldsymbol{t}, \varepsilon)$ can be used to find the numerical solution of the three-term system, where $\boldsymbol{a} = [a_0, a_1, a_2]$, and $\boldsymbol{b} = [\beta_1, \beta_2]$. The argument ε is the error tolerance. `ml_step` code list can be found in [207].

Appendix B
Application of Numerical Inverse Laplace Transform Algorithms in Fractional-Order Signal Processing

B.1 Introduction

Laplace transform has been considered as a useful tool to solve integer-order or some simple fractional-order differential equations [237, 302]. Inverse Laplace transform is an important but difficult step in the application of Laplace transform technique in solving differential equations. The inverse Laplace transformation can be accomplished analytically according to its definition, or by using Laplace transform tables. For a complicated differential equation, however, it is difficult to analytically calculate the inverse Laplace transformation. So, the numerical inverse Laplace transform algorithms are often used to obtain the numerical results. Motivated by taking advantages of numerical inverse Laplace transform algorithms in fractional calculus, we investigate the validity of applying these numerical algorithms in solving fractional-order differential equations.

Many numerical inverse Laplace transform algorithms have been proposed to solve the Laplace transform inversion problems. *Weeks* numerical inversion of Laplace transform algorithm was provided using the Laguerre expansion and bilinear transformations [319]. *Direct* numerical inversion of Laplace transform algorithm, which is based on the trapezoidal approximation of the Bromwich integral, was introduced in [294]. Based on accelerating the convergence of the Fourier series using the trapezoidal rule, *Invlap* method for numerical inversion of Laplace transform was proposed in [79]. *Gavsteh* numerical inversion of Laplace transform algorithm was introduced in [282], and the *NILT* fast numerical inversion of Laplace transforms algorithm was proposed in [33]. The *NILT* method is based on the application of fast Fourier transformation followed by the so-called ε-algorithm to speed up the convergence of infinite complex Fourier series. The algorithm was improved using a quotient-difference algorithm in [34]. The quotient-difference algorithm based *NILT* method is numerically more stable in producing the same results in a practical way. Furthermore, some efforts have been made to evaluate the performances of these numerical inverse Laplace transform algorithms [75, 84, 151]. However, there is a lack of good assessment for applying numerical inverse Laplace transform algorithms in solving fractional-order differential equations. In this appendix, *Invlap*, *Gavsteh* and improved *NILT*, which is simply called *NILT* in this

H. Sheng et al., *Fractional Processes and Fractional-Order Signal Processing*, Signals and Communication Technology,
DOI 10.1007/978-1-4471-2233-3,

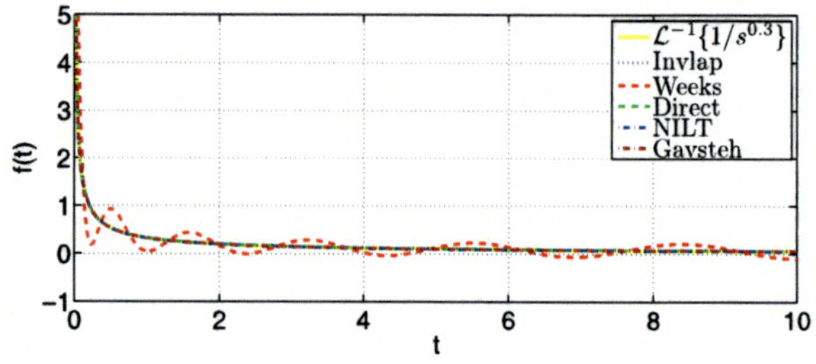

Fig. B.1 Inverse Laplace transform of $F(s) = 1/s^{0.3}$

paper, are tested using Laplace transform of simple and complicated fractional-order differential equations.

Fractional calculus is a part of mathematics dealing with derivatives of arbitrary order [139, 203, 209, 218, 237]. A growing number of fractional-order differential equation based models are proposed to describe physical phenomena and complex dynamic systems [47, 228, 266]. Moreover, some variable-order fractional models and distributed-order fractional models were proposed to understand or describe the nature of complex phenomena in a better way [46, 180]. The rapid growth of fractional-order models leads to the emergence of complicated fractional-order differential equations, and brings forward challenges for solving these complicated equations [196, 323]. In this Appendix, we will investigate the validity of numerical inverse Laplace transform algorithms to overcome these difficulties.

B.2 Numerical Inverse Laplace Transform Algorithms

The Laplace transform and the inverse Laplace transform have been introduced in Chap. 1. Many numerical methods have been proposed to calculate the inversion of Laplace transforms. In order to take advantages of these numerical inverse Laplace transform algorithms, some efforts have been made to test and evaluate the performances of these numerical methods [75, 84, 151]. It has been concluded that the choice of right algorithm depends upon the problem to be solved [151]. So, we tested these numerical algorithms using a simple fractional-order integrator with transfer function $F(s) = 1/s^{\alpha}$. Figure B.1 shows the comparison results of *Invlap*, *Weeks*, *Direct*, *Gavsteh* and *NILT*, these five numerical inversion Laplace transform algorithms for the fractional-order filter $F(s) = 1/s^{0.3}$. It can be seen that, except *Weeks* method, other four methods generate acceptable numerical results. In some cases, however, the *Direct* method did not converge in our tests. In contrast, *Invlap*, *Gavsteh* and *NILT* numerical algorithms performed better, so in this study, we concentrate on the validity of *Invlap*, *Gavsteh* and *NILT* numerical inverse Laplace transform algorithms based on the MATLAB codes in [35, 119, 279]. The theories of these three numerical inverse Laplace transform algorithms can be found in [33, 79, 282].

Fig. B.2 Inverse Laplace transform of $F(s) = 1/(s+1)^{0.8}$

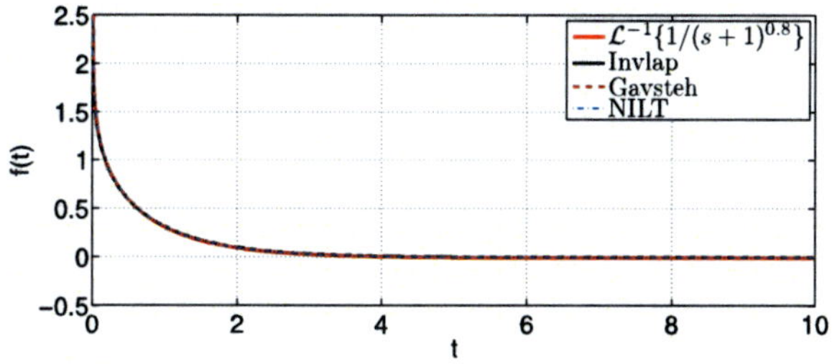

(a) Inverse Laplace transform

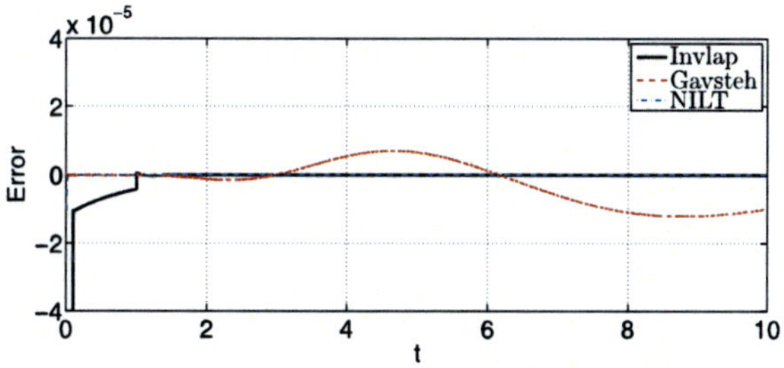

(b) Error

B.3 Some Application Examples of Numerical Inverse Laplace Transform Algorithms in Fractional Order Signal Processing

In this section, we present some application examples of numerical inverse Laplace transform algorithms for some Laplace transforms of fractional-order differential equations. In all the figures below, the red line represents the analytical inverse Laplace transform $f(t)$, the black line represents the *Invlap* result $\hat{f}_{Invlap}(t)$, the brown line represents the *Gavsteh* result $\hat{f}_{Gavsteh}(t)$, and the blue line represents the *NILT* result $\hat{f}_{NILT}(t)$.

B.3.1 Example A

In this example, *Invlap*, *Gavsteh* and *NILT* numerical inverse Laplace transform algorithms were used to calculate the inverse Laplace transformation of a fractional-order low-pass filter

$$\mathcal{L}\{f(t)\} = F(s) = \frac{1}{(s+1)^{\alpha}}. \tag{B.1}$$

The analytical inverse Laplace transform of (B.1) is

$$f(t) = t^{\alpha-1} E_{1,\alpha}^{\alpha}(-t) = \mathrm{e}^{-t} t^{\alpha-1}, \tag{B.2}$$

where $E_{\rho,\mu}^{\gamma}$ is a generalized Mittag-Leffler function (A.3).

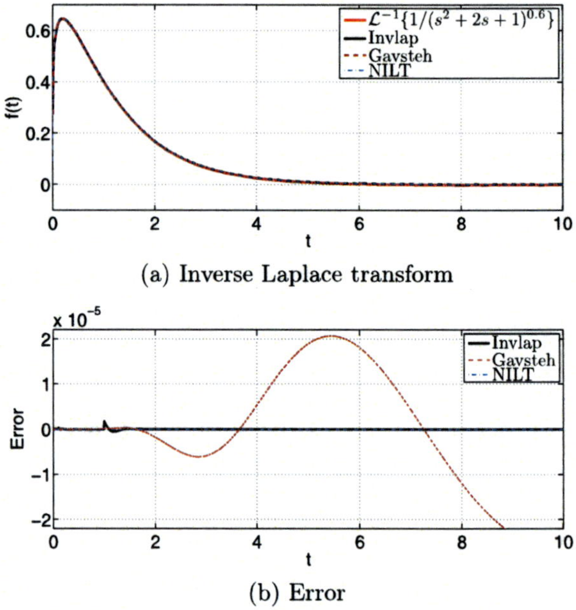

Fig. B.3 Inverse Laplace transform of $F(s) = 1/(s^2 + 2s + 1)^{0.6}$

Figure B.2(a) shows the inverse Laplace transform comparison of $F(s) = \frac{1}{(s+1)^{0.8}}$. Figure B.2(b) illustrates the absolute errors of three numerical inverse Laplace transform algorithms for this example. It can be seen that the numerical result of *NILT* algorithm has the minimum error. Compared with the *NILT* algorithm, the *Invlap* algorithm converges slowly. The *Gavsteh* algorithm has divergent tendency with the elapse of time.

B.3.2 Example B

In this example, we discuss the numerical inverse Laplace transform of the fractional second-order filter (5.63) [171], which has been discussed in Chap. 5.

When $a^2 - 4b = 0$

The analytical inverse Laplace transform of (5.63) is (5.70). Figure B.3(a) presents the inverse Laplace transform comparison of (5.63) for $a = 2$, $b = 1$, and $\gamma = 0.6$. Figure B.3(b) illustrates the absolute errors of the three algorithms for this example. It can be seen that the numerical result of *NILT* algorithm has the minimum error. The *Invlap* algorithm has small errors around $t = 1$, and the *Gavsteh* algorithm has divergent tendency with the elapse of time.

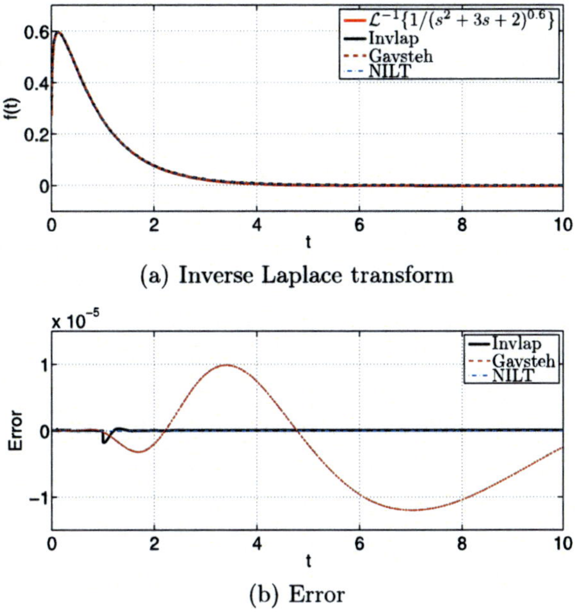

Fig. B.4 Inverse Laplace transform of $F(s) = 1/(s^2 + 3s + 2)^{0.6}$

When $a^2 - 4b > 0$

The two zeroes of $s^2 + as + b$ are $s_1 = \frac{-a-\sqrt{\Delta}}{2}$ and $s_2 = \frac{-a+\sqrt{\Delta}}{2}$, where $\Delta = a^2 - 4b > 0$. The analytical inverse Laplace transform of (5.63) is (5.74) [171]. Figure B.4(a) presents the inverse Laplace transform comparison of (5.63) for $a = 3$, $b = 2$, and $\gamma = 0.6$. Figure B.4(b) illustrates the absolute errors of the three algorithms for this example. The error fluctuations of these three algorithms are similar to the above example.

When $a^2 - 4b < 0$

In this case there are three branch points: $s = s_1 = \frac{-a-i\sqrt{-\Delta}}{2}$, $s = s_2 = \frac{-a+i\sqrt{-\Delta}}{2}$ and $s = \infty$, where $\Delta = a^2 - 4b < 0$. The analytical inverse Laplace transform of (5.63) is (5.75). Figure B.5(a) presents the inverse Laplace transform comparison of (5.63) for $a = 0$, $b = 1$, and $\gamma = 0.6$. Figure B.5(b) illustrates the absolute errors of the three algorithms for this example. It can be seen that the numerical results of *NILT* algorithm and *Invlap* algorithm have small errors, but the *Gavsteh* algorithm has big error for this example.

B.3.3 Example C

This example concentrates on the inverse Laplace transformation of distributed order integrator/differentiator of the form (7.5) [168]. There are two branch points

Fig. B.5 Inverse Laplace transform of $F(s) = 1/(s^2+1)^{0.6}$

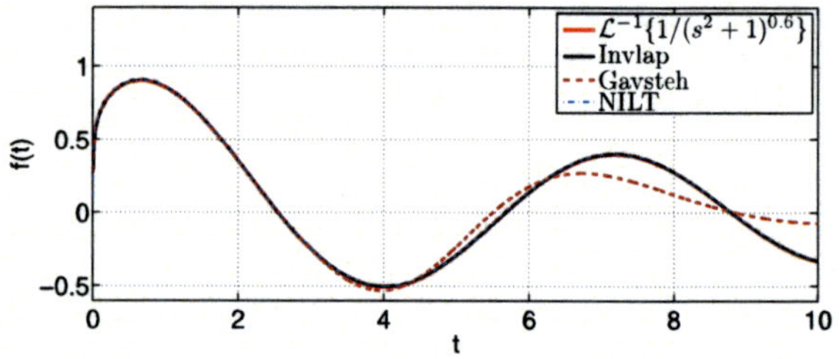

(a) Inverse Laplace transform

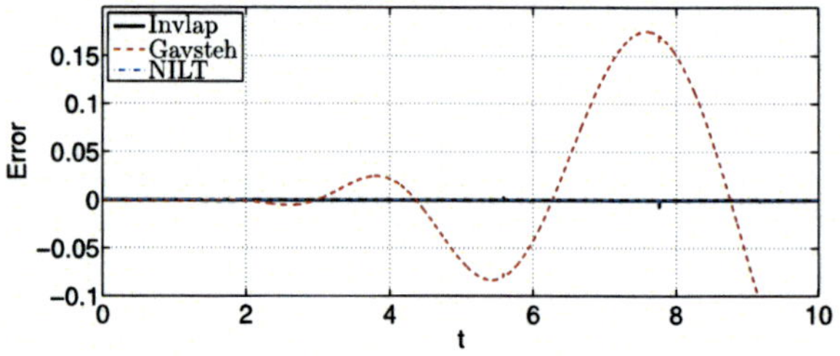

(b) Error

Fig. B.6 Inverse Laplace transform of $F(s) = \int_{0.5}^{0.8} \frac{1}{s^\alpha}\, \mathrm{d}\alpha$

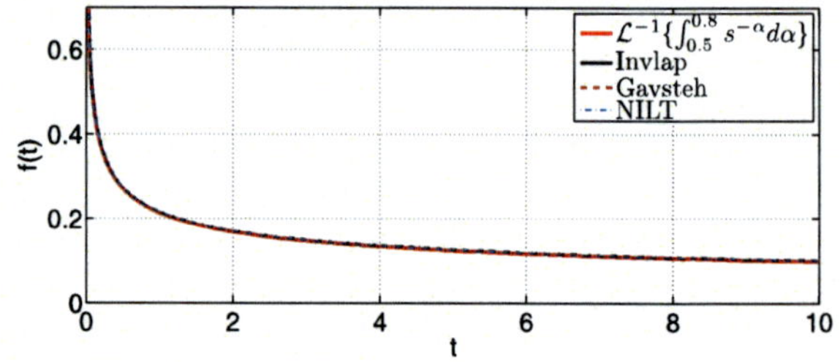

(a) Inverse Laplace transform

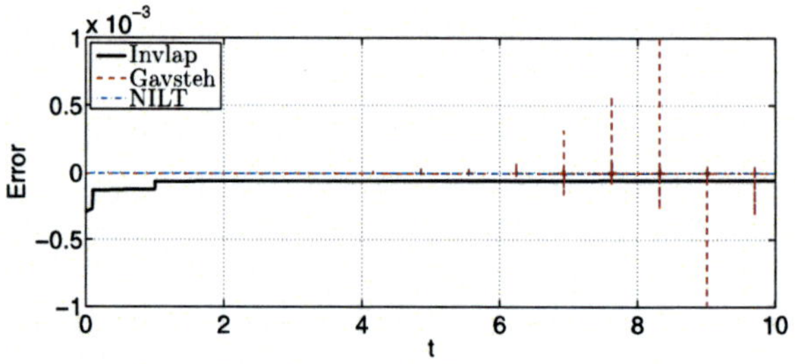

(b) Error

$s = 0$ and $s = \infty$. The analytical inverse Laplace transform of (7.5) is (7.8) [168]. Two cases of the inverse Laplace transform comparison for (7.5) are provided in Figs. B.6 and B.7. Figure B.6(a) presents the inverse Laplace transform comparison of (7.5) for $a = 0.5$, $b = 0.8$, and Fig. B.7(a) presents the inverse Laplace transform comparison of (7.5) for $a = -0.8$, $b = -0.5$. Figures B.6(b) and B.7(b) illustrate

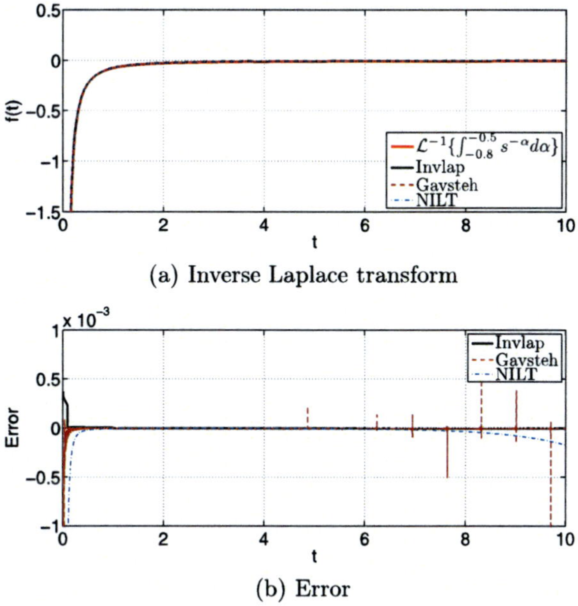

Fig. B.7 Inverse Laplace transform of $F(s) = \int_{-0.8}^{-0.5} \frac{1}{s^\alpha} \mathrm{d}\alpha$

the absolute errors of three numerical inverse Laplace transform algorithms for this example, respectively. It can be seen that the numerical result of *NILT* algorithm has the minimum error for $a = 0.5$, $b = 0.8$, and *Invlap* algorithm has the minimum error for $a = -0.8$, $b = -0.5$.

B.3.4 Example D

This example concentrates on the inverse Laplace transformation of distributed-order low-pass filter with form (7.17) [169]. The analytical inverse Laplace transform of (7.17) is (7.18). Figure B.8(a) presents the inverse Laplace transform comparison of (7.17) for $a = 0.6$, $b = 1$, and $\lambda = 1$. Figure B.8(b) illustrates the absolute errors of three numerical inverse Laplace transform algorithms for this example. All these three numerical algorithms have almost the same error curve.

B.3.5 Example E

In this example, we discuss the inverse Laplace transform of fractional-order distributed parameter low-pass filter with the form (7.24) [170]. The analytical inverse Laplace transform of (7.24) is (7.28). Figure B.9(a) presents the inverse Laplace transform comparison of (7.24) for $a = 0$, $b = 1$, and $\alpha = 0.9$. Figure B.9(b) illustrates the absolute errors of three numerical inverse Laplace transform algorithms

Fig. B.8 Inverse Laplace transform of $\frac{1}{0.4}\int_{0.6}^{1}\frac{1}{(s+\lambda)^{\alpha}}\,\mathrm{d}\alpha$

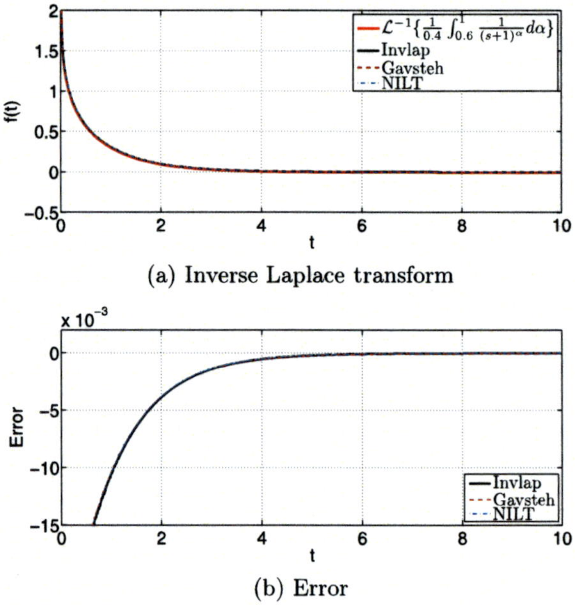

(a) Inverse Laplace transform

(b) Error

Fig. B.9 Inverse Laplace transform of $\int_{0}^{1}\frac{1}{Ts^{0.9}+1}\,\mathrm{d}T$

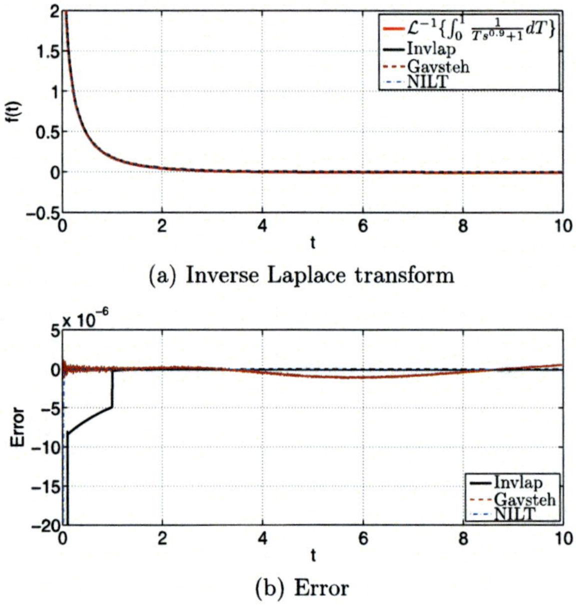

(a) Inverse Laplace transform

(b) Error

for this example. It can be seen that the numerical result of *NILT* algorithm has the smallest error. Compared with *NILT* method, the *Invlap* algorithm converges slowly. The *Gavsteh* algorithm obviously has fluctuating error.

Table B.1 Standard errors for Examples A–E

	$T = 5$	$T = 10$	$T = 15$
Invlap			
Example A	8.6414×10^{-5}	6.1098×10^{-5}	4.9884×10^{-5}
Example B (Case 1)	1.1639×10^{-8}	1.1070×10^{-7}	9.0384×10^{-8}
Example B (Case 2)	2.4064×10^{-8}	1.7773×10^{-7}	1.4511×10^{-7}
Example B (Case 3)	2.1739×10^{-5}	1.2662×10^{-4}	1.7414×10^{-4}
Example C (Case 1)	9.6377×10^{-5}	7.5835×10^{-5}	6.9281×10^{-5}
Example C (Case 2)	3.4413×10^{-5}	2.4330×10^{-5}	1.9865×10^{-5}
Example D	8.9442×10^{-3}	6.3226×10^{-3}	5.1622×10^{-3}
Example E	7.1917×10^{-5}	5.0848×10^{-5}	4.1516×10^{-5}
Gavsteh			
Example A	3.4267×10^{-6}	6.4948×10^{-6}	6.0408×10^{-6}
Example B (Case 1)	6.6759×10^{-6}	1.3451×10^{-5}	1.4470×10^{-5}
Example B (Case 2)	5.0929×10^{-6}	7.2318×10^{-6}	6.5926×10^{-6}
Example B (Case 3)	1.5884×10^{-2}	8.8690×10^{-2}	1.4788×10^{-1}
Example C (Case 1)	5.7844×10^{-7}	2.3653×10^{-5}	1.9313×10^{-5}
Example C (Case 2)	3.8865×10^{-5}	3.4677×10^{-5}	2.8328×10^{-5}
Example D	8.7280×10^{-3}	6.1711×10^{-3}	5.0385×10^{-3}
Example E	3.5116×10^{-7}	5.4424×10^{-7}	7.1124×10^{-7}
NILT			
Example A	2.6442×10^{-7}	1.8459×10^{-7}	1.5008×10^{-7}
Example B (Case 1)	1.1489×10^{-8}	8.0071×10^{-9}	6.5061×10^{-9}
Example B (Case 2)	2.2618×10^{-8}	1.5955×10^{-8}	1.3016×10^{-8}
Example B (Case 3)	2.1739×10^{-5}	1.2662×10^{-4}	1.7414×10^{-4}
Example C (Case 1)	3.1290×10^{-7}	2.2589×10^{-7}	1.8808×10^{-7}
Example C (Case 2)	4.7214×10^{-3}	3.2335×10^{-3}	2.6146×10^{-3}
Example D	8.7280×10^{-3}	6.1712×10^{-3}	5.0386×10^{-3}
Example E	8.4523×10^{-7}	5.9010×10^{-7}	4.7974×10^{-7}

From Figs. B.2–B.9 we can see that these three numerical inverse Laplace transform algorithms perform very well on most of fractional-order differential equations. In some cases, however, *Gavsteh* numerical algorithm leads to some errors. In order to quantitatively evaluate the performance of these three numerical Laplace transform algorithms, we calculate the standard errors S of these algorithms for different fractional-order differential equations. The standard errors S is defined as

$$S = \sqrt{\frac{\sum_{i=1}^{n}(x_i - u_i)^2}{n-1}}, \tag{B.3}$$

where u_i is the analytical result, and x_i is the result of numerical Laplace transform algorithm. Table B.1 presents the standard errors of these three numerical Laplace transform algorithms for Examples A–E. From Table B.1 we can see that the accuracy of the numerical algorithm depends upon the fractional-order differential equations to be solved. *NILT* numerical algorithm has the best accuracy in most of the cases, and *Invlap* numerical algorithm can provide acceptable results. But *Gavsteh* algorithm failed to calculate the fractional-order differential equations in some cases. For *NILT* numerical algorithm, the error decreases with the increase of t in most of the cases, but the *Gavsteh* algorithm is quite the opposite.

B.4 Conclusion

In this appendix, we investigated the application potential of numerical Laplace transform algorithms in fractional-order signal processing. Three numerical algorithms *Invlap*, *Gavsteh* and *NILT* are studied. These three numerical inverse Laplace transform algorithms are applied in some Laplace transforms of fractional-order differential equations. The analytical results show that these three algorithms perform well for most of the Laplace transform of fractional-order fractional differential equations. The *NILT* numerical algorithm performed the best, and the *Gavsteh* algorithm failed in some cases. In summary, *Invlap* and *NILT* numerical inverse Laplace transform algorithms are effective and reliable for fractional-order differential equations. So, these numerical inverse Laplace transform algorithms provide an easy way to numerically solve some complicated fractional-order differential equations.

Appendix C
Some Useful Webpages

C.1 Useful Homepages

- Dr. YangQuan Chen
  http://mechatronics.ece.usu.edu/yqchen/
  http://sites.google.com/site/yangquanchen/
- Dr. Igor Podlubny
  http://people.tuke.sk/igor.podlubny/
- Dr. Manuel Duarte Ortigueira
  http://www.uninova.pt/~mdo/
- Dr. Tom T. Hartley
  http://coel.ecgf.uakron.edu/hartley/index.html
- Dr. Blas Vinagre
  http://eii.unex.es/profesores/bvinagre/
- Dr. J.A. Tenreiro Machado
  http://ave.dee.isep.ipp.pt/~jtm/
- Dr. Hu Sheng
  https://sites.google.com/site/hushenghomepage/
- Dr. Ming Li
  http://www.ee.ecnu.edu.cn/teachers/mli/js_lm(Eng).htm

C.2 Useful Codes

- Hurst estimator:
  http://www.mathworks.com/matlabcentral/fileexchange/21028
- Hurst parameter estimators:
  http://www.mathworks.com/matlabcentral/fileexchange/19148
- Mittag-Leffler function:
  http://www.mathworks.com/matlabcentral/fileexchange/8738
- Generalized Mittag-Leffler function:
  http://www.mathworks.com/matlabcentral/fileexchange/20849

H. Sheng et al., *Fractional Processes and Fractional-Order Signal Processing*,
Signals and Communication Technology,
DOI 10.1007/978-1-4471-2233-3, © Springer-Verlag London Limited 2012

- Generalized generalized Mittag-Leffler function:
  http://www.mathworks.com/matlabcentral/fileexchange/21454
- Impulse response invariant discretization of fractional-order low-pass filters:
  http://www.mathworks.com/matlabcentral/fileexchange/21365
- Step response invariant discretization of fractional-order integrators or differentiators:
  http://www.mathworks.com/matlabcentral/fileexchange/21363
- Impulse response invariant discretization of fractional-order integrators or differentiators:
  http://www.mathworks.com/matlabcentral/fileexchange/21342
- Hybrid symbolic and numerical simulation studies of time-fractional order:
  http://www.mathworks.com/matlabcentral/fileexchange/10532
- Oustaloup-recursive-approximation for fractional-order differentiators:
  http://www.mathworks.com/matlabcentral/fileexchange/3802
- A new IIR-type digital fractional order differentiator:
  http://www.mathworks.com/matlabcentral/fileexchange/3518
- Low-pass FIR digital differentiator design:
  http://www.mathworks.com/matlabcentral/fileexchange/3516
- Predictor-corrector method for variable-order, random-order fractional relaxation equation:
  http://www.mathworks.com/matlabcentral/fileexchange/26407
- Impulse response invariant discretization of distributed-order low-pass filter:
  http://www.mathworks.com/matlabcentral/fileexchange/26868
- Impulse response invariant discretization of fractional second-order filter:
  http://www.mathworks.com/matlabcentral/fileexchange/26442
- Impulse response invariant discretization of distributed-order integrator:
  http://www.mathworks.com/matlabcentral/fileexchange/26380
- Impulse response invariant discretization of BICO (Bode's Ideal Cut-Off) transfer function:
  http://www.mathworks.com/matlabcentral/fileexchange/28398
- Step response invariant discretization of BICO (Bode's Ideal Cut-Off) transfer function:
  http://www.mathworks.com/matlabcentral/fileexchange/28399
- Matrix approach to discretization of ODEs and PDEs of arbitrary real order:
  http://www.mathworks.com/matlabcentral/fileexchange/22071

Appendix D
MATLAB Codes of Impulse Response Invariant Discretization of Fractional-Order Filters

D.1 Impulse Response Invariant Discretization of Distributed-Order Integrator

```
%*****************************************************************
function [sr]=irid_doi(Ts,a,b,p,q)
%
% irid_doi() is prepared to compute a discrete-time finite
% dimensional (z) transfer function to approximate a distributed
% order integrator int(1/(s^r),r,a,b), where "s" is the Laplace
% transform variable, where 'a', 'b' are any real numbers in the
% range of (0.5,1), and a<b. 'p' and 'q' are integer and p>=q.
%
% The approximation keeps the impulse response "invariant"
%*****************************************************************
% IN:
%    Ts: The sampling period
%    a : Lower limit of integral
%    b : Upper limit of integral
%    p : Denominator order of the approx. z-transfer function
%    q : Numerator order of the approximate z-transfer function
%*****************************************************************
% OUT:
%    sr: returns the LTI object that approximates the
%    int(1/s^r,r,a,b) with invariant impulse response.
%*****************************************************************
% TEST CODE
%    doi=irid_doi(0.001,0.75,1,5,5);
%*****************************************************************
%Written by Hu Sheng, Yan Li and YangQuan Chen
%*****************************************************************
```

H. Sheng et al., *Fractional Processes and Fractional-Order Signal Processing*, Signals and Communication Technology,
DOI 10.1007/978-1-4471-2233-3, © Springer-Verlag London Limited 2012

```
if nargin<4; p=5,q=5; end

if p<3 | q<3
   sprintf('%s','The order of the approximate transfer
   function should be greater than 2')
return,end

if p<q,
   sprintf('%s','The denominator order of the approx. z-transfer
   function should be greater than or equal to the Numerator')
   return,end

if Ts<=0 , sprintf('%s','Sampling period has to be positive')
return, end

if a<=0.5 | a>1 | b<=0.5 | b>1
   sprintf('%s','The fractional order should be in (0.5,1)')
return,end

if a>=b
   sprintf('%s','The Upper limit of integral should be
   greater than the Lower limit of integral')
return, end

close all;
wmax0=2*pi/Ts/2;
L=1/Ts;
t=[0:L-1]*Ts; ht=[ ];
for k=1:length(t)
   ht(k)=quadgk(@(x)integrand(a,b,t(k),x),0,inf);
end
h=[ht(2:end).*Ts];
[B,A]=stmcb((h),q,p);
sprintf('IRI discrete approx. transfer function:')
sr = tf(B,A,Ts)
hht=impulse(sr,t);

wmax=floor(log10(wmax0))+1; wmin=wmax-5;
w=logspace(wmin,wmax,1000);
j=sqrt(-1);
srfr=((j.*w).(-a)-(j.*w).^(-b))./log(j.*w);
srfr1=freqresp(sr,w);
```

```
figure;
subplot(3,1,1)
plot(t,ht,'b'); hold on plot(t,hht./Ts,'r-.')
axis([Ts,Ts.*L,0,1]);
xlabel('Time');ylabel('Impulse response'); grid on;
legend('impulse response of \int_a^b{s^{-\alpha}}d\alpha', ...
'approximated impulse response');

subplot(3,1,2)
semilogx(w,20*log10(abs(srfr)),'b');hold on;
semilogx(w,20*log10(abs(reshape(srfr1, 1000, 1))),'r-.');
legend('mag. Bode of \int_a^b{s^-\alpha}}d\alpha', ...
'approximated mag. Bode');
xlabel('Frequency (Hz)');ylabel('Magnitude(dB)');grid on;

subplot(3,1,3)
semilogx(w,(180/pi) * (angle(srfr)),'b');hold on;
semilogx(w,(180/pi) * (angle(reshape(srfr1, 1000, 1))),'r-.');
grid on;
xlabel('Frequency (Hz)');ylabel('Phase (degrees)');
legend('phase Bode of \int_a^bs^{-\alpha}}d\alpha',...
'approximated phase Bode')

end
%*****************************************************************



%*****************************************************************
function y=integrand(a,b,t,x)
%*****************************************************************
i=sqrt(-1);
y=(exp(-x.*t).*(x.^(b-a).*exp(-i*(b-a)*pi)-1)./
((x.^b).*exp(-i*b*pi).*(log(x)-i*pi))-exp(-x.*t).*
(x.^(b-a).*exp(i*(b-a)*pi)-1)./
((x.^b).*exp(i*b*pi).*(log(x)+i*pi)))/(2*pi*i);

end
%*****************************************************************
```

D.2 Impulse Response Invariant Discretization of Fractional Second-Order Filter

```
%***************************************************************
function [sr]=irid_fsof(Ts,a,b,r,norder)
%
% irid_fsof function is to compute a discrete-time finite
% dimensional (z) transfer function to approx. a continuous-time
% fractional second order low-pass filter [1/(s^2 +a*s +b)]^r,
% where "s" is the Laplace transform variable;
% "r" is a real number in the range of (0,1);
% a and b are the time constant of LPF
% [1/(s^2 + a*s + b)]^r,
% where a, b >= 0.
%
% The approximation keeps the impulse response "invariant"
%***************************************************************
% IN:
%    a, b: the time constant of (the first order) LPF
%     (a and b are arbitrary positive real numbers)
%    r: the fractional order in (0,1)
%    Ts: the sampling period
%    norder: the finite order of the approx. z-transfer function
%    (the orders of den. and num. z-polynomials are the same)
%***************************************************************
% OUT:
%    sr: returns LTI object that approx. [1/(s^2 + a*s + b)]^r
%     in the sense of invariant impulse response.
%***************************************************************
% TEST CODE
%    [sr]=irid_fsof(0.01,3,2,.8,5);
%***************************************************************
% Written by Hu Sheng, Yan Li and YangQuan Chen
%***************************************************************

if nargin<5; norder=5; end

if a < 0 | b < 0
    sprintf('%s','a and b constant has to be positive')
return, end

if Ts < 0
    sprintf('%s','Sampling period has to be positive'),
return, end
```

```
if r>=1 | r<= 0
    sprintf('%s','The fractional order should be in (0,1)')
return, end
if norder<2
    sprintf('%s','The order of the approx. transfer function
   has to be greater than 1')
return, end

close all;
wmax0=2*pi/Ts/2;
wmax=floor(1+ log10(wmax0) );
wmin=wmax-5; w=logspace(wmin,wmax,1000);
j=sqrt(-1); L=10/Ts;
t=[1:L]*Ts; y=[ ]; ht=[ ]; y1=[ ]; y2=[ ];
if a^2-4*b<0
    for k=1:length(t)
       y1(k)=quad(@(tau)realconvolution((-a)/2,sqrt(-a^2+4*b)/2,
r,tau,t(k)),0,t(k));
       y2(k)=quad(@(tau)imconvolution((-a)/2,sqrt(-a^2+4*b)/2,
r,tau,t(k)),0,t(k));
    end
    ht=y1+y2;
elseif a^2-4*b==0
    ht = exp(-sqrt(b).*t).*t.^(2*r-1)/gamma(2*r);
else
    for k=1:length(t)
       ht(k)=quadgk(@(x)integration2(x,a,b,r,t(k)),0,t(k));
    end
    s=(-a+sqrt(abs(a^2-4*b)))/2;
    ht= (exp(s.*t)/gamma(r)/gamma(r)).*ht;
end
h = [ht.*Ts];
q=norder;p=norder; [B,A]=stmcb((h),q,p);
sprintf('IRI discrete approx. transfer function:')
sr=tf(B,A,Ts)
hht=impulse(sr,t);
srfr=(1./((j*w).^2 +a*j*w+b)).^(r);
srfr1=freqresp(sr,w);

figure;
subplot(3,1,1)
plot(t,ht,'b'); hold on; plot(t,hht./Ts,'r-.')
axis([Ts,Ts.*L,-0.5,1]); xlabel('Time');
ylabel('Impulse response');
grid on;
```

```
legend(['impulse response of 1/(s^2 + ',num2str(a), '* s +',...
num2str(b),' )^{',num2str(abs(r)),'}'],
'approx. impulse response');

subplot(3,1,2)
semilogx(w,20*log10(abs(srfr)),'b');hold on;
semilogx(w,20*log10(abs(reshape(srfr1, 1000, 1))),'r-.');
legend(['mag. Bode of 1/(s^2 + ',num2str(a), '* s +',...
num2str(b),' )^{',num2str(abs(r)),'}'],'approx. mag. Bode');
xlabel('Frequency (Hz)');ylabel('Magnitude (dB)');
grid on;

subplot(3,1,3)
semilogx(w,(180/pi) * (angle(srfr)),'b');hold on;
semilogx(w,(180/pi) * (angle(reshape(srfr1, 1000, 1))),'r-.');
grid on;
xlabel('Frequency (Hz)');ylabel('Phase (degrees)');
legend(['phase Bode of 1/(s^2 + ',...
num2str(a), '* s +',num2str(b),' )^{',num2str(abs(r)),'}'],
'approx. phase Bode')

end
%**************************************************************



%**************************************************************
function y=imconvolution(reroot,imroot,gammac,tau,t)
%**************************************************************
y=(1/gamma(gammac)/gamma(gammac)).*
(tau.^(gammac-1).*exp(reroot.*tau).*sin(imroot.*tau)).*
((t-tau).^(gammac-1).*exp(reroot.*(t-tau)).*sin(imroot.*(t-tau)));

end
%**************************************************************
%**************************************************************
```

```
%****************************************************************
function y=realconvolution(reroot,imroot,gammac,tau,t)
%****************************************************************
y=(1/gamma(gammac)/gamma(gammac)).*
(tau.^(gammac-1).*exp(reroot.*tau).*cos(imroot.*tau)).*
((t-tau).^(gammac-1).*exp(reroot.*(t-tau)).*cos(imroot.*(t-tau)));

end
%****************************************************************
```

```
%****************************************************************
function y = integration2(x,a,b,gamma,t)
%****************************************************************
sqrtDelta = sqrt(abs(a^2-4*b));
s1=(-a-sqrtDelta)/2;
s2=(-a+sqrtDelta)/2;
y = exp((s1-s2).*x).*(x.^(gamma-1)).*((t-x).^(gamma-1));

end
%****************************************************************
```

D.3 Impulse Response Invariant Discretization of Distributed-Order Low-Pass Filter

```
%****************************************************************
function [sr]=irid_dolpf(Ts,a,b,c,p,q)
%
% irid_dolpf function is to compute a discrete-time finite
% dim. (z) transfer function to approx. a distributed order
% integrator ((c^r)/(b-a))*int(1/(s+c)^r,r,a,b),
% where "s" is the Laplace transform variable.
% Where 'a' and 'b' are arbitrary real numbers in the range of
% (0.5,1), and a<b. c>=0, 'p' and 'q' are integer and p>=q.
%
% The proposed approx. keeps the impulse response "invariant"
%****************************************************************
% IN:
%    Ts: The sampling period
%    a : Lower limit of integral
%    b : Upper limit of integral
%    c : A constant and c>=0
%    p : Denominator order of the approx. z-transfer function
%    q : Numerator order of the approximate z-transfer function
```

```
%*****************************************************************
% OUT:
%    sr: returns the LTI object that approximates the
%    ((c^r)/(b-a))*int(1/(s+c)^r,r,a,b)
%    in the sense of invariant impulse response.
%*****************************************************************
% TEST CODE
%    [doi]=irid_dolpf(0.001,0.6,1,1,5,5)
%*****************************************************************
% Written by Hu Sheng, Yan Li and YangQuan Chen
%*****************************************************************

if nargin<4; p=5,q=5; end

if p<3 | q<3
sprintf('%s','The order of the approx. transfer function ...
should be greater than 2')
return,end

if p<q
sprintf('%s','The denom. order of the approx. z-transfer function
... should be greater than or equal to the Numerator')
return, end

if Ts <= 0
sprintf('%s','Sampling period has to be positive')
return, end

if a>=b
sprintf('%s','The Upper limit of integral should be greater...
than the Lower limit of integral'),
return, end

close all;
wmax0=2*pi/Ts/2;
L=1/Ts;
t=[0:L-1]*Ts; ht=[ ];
for k=1:length(t)
    ht(k)=quadgk(@(x)integrand_dolpf(a,b,c,t(k),x),0,inf)/(b-a);
end
h=[ht(2:end).*Ts];
[B,A]=stmcb((h),q,p);
sprintf('IRI discrete approximated transfer function:')

sr = tf(B,A,Ts)
hht=impulse(sr,t);

wmax=floor(log10(wmax0))+1;
wmin=wmax-5;
w=logspace(wmin,wmax,1000);
j=sqrt(-1);
srfr=((j.*w+c).^(-a)-(j.*w+c).^(-b))./log(j.*w+c)./(b-a);
srfr1=freqresp(sr,w);
```

```
figure;
subplot(3,1,1)
plot(t,ht,'b'); hold on; plot(t,hht./Ts,'r-.');
axis([Ts,Ts.*L,0,5]);
xlabel('Time');ylabel('Impulse response');
legend('impulse response of c^\alpha/(b-a)*\int_a^b(s+c)^...
-\alphad\ alpha','approximated impulse response');
grid on;

subplot(3,1,2)
semilogx(w,20*log10(abs(srfr)),'b');hold on;
semilogx(w,20*log10(abs(reshape(srfr1, 1000, 1))),'r-.');
legend('mag. Bode of c^\alpha/(b-a)*\int_a^b(s+c)^...
-\alphad\alpha','approximated mag. Bode');
xlabel('Frequency (Hz)');
ylabel('Magnitude (dB)');grid on;

subplot(3,1,3)
semilogx(w,(180/pi) * (angle(srfr)),'b');hold on;
semilogx(w,(180/pi) * (angle(reshape(srfr1, 1000, 1))),'r-.');
grid on
xlabel('Frequency (Hz)');ylabel('Phase (degrees)');
legend('phase Bode of c^\alpha/(b-a)*\int_a^b(s+c)^...
-\alphad\alpha','approximated phase Bode')

end
%**************************************************************




%**************************************************************
function y=integrand_dolpf(a,b,lambdac,t,x)
%**************************************************************
i=sqrt(-1);
y =exp(-lambdac*t).*(exp(-x.*t).*(x.^(b-a).* ...
exp(-i*(b-a)*pi)-1)./
((x.^b).*exp(-i*b*pi).*(log(x)-i*pi))-exp(-x.*t).* ...
(x.^(b-a).*exp(i*(b-a)*pi)-1)./
((x.^b).*exp(i*b*pi).*(log(x)+i*pi)))/(2*pi*i);

end
%**************************************************************
```

References

1. Abry, P., Veitch, D.: Wavelet analysis of long-range-dependent traffic. IEEE Trans. Inf. Theory **44**(1), 2–15 (1998)
2. Adams, J.L., Hartley, T.T., Lorenzo, C.F.: Identification of complex order-distributions. J. Vib. Control **14**(9–10), 1375–1388 (2008)
3. Al-Alaoui, M.A.: Novel digital integrator and differentiator. Electron. Lett. **29**(4), 376–378 (1993)
4. Alcor, D., Gouzer, G., Triller, A.: Single-particle tracking methods for the study of membrane receptors dynamics. Eur. J. Neurosci. **30**(6), 987–997 (2009)
5. Antoniou, A.: Digital Filters: Analysis, Design and Applications, 2nd edn. McGraw-Hill, New York (1993)
6. Arnow, T.: Water-level and water-quality changes in Great Salt Lake, Utah, 1847–1983. U.S. Dept. of the Interior (1984)
7. Arnow, T.: Hydrologic characteristics of the Great Salt Lake, Utah, 1847–1986. U.S. G.P.O. Washington, DC (1990)
8. Asmussen, S.: Applied Probability and Queues (Stochastic Modelling and Applied Probability), 2nd edn. Springer, Berlin (2003)
9. Atanackovic, T.M., Budincevic, M., Pilipovic, S.: On a fractional distributed-order oscillator. J. Phys. A, Math. Gen. **38**(30), 6703–6713 (2005)
10. Atanackovic, T.M., Pilipovic, S., Zorica, D.: Existence and calculation of the solution to the time distributed order diffusion equation. Phys. Scr. **T136**, 014012 (2009) (6 pp.)
11. Atanackovic, T.M., Oparnica, L., Pilipovic, S.: On a nonlinear distributed order fractional differential equation. J. Math. Anal. Appl. **328**(1), 590–608 (2007)
12. Atanackovic, T.M., Pilipovic, S., Zorica, D.: Time distributed-order diffusion-wave equation. I. Volterra-type equation. Proc. R. Soc. Lond. Ser. A **465**, 1869–1891 (2009)
13. Atanackovic, T.M., Pilipovic, S., Zorica, D.: Time distributed-order diffusion-wave equation. II. Applications of Laplace and Fourier transformations. Proc. R. Soc. Lond. Ser. A **465**, 1893–1917 (2009)
14. Babcock, H.P., Chen, C., Zhuang, X.: Using single-particle tracking to study nuclear trafficking of viral genes. Biophys. J. **87**(4), 2749–2758 (2004)
15. Bagley, R.L., Torvik, P.J.: On the existence of the order domain and the solution of distributed order equations (Parts I, II). Int. J. Appl. Math. **2**, 865–882, 965–987 (2000)
16. Bailey, D.H., Swarztrauber, P.N.: The fractional Fourier transform and applications. SIAM Rev. **33**(3), 389–404 (1991)
17. Barbosa, R.S., Machado, J.A.T.: Implementation of discrete-time fractional-order controllers based on LS approximations. Acta Polytech. Hung. **3**(4), 5–22 (2006)
18. Bardet, J.M., Lang, G., Oppenheim, G., Phillipe, A., Taqqu, M.S.: Generators of long-range dependent processes: a survey. In: Doukhan, P., Oppenheim, G., Taqqu, M.S. (eds.) Theory and Applications of Long-Range Dependence, pp. 579–623 (2003)

H. Sheng et al., *Fractional Processes and Fractional-Order Signal Processing*, Signals and Communication Technology,
DOI 10.1007/978-1-4471-2233-3,

19. Barndorff-Nielsen, O.E.: Lévy Processes: Theory and Applications, 1st edn. Birkhauser, Basel (2001)
20. Barnes, J.A., Allan, D.W.: A statistical model of flicker noise. Proc. IEEE **54**(2), 176–178 (1996)
21. Beine, M., Laurent, S.: Structural change and long memory in volatility: new evidence from daily exchange rates. In: Econometric Society World Congress 2000 Contributed Papers 2013/10473, ULB—Universite Libre de Bruxelles (2000)
22. Beran, J.: Statistics for Long-Memory Processes, 1st edn. CRC Press, Boca Raton (1994)
23. Bertocci, U., Gabrielli, C., Huet, F., Keddam, M.: Noise resistance applied to corrosion measurements. J. Electrochem. Soc. **144**(1), 31–37 (1997)
24. Biagini, F., Hu, Y., Oksendal, B., Zhang, T.: Stochastic Calculus for Fractional Brownian Motion and Applications, 1st edn. Springer, Berlin (2008)
25. Blu, T., Unser, M.: The fractional spline wavelet transform: definition and implementation. In: Proceedings of the 25th IEEE International Conference on Acoustics, Speech, and Signal Processing (ICASSP'00), Istanbul, Turkey, vol. I, pp. 512–515 (2000)
26. Bohannan, G.: Application of fractional calculus to polarization dynamics in solid dielectric materials. Ph.D. Dissertation, Montana State University (November 2000)
27. Bohannan, G.W.: Analog realization of a fractional control element—revisited. In: IEEE CDC2002 Tutorial Workshop, Las Vegas, NE, USA (2002). http://mechatronics.ece.usu.edu/foc/cdc02tw/
28. Bohannan, G.W.: Analog fractional order controller in temperature and motor control applications. J. Vib. Control **14**(9–10), 1487–1498 (2008)
29. Bollerslev, T., Mikkelsen, H.O.: Modeling and pricing long memory in stock market volatility. J. Econom. **73**(1), 151–184 (1996)
30. Boutahar, M., Dufrénot, G., Péguin-Feissolle, A.: A simple fractionally integrated model with a time-varying long memory parameter d_t. Comput. Econ. **31**(3), 225–241 (2008)
31. Box, G., Jenkins, G.: Time Series Analysis: Forecasting and Control, 3rd edn. Prentice Hall, New York (1976)
32. Bracewell, R.N.: The Fourier Transform and Its Applications, 3rd edn. McGraw-Hill, New York (1999)
33. Brančík, L.: Programs for fast numerical inversion of Laplace transforms in MATLAB language environment. In: Proceedings of the 7th Conference MATLAB'99, Prague, Czech Republic, Nov. 1999, pp. 27–39 (1999)
34. Brančík, L.: Utilization of quotient-difference algorithm in FFT-based numerical ILT method. In: Proceedings of the 11th International Czech-Slovak Scientific Conference Radioelektronika 2001, Brno, Czech Republic, May 2001, pp. 352–355 (2001)
35. Brančík, L.: Matlab oriented matrix Laplace transforms inversion for distributed systems simulation. In: Proceedings of the 12th International Czech-Slovak Scientific Conference Radioelektronika, Slovakia (2002)
36. Breiman, L.: Probability. Addison-Wesley, Reading (1968)
37. Brockwell, P.J., Davis, R.A.: Time Series: Theory and Methods, 2nd edn. Springer, Berlin (1998)
38. Brzeźniak, Z., Zabczyk, J.: Regularity of Ornstein-Uhlenbeck processes driven by a Lévy white noise. Potential Anal. **32**(2), 153–188 (2010)
39. Cajueiro, D.O., Tabak, B.M.: Time-varying long-range dependence in US interest rates. Chaos Solitons Fractals **34**(2), 360–367 (2007)
40. Cambanis, S., Maejima, M., Samorodnitsky, G.: Characterization of linear and harmonizable fractional stable motions. Stoch. Process. Appl. **42**, 91–110 (1992)
41. Candan, C., Kutay, M.A., Ozaktas, H.M.: The discrete fractional Fourier transform. IEEE Trans. Signal Process. **48**(5), 1329–1337 (2000)
42. Cao, Y.: Correcting the minimum ITAE standard forms of zero-displacement-error systems. J. Zhejiang Univ. Sci. **23**(4), 550–559 (1989)
43. Cappé, O., Moulines, E., Pesquet, J.-C., Petropulu, A., Yang, X.: Long-range dependence and heavy-tail modeling for teletraffic data. IEEE Signal Process. Mag. **19**(3), 14–27 (2002)

44. Capps, G.G., Pine, S., Edidin, M., Zuniga, M.C.: Short class I major histocompatibility complex cytoplasmic tails differing in charge detect arbiters of lateral diffusion in the plasma membrane. Biophys. J. **86**(5), 2896–2909 (2004)
45. Caputo, M.: Elasticità e Dissipazione. Zanichelli, Bologna (1969)
46. Caputo, M.: Mean fractional-order-derivatives differential equations and filters. Ann. Univ. Ferrara **41**(1), 73–84 (1995)
47. Caputo, M.: Distributed order differential equations modelling dielectric induction and diffusion. Fract. Calc. Appl. Anal. **4**(4), 421–442 (2001)
48. Carlson, G., Halijak, C.: Approximation of fractional capacitors $(1/s)^{(1/n)}$ by a regular Newton process. IEEE Trans. Circuit Theory **11**(2), 210–213 (1964)
49. Çenesiz, Y., Keskin, Y., Kurnaz, A.: The solution of the Bagley-Torvik equation with the generalized Taylor collocation method. J. Franklin Inst. **347**(2), 452–466 (2010)
50. Charef, A.: Analogue realization of fractional-order integrator, differentiator and fractional $PI^{\lambda}D^{\mu}$ controller. In: IEE Proceedings-Control Theory and Applications, vol. 153, pp. 714–720 (2006)
51. Chen, C.: Hurst parameter estimate. http://www.mathworks.com/matlabcentral/fileexchange/19148 (Mar 2008)
52. Chen, J.F., Bogaerts, W.F.: The physical meaning of noise resistance. Corros. Rev. **37**(11), 1839–1842 (1995)
53. Chen, W., Sun, H., Zhang, X., Korosak, D.: Anomalous diffusion modeling by fractal and fractional derivatives. Comput. Math. Appl. **59**(5), 1754–1758 (2009)
54. Chen, Y.Q.: Low-pass IIR digital differentiator design. http://www.mathworks.com/matlabcentral/fileexchange/3517 (2003)
55. Chen, Y.Q.: A new IIR-type digital fractional order differentiator. http://www.mathworks.com/matlabcentral/fileexchange/3518 (2003)
56. Chen, Y.Q.: Fractional order signal processing in biology/biomedical signal analysis. In: Fractional Order Calculus Day at Utah State University (2005). http://mechatronics.ece.usu.edu/foc/event/FOCDay@USU/
57. Chen, Y.Q.: Impulse response invariant discretization of fractional order integrators or differentiators. http://www.mathworks.com/matlabcentral/fileexchange/21342 (2008)
58. Chen, Y.Q.: Impulse response invariant discretization of fractional order low-pass filters. http://www.mathworks.com/matlabcentral/fileexchange/21365 (2008)
59. Chen, Y.Q., Moore, K.L.: Discretization schemes for fractional-order differentiators and integrators. IEEE Trans. Circuits Syst. I, Fundam. Theory Appl. **49**(3), 363–367 (2002)
60. Chen, Y.Q., Sun, R., Zhou, A.: An improved Hurst parameter estimator based on fractional Fourier transform. Telecommun. Syst. **43**(3–4), 197–206 (2010)
61. Chen, Y.Q., Sun, R., Zhou, A., Zaveri, N.: Fractional order signal processing of electrochemical noises. J. Vib. Control **14**(9–10), 1443–1456 (2008)
62. Chen, Y.Q., Vinagre, B.M.: A new IIR-type digital fractional order differentiator. Signal Process. **83**(11), 2359–2365 (2003)
63. Chen, Y.Q., Vinagre, B.M., Podlubny, I.: Continued fraction expansion approaches to discretizing fractional order derivatives—an expository review. Nonlinear Dyn. **38**(16), 155–170 (2004)
64. Cheng, C.F., Tsay, Y.T., Wu, T.T.: Walsh operational matrices for fractional calculus and their application to distributed systems. J. Franklin Inst. **303**(3), 267–284 (1977)
65. Clegg, R.G.: A practical guide to measuring the Hurst parameter. Int. J. Simul. Syst. Sci. Technol. **7**(2), 3–14 (2006)
66. Coeurjolly, J.-F.: Identification of multifractional Brownian motion. Bernoulli **11**(6), 987–1008 (2005)
67. Cohen, S., Marty, R.: Invariance principle, multifractional Gaussian processes and long-range dependence. Ann. Inst. Henri Poincaré B, Probab. Stat. **44**(3), 475–489 (2008)
68. Coimbra, C.F.M.: Mechanics with variable-order differential operators. Ann. Phys. **12**(11–12), 692–703 (2003)
69. Connolly, J.A.: The numerical solution of fractional and distributed order differential equations. Thesis, University of Liverpool (December 2004)

70. Cont, R.: Long range dependence in financial markets. In: Lévy-Véhel, J., Lutton, E. (eds.) Fractals in Engineering: New Trends in Theory and Applications. Springer, Berlin (2005)
71. Cottone, G., Di Paola, M.: On the use of fractional calculus for the probabilistic characterization of random variables. Probab. Eng. Mech. **24**(3), 321–330 (2009)
72. Cottone, G., Di Paola, M., Metzler, R.: Fractional calculus approach to the statistical characterization of random variables and vectors. Phys. A, Stat. Mech. Appl. **389**(5), 909–920 (2010)
73. Crovella, M.E., Bestavros, A.: Self-similarity in world wide web traffic evidence and possible causes. IEEE/ACM Trans. Netw. **5**(6), 835–846 (1997)
74. Dalir, M., Bashour, M.: Applications of fractional calculus. Appl. Math. Sci. **4**(21–24), 1021–1032 (2010)
75. Davies, B., Martin, B.: Numerical inversion of the Laplace transform: a survey and comparison of methods. J. Comput. Phys. **33**(1), 1–32 (1979)
76. D'Azzo, J.J., Houpis, C.H., Sheldon, S.N.: Linear Control System Analysis and Design, 5th edn. CRC Press, Boca Raton (2003)
77. De Espíndola, J.J., Bavastri, C.A., De Oliveira Lopes, E.M.: Design of optimum systems of viscoelastic vibration absorbers for a given material based on the fractional calculus model. J. Vib. Control **14**(9–10), 1607–1630 (2008)
78. de Espíndola, J.J., Bavastri, C.A., Lopes, E.M.O.: On the passive control of vibrations with viscoelastic dynamic absorbers of ordinary and pendulum types. J. Franklin Inst. **347**(1), 102–115 (2010)
79. de Hoog, F.R., Knight, J.H., Stokes, A.N.: An improved method for numerical inversion of Laplace transforms. SIAM J. Sci. Comput. **3**(3), 357–366 (1982)
80. Dexter, S.C., Duquette, D.J., Siebert, O.W., Videla, H.A.: Use and limitations of electrochemical techniques for investigating microbiological corrosion. Corros. Sci. **47**(4), 308–318 (1991)
81. Diethelm, K., Ford, N.J.: Numerical analysis for distributed-order differential equations. J. Comput. Appl. Math. **225**(1), 96–104 (2009)
82. Doornik, J.A.: An Object-Oriented Matrix Programming Language OxMetrics, 1st edn. Timberlake Consultants, London (1998)
83. Dorf, R.C.: Modern Control Systems. Addison-Wesley/Longman, Reading/Harlow (1989)
84. Duffy, D.G.: On the numerical inversion of Laplace transforms: comparison of three new methods on characteristic problems from applications. ACM Trans. Math. Softw. **19**(3), 333–359 (1993)
85. Eden, D.A.: Electrochemical noise—the first two octaves. In: Corrosion, vol. 98, San Diego (March 1998)
86. Edidin, M.: Shrinking patches and slippery rafts: scales of domains in the plasma membrane. Trends Cell Biol. **11**(12), 492–496 (2001)
87. Erseghe, T., Kraniauskas, P., Cariolaro, G.: Unified fractional Fourier transform and sampling theorem. IEEE Trans. Signal Process. **47**(12), 3419–3423 (1999)
88. Etter, D.M., Stearns, S.D.: Adaptive estimation of time delay in sampled system. IEEE Trans. Acoust. Speech Signal Process. **29**(3), 582–587 (1981)
89. Ewing, R.L., Abdel-Aty-Zohdy, H.S., Hollenbeck, M.C., Stevens, K.S.: Fractional-order signal processing using a polymer-electrolyte transistor. In: 51st Midwest Symposium on Circuits and Systems, MWSCAS 2008, Knoxville, USA, pp. 601–604 (2008)
90. Falconer, K.J.: The local structure of random processes. J. Lond. Math. Soc. **67**(3), 657–672 (2003)
91. Falconer, K.: Fractal Geometry: Mathematical Foundations and Applications, 1st edn. Wiley, New York (1990)
92. Fama, E.F., Roll, R.: Parameter estimates for symmetric stable distributions. J. Am. Stat. Assoc. **66**(334), 331–338 (1971)
93. Feller, W.: The fundamental limit theorems in probability. Bull. Am. Math. Soc. **51**(11), 800–832 (1945)
94. Feller, W.: An Introduction to Probability Theory and Its Applications, 2nd edn. Wiley, New York (1966)

95. Ferdi, Y.: Impulse invariance-based method for the computation of fractional integral of order $0 < \alpha < 1$. Comput. Electr. Eng. **35**(5), 722–729 (2009)
96. Fukami, T., Chen, R.-H.: Crystal structure and electrical conductivity of $LiN_2H_5SO_4$ at high temperature. Jpn. J. Appl. Phys. **37**(3A), 925–929 (1998)
97. Geweke, J., Porter-Hudak, S.: The estimation and application of long memory time series models. J. Time Ser. Anal. **4**, 221–238 (1983)
98. Glöckle, W.G., Nonnenmacher, T.F.: A fractional calculus approach to self-similar protein dynamics. Biophys. J. **68**(1), 46–53 (1995)
99. Gnedenko, B.V., Kolmogorov, A.N.: Limit Distributions for Sums of Independent Random Variables, 1st edn. Wiley-Interscience, New York (1968)
100. Goldberger, A.L., Amaral, L.A.N., Glass, L., Hausdorff, J.M., Ivanov, P.C., Mark, R.G., Mietus, J.E., Moody, G.B., Peng, C.-K., Stanley, H.E.: PhysioBank, PhysioToolkit, and PhysioNet: Components of a new research resource for complex physiologic signals. Circulation **101**(23), e215–e220 (2000)
101. Gonzáez-Núñez, M.A., Uruchurtu-Chavarín, J.: R/S fractal analysis of electrochemical noise signals of three organic coating samples under corrosion condition. Corros. Sci. Eng. **6**(2003) (2003)
102. Gouyet, J.-F., Mandelbrot, B.: Physics and Fractal Structures, 1st edn. Springer, Berlin (1996)
103. Granger, W., Orr, D.: Infinite variance and research strategy in time series analysis. J. Am. Stat. Assoc. **67**(338), 275–285 (1972)
104. Gray, H.L., Zhang, N.-F., Woodward, W.A.: On generalized fractional processes. J. Time Ser. Anal. **10**(3), 233–257 (1989)
105. Grigolini, P., Palatella, L., Raffaelli, G.: Asymmetric anomalous diffusion: an efficient way to detect memory in time series. Fractals **9**(4), 439–449 (2001)
106. Grossmann, A., Morlet, J.: Decomposition of Hardy functions into square integrable wavelets of constant shape. SIAM J. Math. Anal. **15**(4), 723–736 (1984)
107. Gubner, J.A.: Probability and Random Processes for Electrical and Computer Engineers, 1st edn. Cambridge University Press, Cambridge (2006)
108. Gupta, L., Molfese, D.L., Tammana, R., Simos, P.G.: Nonlinear alignment and averaging for estimating the evoked potential. IEEE Trans. Biomed. Eng. **43**(4), 348–356 (1996)
109. Hahn, S.L.: Hilbert Transforms in Signal Processing, 1st edn. Artech Print on Demand, Norwood (1996)
110. Harmantzis, F., Hatzinakos, D.: Heavy network traffic modeling and simulation using stable FARIMA processes. In: International Teletraffic Congress (ITC-19) (2005)
111. Hartley, T.T., Lorenzo, C.F.: Fractional-order system identification based on continuous order-distributions. Signal Process. **83**(11), 2287–2300 (2003)
112. Hartley, T.T., Lorenzo, C.F.: A frequency-domain approach to optimal fractional-order damping. Nonlinear Dyn. **38**(1–2), 69–84 (2004)
113. Hassler, U.: Regression of spectral estimators with fractionally integrated time series. J. Time Ser. Anal. **14**(4), 369–380 (1993)
114. Hayes, M.H.: Statistical Digital Signal Processing and Modeling, 1st edn. Wiley, New York (1996)
115. Hazarika, N., Tsoi, A.C., Sergejew, A.A.: Nonlinear considerations in EEG signal classification. IEEE Trans. Signal Process. **45**(4), 829–836 (1997)
116. Higuchi, T.: Approach to an irregular time series on the basis of the fractal theory. Physica D, Nonlinear Phenom. **31**(2), 277–283 (1988)
117. Hilfer, R.: Applications of Fractional Calculus in Physics, 1st edn. World Scientific, Singapore (2000)
118. Holcomb, G.R. Jr., Covino, B.S., Eden, D.: State-of-the-art Review of Electrochemical Noise Sensors. product of Department of Energy's Information Bridge: DOE Scientific and Technical Information (2001). http://www.netl.doe.gov/scng/publications/ENStateoftheArt.pdf
119. Hollenbeck, K.J.: INVLAP.M: A MATLAB function for numerical inversion of Laplace transforms by the de Hoog algorithm (1998). http://www.isva.dtu.dk/staff/karl/invlap.htm

120. Hsu, H.: Schaum's Outline of Probability, Random Variables, and Random Processes, 2nd edn. McGraw-Hill, New York (2010)
121. Huang, Z., Li, C.: On fractional stable processes and sheets: White noise approach. J. Math. Anal. Appl. **325**(1), 624–635 (2006)
122. Humbert, P., Agarwal, R.P.: Sur la fonction de Mittag-Leffler et quelques-unes de ses généralisations. Bull. Sci. Math. Ser. II **77**, 180–185 (1953)
123. Hurst, H.E.: Long-term storage capacity of reservoirs. Trans. Am. Soc. Civ. Eng. **116**(3), 770–799 (1951)
124. Hwa, R.C., Ferree, T.C.: Scaling properties of fluctuations in the human electroencephalogram. Phys. Rev. E **66**(2), 02190 (2002), 1–18
125. Iber, C., Ancoli-Israel, S., Chesson, A., Quan, S.F.: The AASM Manual for the Scoring of Sleep and Associated Events: Rules, Terminology and Technical Specifications. American Academy of Sleep Medicine, Darien (2007)
126. Ichimaru, Y., Moody, G.B.: Development of the polysomnographic database on CD-ROM. Psychiatry Clin. Neurosci. **53**(2), 175–177 (1999)
127. Ignaccolo, M., Latka, M., Jernajczyk, W., Grigolini, P., West, B.J.: The dynamics of EEG entropy. J. Biol. Phys. **36**(2), 185–196 (2010)
128. Ingman, D., Suzdalnitsky, J.: Application of differential operator with servo-order function in model of viscoelastic deformation process. J. Eng. Mech. **131**(7), 763–767 (2005)
129. Ingman, D., Suzdalnitsky, J., Zeifman, M.: Constitutive dynamic-order model for nonlinear contact phenomena. J. Appl. Mech. **67**(2), 383–390 (2000)
130. Jach, A., Kokoszka, P.: Robust wavelet-domain estimation of the fractional difference parameter in heavy-tailed time series: an empirical study. Methodol. Comput. Appl. Probab. **12**(1), 177–197 (2010)
131. Jesus, I.S., Machado, J.A.T.: Development of fractional order capacitors based on electrolyte processes. Nonlinear Dyn. **56**(1–2), 45–55 (2009)
132. Jin, S., Haggie, P.M., Verkman, A.S.: Single-particle tracking of membrane protein diffusion in a potential: simulation, detection, and application to confined diffusion of CFTR Cl^- channels. Biophys. J. **93**(3), 1079–1088 (2007)
133. Jin, S., Verkman, A.S.: Single particle tracking of complex diffusion in membranes: simulation and detection of barrier, raft, and interaction phenomena. J. Phys. Chem. **111**(14), 3625–3632 (2007)
134. Navarro, R. Jr., Tamangan, R., Guba-Natan, N., Ramos, E., Guzman, A.D.: The identification of long memory process in the Asean-4 stock markets by fractional and multifractional Brownian motion. Philipp. Stat. **55**(1–2), 65–83 (2006)
135. Jumarie, G.: Fractional multiple birth-death processes with birth probabilities $\lambda_i(\Delta t)^\alpha + o((\Delta t)^\alpha)$. J. Franklin Inst. **347**(10), 1797–1813 (2010)
136. Jury, E.I.: Theory and Application of the Z-Transform Method, 1st edn. Krieger, Melbourne (1973)
137. Kawaba, K., Nazri, W., Aun, H.K., Iwahashi, M., Kambayashi, N.: A realization of fractional power-law circuit using OTAs. In: The 1998 IEEE Asia-Pacific Conference on Circuits and Systems (IEEE APCCAS'98), Chiangmai, Thailand, pp. 249–252 (1998)
138. Kettani, H., Gubner, J.A.: A novel approach to the estimation of the long-range dependence parameter. IEEE Trans. Circuits Syst. **53**(6), 463–467 (2006)
139. Kilbas, A.A., Srivastava, H.M., Trujillo, J.J.: Theory and Applications of Fractional Differential Equations, vol. 204. Elsevier, New York (2006)
140. Kilbas, A.A., Saigo, M., Saxena, R.K.: Generalized Mittag-Leffler function and generalized fractional calculus operators. Integral Transforms Spec. Funct. **15**(1), 31–49 (2004)
141. Kochubei, A.N.: Distributed order calculus and equations of ultraslow diffusion. J. Math. Anal. Appl. **340**(1), 252–281 (2008)
142. Koeller, R.C.: Applications of fractional calculus to the theory of viscoelasticity. J. Appl. Mech. **51**(2), 299–307 (1984)
143. Kokoszka, P.S., Taqqu, M.S.: Fractional ARIMA with stable innovations. Stoch. Process. Appl. **60**, 19–47 (1995)

144. Kolmogorov, A.N.: Wienersche Spiralen und einige, andere interessante Kurven in Hilbertschen Raum. Dokl. Akad. Nauk SSSR **26**, 115–118 (1940)
145. Komkov, V.: Optimal Control Theory for the Damping of Vibrations of Simple Elastic Systems. Springer, Berlin (1972)
146. Kong, X., Qiu, T.: Adaptive estimation of latency change in evoked potentials by direct least mean p-norm time-delay estimation. IEEE Trans. Biomed. Eng. **46**(8), 994–1003 (1999)
147. Kong, X., Qiu, T.: Latency change estimation for evoked potentials via frequency selective adaptive phase spectrum analyzer. IEEE Trans. Biomed. Eng. **46**(8), 1004–1012 (1999)
148. Kong, X., Qiu, T.: Latency change estimation for evoked potentials: a comparison of algorithms. Med. Biomed. Eng. Comput. **39**(2), 208–224 (2001)
149. Kong, X., Thakor, N.V.: Adaptive estimation of latency changes in evoked potentials. IEEE Trans. Biomed. Eng. **43**(2), 189–197 (1996)
150. Kono, N., Maejima, M.: Self-similar stable processes with stationary increments. In: Cambanis, S., Samorodnitsky, G., Taqqu, M. (eds.) Stable Processes and Related Topics, 1st edn., pp. 275–295. Birkhauser, Boston (1991)
151. Kotyk, J.: Matlab algorithms for the Laplace transform inversion. In: Proceedings of MATLAB Conference 2005 (2005). http://dsp.vscht.cz/konference_matlab/matlab05/prispevky/kotyk/kotyk.pdf
152. Koutsoyiannis, D.: Coupling stochastic models of different time scales. Water Resour. Res. **37**(2), 379–392 (2001)
153. Koutsoyiannis, D.: Climate change, the Hurst phenomenon, and hydrological statistics. Hydrol. Sci. J. **48**(1), 3–24 (2003)
154. Krishna, B.T.: Studies on fractional order differentiators and integrators: a survey. Signal Process. **91**(3), 386–426 (2011)
155. Krishna, B.T., Reddy, K.V.V.S.: Active and passive realization of fractance device of order 1/2. Act. Passive Electron. Compon. (2008). doi:10.1155/2008/369421
156. Kusumi, A., Ike, H., Nakada, C., Murase, K., Fujiwara, T.: Single-molecule tracking of membrane molecules: plasma membrane compartmentalization and dynamic assembly of raft-philic signaling molecules. Semin. Immunol. **17**(1), 3–21 (2005)
157. Lahiri, A., Rawat, T.K.: Noise analysis of single stage fractional-order low-pass filter using stochastic and fractional calculus. ECTI Trans. Electr. Eng. Electron. Commun. **7**(2), 136–143 (2009)
158. Lall, U., Moon, Y.I., Kwon, H.H., Bosworth, K.: Locally weighted polynomial regression: parameter choice and application to forecasts of the Great Salt Lake. Water Resour. Res. **42**(5), W05422.1–W05422.11 (2006)
159. Lall, U., Sangoyomi, T., Abarbanel, H.D.I.: Nonlinear dynamics of the Great Salt Lake: nonparametric short-term forecasting. Water Resour. Res. **32**(4), 975–985 (1996)
160. Lambourne, A.: Our Inland Sea: The Story of A Homestead. Deseret News, Salt Lake City (1990)
161. Laurent, S., Peters, J.P.: Estimating and Forecasting ARCH Models Using G@RCH, 1st edn. London: Timberlake Consultants Press, London (2004)
162. Li, M., Lim, S.C.: A rigorous derivation of power spectrum of fractional Gaussian noise. Fluct. Noise Lett. **6**(4), C33–C36 (2006)
163. Li, M., Lim, S.C.: Modeling autocorrelation functions of long-range dependent teletraffic series based on optimal approximation in Hilbert space-a further study. Appl. Math. Model. **31**(3), 625–631 (2007)
164. Li, M.: Fractal time series—a tutorial review. Math. Probl. Eng. (2010). doi:10.1155/2010/157264
165. Li, M., Li, J.-Y.: On the predictability of long-range dependent series. Math. Probl. Eng. (2010). doi:10.1155/2010/397454
166. Li, Q., Tricaud, C., Sun, R., Chen, Y.Q.: Great Salt Lake surface level forecasting using FIGARCH modeling. In: IDETC/CIE 2007, Las Vegas, September 2007
167. Li, Y., Sheng, H., Chen, Y.Q.: Impulse response invariant discretization of distributed order low-pass filter. http://www.mathworks.com/matlabcentral/fileexchange/authors/82211 (2010)

168. Li, Y., Sheng, H., Chen, Y.Q.: On distributed order integrator/differentiator. Signal Process. **91**(5), 1079–1084 (2010)
169. Li, Y., Sheng, H., Chen, Y.Q.: On distributed order low pass filter. In: 2010 IEEE/ASME International Conference on Mechatronic and Embedded Systems and Applications, Qingdao, ShanDong, China (2010)
170. Li, Y., Sheng, H., Chen, Y.Q.: On the fractional-order distributed parameter low-pass filter. In: The 49th IEEE Conference on Decision and Control, Atlanta, Georgia, USA (2010)
171. Li, Y., Sheng, H., Chen, Y.Q.: Analytical impulse response of a fractional second order filter and its impulse response invariant discretization. Signal Process. **91**(3), 498–507 (2011)
172. Lim, S.C.: Fractional Brownian motion and multifractional Brownian motion of Riemann-Liouville type. J. Phys. A, Math. Gen. **34**, 1301–1310 (2001)
173. Lim, S.C., Teo, L.P.: The fractional oscillator process with two indices. J. Phys. A, Math. Theor. **42**(6) (2009)
174. Linkenkaer-Hansen, K., Nikouline, V.V., Palva, J.M., Ilmoniemi, R.J.: Long-range temporal correlations and scaling behavior in human brain oscillations. J. Neurosci. **21**(4), 1370–1377 (2001)
175. Lion, A.: On the thermodynamics of fractional damping elements. Contin. Mech. Thermodyn. **9**(2), 83–96 (1997)
176. Lohmann, A.W., Mendlovic, D., Zalevsky, Z.: Fractional Hilbert transform. Opt. Lett. **21**(4), 281–283 (1996)
177. Loomis, A.L., Harvey, E.N., Hobart, G.A.: Cerebral states during sleep, as studied by human brain potentials. J. Exp. Psychol. **21**(2), 127–144 (1937)
178. López, T., Martínez-González, C.L., Manjarrez, J., Plascencia, N., Balank, A.S.: Fractal analysis of EEG signals in the brain of epileptic rats, with and without biocompatible implanted neuroreservoirs. Appl. Mech. Mater. **15**, 127–136 (2009)
179. Lorenzo, C.F., Hartley, T.T.: Initialization, conceptualization, and application in the generalized fractional calculus. NASA technical paper, NASA/TP 1998-208415 (1998)
180. Lorenzo, C.F., Hartley, T.T.: Variable order and distributed order fractional operators. Nonlinear Dyn. **29**(1–4), 57–98 (2002)
181. Loverro, A.: Fractional calculus: History, definitions and applications for the engineer. Report, Department of Aerospace and Mechanical Engineering, Notre Dame, IN, USA (May 2004)
182. Lubich, C.: Discretized fractional calculus. SIAM J. Math. Anal. **17**(3), 704–719 (1986)
183. Lucio-Garcia, M.A., Gonzalez-Rodriguez, J.G., Martinez-Villafañe, A., Dominguez-Patiño, G., Neri-Flores, M.A., Chacon-Nava, J.G.: A study of hydroxyethyl imidazoline as H_2S corrosion inhibitor using electrochemical noise and electrochemical impedance spectroscopy. J. Appl. Electrochem. **40**(2), 393–399 (2010)
184. Ma, X., Nikias, C.L.: Joint estimation of time delay an frequency delay in impulsive noise using fractional lower order statistics. IEEE Trans. Signal Process. **44**(11), 2669–2687 (1996)
185. Machado, J.A.T.: Analysis and design of fractional-order digital control systems. Syst. Anal. Model. Simul. **27**(2–3), 107–122 (1997)
186. Machell, F.W., Nikias, C.L.: Statistical characteristic of ocean acoustic noise processes. In: Wegman, E.J., Schwartz, S.C., Thomas, J.B. (eds.) Topics in Non-Gaussian Signal Processing, pp. 29–57. Springer, New York (1989). Chap. 3
187. Mainardi, F., Paradisi, P., Gorenflo, R.: Probability distributions generated by fractional diffusion equations. Comput. Econom. (2007). http://arxiv.org/abs/0704.0320v1
188. Mainardi, F.: Fractional Calculus and Waves in Linear Viscoelasticity: An Introduction to Mathematical Models, 1st edn. World Scientific, Singapore (2010)
189. Mainardi, F., Mura, A., Gorenflo, R., Stojanovic, M.: The two forms of fractional relaxation of distributed order. J. Vib. Control **9**, 1249–1268 (2007)
190. Mainardi, F., Mura, A., Pagnini, G., Gorenflo, R.: Time-fractional diffusion of distributed order. http://www.citebase.org/abstract?id=oai:arXiv.org:cond-mat/0701132 (2007)
191. Mainardi, F., Pagnini, G.: The role of the Fox-Wright functions in fractional sub-diffusion of distributed order. J. Comput. Appl. Math. **207**(2), 245–257 (2007)

192. Mandelbrot, B.B.: The Fractal Geometry of Nature, 1st edn. Freeman, New York (1982)
193. Mandelbrot, B.B., Van Ness, J.W.: Fractional Brownian motion, fractional noises and applications. SIAM Rev. **10**(4), 422–437 (1968)
194. Manderlbrot, B.B., Wallis, J.R.: Computer experiments with fractional Gaussian noises. Water Resour. Res. **5**(1), 228–267 (1969)
195. Mansfeld, F., Little, B.: A technical review of electrochemical techniques applied to microbiologically influenced corrosion. Corros. Sci. **32**(3), 247–272 (1991)
196. Marszaek, W.: Block-pulse functions method of the inverse Laplace transform for irrational and transcendental transfer functions. J. Franklin Inst. **318**(3), 193–200 (1984)
197. Matos, C., Ortigueira, M.D.: Fractional filters: an optimization approach. In: Camarinha-Matos, L.M., Pereira, P., Ribeiro, L. (eds.) Emerging Trends in Technological Innovation, IFIP Advances in Information and Communication Technology. IFIP International Federation for Information Processing 2010, vol. 314, pp. 361–366. Springer, Berlin (2010)
198. Matson, D.L., Weiss, M.S.: Evoked potential analysis of impact acceleration experiments. In: AGARD, Electric and Magnetic Activity of the Central Nervous System: Research and Clinical Applications in Aerospace Medicine. 28, vol. 432, pp. 1–13. Neuilly Sur Seine, France (1998)
199. McCulloch, J.H.: Simple consistent estimators of stable distribution parameters. Commun. Stat., Simul. Comput. **15**(4), 1109–1136 (1986)
200. Mcewen, J.A., Anderson, G.B.: Modeling the stationarity and Gaussianity of spontaneous electroencephalographic activity. IEEE Trans. Biomed. Eng. **22**(5), 361–369 (1975)
201. McQuarrie, D.A.: Statistical Mechanics, 2nd edn. University Science Books, Sausalito (2000)
202. Middleton, D.: Statistical-physical models of man-made and natural radio noise, part II. Report 76-86, Institute for Telecommunication Sciences (April 1976)
203. Miller, K.S., Ross, B.: An Introduction to the Fractional Calculus and Fractional Differential Equations. Wiley, New York (1993)
204. Miller, S., Childers, D.: Probability and Random Processes: With Applications to Signal Processing and Communications, 2nd edn. Academic Press, San Diego (2004)
205. Mittag-Leffler, G.M.: Sur la nouvelle fonction $E_\alpha(x)$. C. R. Acad. Sci. Paris **137**, 554–558 (1903)
206. Mitzenmacher, M.: A brief history of generative models for power law and lognormal distributions. Internet Math. **1**(2), 226–251 (2003)
207. Monje, C.A., Chen, Y., Vinagre, B., Xue, D., Feliu, V.: Fractional Order Systems and Control—Fundamentals and Applications (Advances in Industrial Control Series). Springer, Berlin (2010)
208. Moon, M., Skerry, B.: Interpretation of corrosion resistance properties of organic paint films from fractal analysis of electrochemical noise data. J. Coat. Technol. **67**(843), 35–44 (1995)
209. Moshrefi-Torbati, M., Hammond, J.K.: Physical and geometrical interpretation of fractional operators. J. Franklin Inst. **335**(6), 1077–1086 (1998)
210. Mukhopadhyay, S.: Fractional order modeling and control: development of analog strategies for plasma position control of the STOR-1M Tokamak. Master's thesis, Utah State University, Logan, Utah, USA (2009)
211. Muniandy, S.V., Lim, S.C.: Modeling of locally self-similar processes using multifractional Brownian motion of Riemann-Liouville type. Physical Review E **63**(4) (2001)
212. Nagiub, A., Mansfeld, F.: Evaluation of microbiologically influenced corrosion inhibition using electrochemical noise analysis. Corros. Sci. **43**(11), 2001–2009 (2001)
213. Namias, V.: The fractional order Fourier transform and its application to quantum mechanics. IMA J. Appl. Math. **25**(3), 241–265 (1980)
214. Natarajan, K., Acharya, R.U., Alias, F., Tiboleng, T., Puthusserypady, S.K.: Nonlinear analysis of EEG signals at different mental states. Biomed. Eng. Online **3**(1) (2004)
215. Nikias, C.L., Shao, M.: Signal Processing with Alpha-Stable Distributions and Applications, 1st edn. Wiley-Interscience, New York (1995)

216. Nurujjaman, M., Narayanan, R., Iyengar, A.N.S.: Comparative study of nonlinear properties of EEG signals of normal persons and epileptic patients. Nonlinear Biomed. Phys. **3**(1) (2009)
217. Ogata, K.: Modern Control Engineering. Prentice-Hall, Englewood Cliffs (1970)
218. Oldham, K.B., Spanier, J.: The Fractional Calculus. Academic Press, New York (1974)
219. Onalan, O.: Financial asset returns modeling using α-stable self-similar processes. In: International Conference on Stochastic Finance, Instituto Superior de Economia e Gestao-ISEG-Technical University of Lisbon, September 2004
220. Ortigueira, M.D.: Introduction to fractional linear systems-Part 2. discrete-time case. IEE Proc., Vis. Image Signal Process. **147**(1), 71–78 (2000)
221. Ortigueira, M.D.: Introduction to fractional linear systems. Part 1: Continuous-time case. IEE Proc., Vis. Image Signal Process. **147**(1), 62–70 (2000)
222. Ortigueira, M.D.: A new symmetric fractional B-spline. Signal Process. **83**(11), 2311–2318 (2003)
223. Ortigueira, M.D., aes Batista, A.G.: On the relation between the fractional Brownian motion and the fractional derivatives. Phys. Lett. A **372**(7), 958–968 (2008)
224. Osorio, I., Frei, M.G.: Hurst parameter estimation for epileptic seizure detection. Commun. Inf. Syst. **7**(2), 167–176 (2007)
225. Oustaloup, A.: Fractional order sinusoidal oscillators: optimization and their use in highly linear FM modulation. IEEE Trans. Circuits Syst. **28**(10), 1007–1009 (1981)
226. Oustaloup, A.: La Commande CRONE. Hermés, Paris (1991) (in French)
227. Oustaloup, A., Levron, F., Mathieu, B., Nanot, F.M.: Frequency-band complex noninteger differentiator: characterization and synthesis. IEEE Trans. Circuits Syst. I, Fundam. Theory Appl. **47**(1), 25–39 (2000)
228. Padovan, J., Sawicki, J.T.: Nonlinear vibrations of fractionally damped systems. Nonlinear Dyn. **16**(4), 321–336 (1998)
229. Padovan, J., Guo, Y.: General response of viscoelastic systems modelled by fractional operators. J. Franklin Inst. **325**(2), 247–275 (1988)
230. Paris, R.B., Kaminski, D.: Asymptotics and Mellin-Barnes Integrals, 1st edn. Cambridge University Press, Cambridge (2001)
231. Parks, T.W., Burrus, C.S.: Digital Filter Design. Wiley, New York (1987)
232. Peltier, R.F., Vehe, J.L.: Multifractional Brownian motion: definition and preliminary results. Technical report 2645, Institut National de Recherche en Informatique et en Automatique (1995)
233. Peng, C.K., Buldyrev, S.V., Havlin, S., Simons, M., Stanley, H.E., Goldberger, A.L.: Mosaic organization of DNA nucleotides. Phys. Rev. E **49**, 1685–1689 (1994)
234. Perdomo, J.J., Singh, P.M.: Electrochemical noise (ECN) measurements as a corrosion monitoring tool: a review. Corros. Rev. **20**(4–5), 359–378 (2002)
235. Pesquet-Popescu, B., Pesquet, J.-C.: Synthesis of bidimensional α-stable models with long-range dependence. Signal Process. **82**(12), 1927–1940 (2002)
236. Podlubny, I., Petráš, I., Vinagre, B.M., O'Leary, P., Dorčák, L.: Analogue realizations of fractional-order controllers. Nonlinear Dyn. **29**(1–4), 281–296 (2002)
237. Podlubny, I.: Fractional Differential Equations. Academic Press, San Diego (1999)
238. Podlubny, I.: Fractional-order systems and $\mathrm{PI}^{\lambda}\mathrm{D}^{\mu}$-controllers. IEEE Trans. Autom. Control **44**(1), 208–214 (1999)
239. Qiu, T., Kong, X.: A new adaptive latency change estimation algorithm for evoked potentials under Non-Gaussian noise condition. In: Proceedings of the IEEE-EMBS Asia-Pacific Conference on Biomedical Engineering, Hangzhou, pp. 135–136 (2000)
240. Quanser: Heat flow experiment system identification and frequency domain design. Heat Flow Experiment System Manuals (2002). http://www.quanser.com/english/downloads/products/Heatflow.pdf
241. Radwan, A.G., Elwakil, A.S., Soliman, A.M.: Fractional-order sinusoidal oscillators: design procedure and practical examples. IEEE Trans. Circuits Syst. I, Regul. Pap. **55**(7), 2051–2063 (2008)

242. Radwan, A.G., Soliman, A.M., Elwakil, A.S.: Design equations for fractional-order sinusoidal oscillators: practical circuit examples. In: Proceedings of the International Conference on Microelectronics, pp. 89–92, December 2007
243. Ray, B.K., Tsay, R.S.: Bayesian methods for change-point detection in long-range dependent processes. J. Time Ser. Anal. **23**(6), 687–705 (2002)
244. Rea, W., Oxley, L., Reale, M., Brown, J.: Estimators for long range dependence: an empirical study. Electron. J. Stat. **3**, 785–798 (2009)
245. Rechtschaffen, A., Kales, A. (eds.): A Manual of Standardized Terminology, Techniques and Scoring System for Sleep Stages of Human Subjects, 1st edn. Public Health Service, US Government Printing Office, Washington (1968)
246. Roberge, P.R., Lenard, D.R.: Characterization of corroding aluminium alloys with electrochemical noise and electrochemical impedance spectroscopy. J. Appl. Electrochem. **28**(4), 405–410 (1998)
247. Rose, O.: Estimation of the Hurst parameter of long-range dependent time series. Technical report, University of Würzburg Institute of Computer Science (1996)
248. Rossikhin, Y.A., Shitikova, M.V.: Application of fractional derivatives to the analysis of damped vibrations of viscoelastic single mass systems. Acta Mech. **120**(1–4), 109–125 (1997)
249. Rüdinger, F.: Tuned mass damper with fractional derivative damping. Eng. Struct. **28**(13), 1774–1779 (2006)
250. Sabatier, J., Agrawal, O.P., Machado, J.A.T. (eds.): Advances in Fractional Calculus—Theoretical Developments and Applications in Physics and Engineering. Springer, Berlin (2007)
251. Samko, S.G.: Fractional integration and differentiation of variable order. Anal. Math. **21**(3), 213–236 (1995)
252. Samko, S.G., Kilbas, A.A., Marichev, O.I.: Fractional Integrals and Derivatives: Theory and Applications. Gordon & Breach, Philadelphia (1993)
253. Samorodnitsky, G., Taqqu, M.S.: Stable Non-Gaussian Random Processes: Stochastic Models with Infinite Variance, 1st edn. Chapman & Hall/CRC Press, London/Boca Raton (1994)
254. Sastry, S., Bodson, M.: Adaptive Control—Stability Convergence and Robustness. Prentice Hall, New York (1989)
255. Saxena, R.K., Mathai, A.M., Haubold, H.J.: On fractional kinetic equations. Astrophys. Space Sci. **282**(1), 281–287 (2004)
256. Saxena, R.K., Mathai, A.M., Haubold, H.J.: On generalized fractional kinetic equations. Phys. A, Stat. Mech. Appl. **344**, 657–664 (2004)
257. Saxena, R., Singh, K.: Fractional Fourier transform: a novel tool for signal processing. J. Indian Inst. Sci. **85**(1), 11–26 (2005)
258. Saxton, M.J., Jacobson, K.: Single-particle tracking: applications to membrane dynamics. Annu. Rev. Biophys. Biomol. Struct. **26**, 373–399 (1997)
259. Schaumann, R., Van Valkenburg, M.E.: Design of Analog Filters, 1st edn. Oxford University Press, London (2001)
260. Schiff, J.L.: The Laplace Transform: Theory and Applications, 1st edn. Springer, Berlin (1999)
261. Schmidt, V.H., Drumheller, J.E.: Dielectric properties of Lithium Hydrazinium Sulfate. Phys. Rev. B, Solid State **4**(2), 4582–4597 (1971)
262. Sejdić, E., Djurović, I., Stanković, L.: Fractional Fourier transform as a signal processing tool: an overview of recent developments. Signal Process. **91**(6), 1351–1369 (2010)
263. Serinaldi, F.: Use and misuse of some Hurst parameter estimators applied to stationary and non-stationary financial time series. Phys. A, Stat. Mech. Appl. **389**(14), 2770–2781 (2010)
264. Shafieezadeh, A., Ryan, K., Chen, Y.Q.: Fractional order filter enhanced LQR for seismic protection of civil structures. J. Comput. Nonlinear Dyn. **3**(2), 020201.1–1021404.7 (2008)
265. Sheng, H.: Impulse Response Invariant Discretization of Distributed Order Integrator. http://www.mathworks.com/matlabcentral/fileexchange/26380 (2010)
266. Sheng, H., Chen, Y.Q.: FARIMA with stable innovations model of Great Salt Lake elevation time series. Signal Process. **91**(3), 553–561 (2011)

267. Sheng, H., Chen, Y.Q., Qiu, T.: On the robustness of Hurst estimators. IET Signal Process. (2011). doi:10.1049/iet-spr.2009.0241
268. Sheng, H., Chen, Y.Q., Qiu, T.: Tracking performance and robustness analysis of Hurst estimators for multifractional processes. IET Signal Process. (2011)
269. Sheng, H., Li, Y., Chen, Y.Q.: Application of numerical inverse Laplace transform algorithms in fractional calculus. J. Franklin Inst. **348**(2), 315–330 (2011)
270. Sheng, H., Sun, H., Chen, Y.Q., Mounteer, L.C., Kmetzsch, V.G., Miller, C.D., Zhou, A.: A fractional order signal processing (FOSP) technique for chemotaxis quantification using video microscope. In: IDETC/CIE 2009, San Diego, USA, September 2009
271. Sheng, H., Sun, H., Chen, Y.Q., Qiu, T.: Synthesis of multifractional Gaussian noises based on variable-order fractional operators. Signal Process. (2011)
272. Shokooh, A.: A comparison of numerical methods applied to a fractional model of damping materials. J. Vib. Control **5**(3), 331–354 (1999)
273. Signal Processing Toolbox 6.12: http://www.mathworks.com/products/signal
274. Smit, W., de Vries, H.: Rheological models containing fractional derivatives. Rheol. Acta **9**(4), 525–534 (1970)
275. Smith, J.O.: Physical Audio Signal Processing (2008). http://ccrma.stanford.edu/jos/pasp/ Online book
276. Sokolov, I.M., Chechkin, A.V., Klafter, J.: Distributed-order fractional kinetics. http://www.citebase.org/abstract?id=oai:arXiv.org:cond-mat/0401146 (2004)
277. Sornette, D.: Critical Phenomena in Natural Sciences: Chaos, Fractals, Selforganization and Disorder: Concepts and Tools, 2nd edn. Springer, Berlin (2004)
278. Spiegel, M.R.: Schaum's Outline of Theory and Problems of Probability and Statistics. McGraw-Hill, New York (1975)
279. Srigutomo, W.: Gaver-Stehfest algorithm for inverse Laplace transform. http://www.mathworks.com/matlabcentral/fileexchange/9987 (2006)
280. Srokowski, T.: Lévy flights in nonhomogeneous media: distributed-order fractional equation approach. Phys. Rev. E **78**(3), 031135 (2008)
281. Stanislavsky, A.A.: Probability interpretation of the integral of fractional order. Theor. Math. Phys. **138**(3), 418–431 (2004)
282. Stehfest, H.: Algorithm 368: Numerical inversion of Laplace transform. Commun. ACM **13**(1), 47–49 (1970)
283. Steiglitz, K., McBride, L.: A technique for the identification of linear systems. IEEE Trans. Autom. Control **10**(4), 461–464 (1965)
284. Stoev, S.: Simulates fractional Gaussian noise (FGN) by using the FFT—`fftfgn`. http://www.mathworks.com/matlabcentral/fileexchange/5702 (2004)
285. Stoev, S., Taqqu, M.S.: Simulation methods for linear fractional stable motion and FARIMA using the fast Fourier transform. Fractals **12**(1), 95–121 (2004)
286. Stoev, S., Taqqu, M.S.: Stochastic properties of the linear multifractional stable motion. Adv. Appl. Probab. **36**(4), 1085–1115 (2004)
287. Stoev, S., Taqqu, M.S., Park, C., Michailidis, G., Marron, J.S.: LASS: a tool for the local analysis of self-similarity. Comput. Stat. Data Anal. **50**(9), 2447–2471 (2006)
288. Stoev, S.A., Taqqu, M.S.: How rich is the class of multifractional Brownian motions? Stoch. Process. Appl. **116**(2), 200–221 (2006)
289. Sun, H.: Predictor-corrector method for variable-order, random-order fractional relaxation equation. MATLAB Central-File Exchange. http://www.mathworks.com/matlabcentral/fileexchange/26407 (2010)
290. Sun, H., Chen, W., Chen, Y.Q.: Variable-order fractional differential operators in anomalous diffusion modeling. Phys. A, Stat. Mech. Appl. **388**(21), 4586–4592 (2009)
291. Sun, H., Chen, W., Sheng, H., Chen, Y.Q.: On mean square displacement behaviors of anomalous diffusions with variable and random orders. Phys. Lett. A **374**(7), 906–910 (2010)
292. Sun, R.: Fractional order signal processing: techniques and applications. M.S. Thesis, Dept. of Electrical and Computer Engineering, Utah State University, Logan, UT, USA (2007)

293. Sun, R., Chen, Y.Q., Li, Q.: The modeling and prediction of Great Salt Lake elevation time series based on ARFIMA. In: Proceedings of the ASME 2007 International Design Engineering Technical Conferences and Computers and Information in Engineering Conference, Las Vegas (2007)
294. Talbot, A.: The accurate numerical inversion of Laplace transforms. J. Appl. Math. **23**(1), 97–120 (1979)
295. Tao, R., Zhang, F., Wang, Y.: Fractional power spectrum. IEEE Trans. Signal Process. **56**(9), 4199–4206 (2008)
296. Taqqu, M.S., Teverovsky, V.: Robustness of Whittle type estimators for time series with long-range dependence. Stoch. Models **13**(4), 723–757 (1997)
297. Taqqu, M.S., Teverovsky, V., Willinger, W.: Estimators for long-range dependence: an empirical study. Fractals **3**(4), 785–788 (1995)
298. Taqqu, M.S., Teverovsky, V.: On estimating the intensity of long-range dependence in finite and infinite variance time series. In: Adler, R.J., Feldman, R.E., Taqqu, M.S.T. (eds.) A Practical Guide to Heavy Tails: Statistical Techniques and Applications, pp. 177–217 (1998)
299. Tavazoei, M.S.: Notes on integral performance indices in fractional-order control systems. J. Process Control **20**(3), 285–291 (2010)
300. Taylor, M.J.: Historical surface levels of the Great Salt Lake. Ph.D. Thesis, Utah State University, Logan, UT (1995)
301. Trivedi, K.S.: Probability and Statistics with Reliability, Queueing, and Computer Science Applications, 2nd edn. Wiley-Interscience, New York (2001)
302. Tsai, P.-Y., Chen, C.-K.: An approximate analytic solution of the nonlinear Riccati differential equation. J. Franklin Inst. **347**(10), 1850–1862 (2010)
303. Tsao, Y.Y.: Fractal concepts in the analysis of dispersion or relaxation processes. Ph.D. Dissertation, Drexel University (June 1987)
304. Tseng, C.C.: Design of fractional order digital FIR differentiator. IEEE Signal Process. Lett. **8**(3), 77–79 (2001)
305. Acharya, R.U., Faust, O., Kannathal, N., Chua, T., Laxminarayan, S.: Non-linear analysis of EEG signals at various sleep stages. Comput. Methods Programs Biomed. **80**(1), 37–45 (2005)
306. U. S. Geological Survey: Great Salt Lake, Utah. http://ut.water.usgs.gov/greatsaltlake/ (2009)
307. Umarov, S., Steinberg, S.: Random walk models associated with distributed fractional order differential equations. In: IMS Lecture Notes Monogr. Ser., vol. 51, pp. 117–127 (2006)
308. Unser, M., Blu, T.: Fractional splines and wavelets. SIAM Rev. **42**(1), 43–67 (2000)
309. Unser, M., Aldroubi, A., Eden, M.: B-spline signal processing I. theory. IEEE Trans. Signal Process. **41**(2), 821–833 (1993)
310. Vaz, C.A., Thakor, N.V.: Adaptive Fourier estimation of time-varying evoked potentials. IEEE Trans. Biomed. Eng. **36**(4), 448–455 (1989)
311. Vinagre, B.M., Chen, Y.Q., Petras, I.: Two direct Tustin discretization methods for fractional-order differentiator/integrator. J. Franklin Inst. **340**(5), 349–362 (2003)
312. Vinagre, B.M., Petras, I., Merchan, P., Dorcak, L.: Two digital realization of fractional controllers: application to temperature control of a solid. In: Proceedings of the European Control Conference (ECC2001), Porto, Portugal, pp. 1764–1767 (2001)
313. Vinagre, B.M., Podlubny, I., Hernández, A., Feliu, V.: Some approximations of fractional order operators used in control theory and applications. Fract. Calc. Appl. Anal. **3**(3), 231–248 (2000)
314. Šušmáková, K.: Human sleep and sleep EEG. Meas. Sci. Rev. **4**(2), 59–74 (2004)
315. Wang, C.-H.: On the generalization of block pulse operational matrices for fractional and operational calculus. J. Franklin Inst. **315**(2), 91–102 (1983)
316. Wang, H.: Application of electrochemical noise technique in multiphase flow. Corrosion (2005). Paper No. 05368
317. Watson, A.B. (ed.): Digital Images and Human Vision, 1st edn. MIT Press, Cambridge (1993)

318. Watters, P.A.: Fractal structure in the electroencephalogram. Complex. Int. **5** (1998). http://www.complexity.org.au/ci/vol05/watters/watters.html
319. Weeks, W.T.: Numerical inversion of Laplace transforms using Laguerre functions. J. ACM **13**(3), 419–429 (1966)
320. Whitcher, B., Jensen, M.J.: Wavelet estimation of a local long memory parameter. Explor. Geophys. **31**(2), 94–103 (2000)
321. Winder, S.: Analog and Digital Filter Design, 2nd edn. Newnes, London (2002)
322. Woodward, W.A., Cheng, Q.C., Gray, H.L.: A *k*-factor GARMA long-memory model. J. Time Ser. Anal. **19**(4), 485–504 (1998)
323. Wu, X., Li, J., Chen, G.: Chaos in the fractional order unified system and its synchronization. J. Franklin Inst. **345**(4), 392–401 (2008)
324. Xiao, H., Mansfeld, F.: Evaluation of coating degradation with electrochemical impedance spectroscopy and electrochemical noise analysis. J. Electrochem. Soc. **141**(9), 2332–2337 (1994)
325. Xu, M., Tan, W.: Intermediate processes and critical phenomena: theory, method and progress of fractional operators and their applications to modern mechanics. Sci. China Ser. G, Phys. Astron. **49**(3), 257–272 (2006)
326. Xue, D., Atherton, D.P.: A suboptimal reduction algorithm for linear systems with a time delay. Int. J. Control **60**(2), 181–196 (1994)
327. Ying, W., Huerta, G., Steinberg, S., Zuniga, M.: Time series analysis of particle tracking data for molecular motion on the cell membrane. Bull. Math. Biol. **71**(8), 1967–2024 (2009)
328. Zaveri, N., Sun, R., Zufelt, N., Zhou, A., Chen, Y.Q.: Evaluation of microbially influenced corrosion with electrochemical noise analysis and signal processing. Electrochim. Acta **52**(9–10), 5795–5807 (2007)
329. Zaveri, N.A.: Biocorrosion studies of surface modified bioimplant material in simulated body fluids. Master of Science Thesis, Department of Biological and Irrigation Engineering, Utah State University, Logan, UT (2007)

Index

H. Sheng et al., *Fractional Processes and Fractional-Order Signal Processing*,
Signals and Communication Technology,
DOI 10.1007/978-1-4471-2233-3,